INTRODUCTION TO ENGINEERING SYSTEMS

This book is part of the

ALLYN AND BACON SERIES IN ENGINEERING

Consulting Editor: FRANK KREITH
UNIVERSITY OF COLORADO

INTRODUCTION TO ENGINEERING SYSTEMS

Raymond G. Jacquot

The University of Wyoming

Francis M. Long

The University of Wyoming

ALLYN AND BACON, INC.

Boston London Sydney Toronto

A division of Simon & Schuster
160 Gould Street
Needham Heights, Massachusetts 02194

Library of Congress Cataloging-in-Publication Data

Jacquot, Raymond G., 1938–
Introduction to engineering systems.

(Allyn and Bacon series in engineering)
Bibliography: p.
Includes index.
1. Engineering—Mathematical models. 2. System analysis. I. Long, Francis M. II. Title. III. Series.
TA342.J33 1988 620.7′2 87-35129
ISBN 0-205-11291-9

Printed in the United States of America

10 9 8 7 6 5 4 3 2 1 92 91 90 89 88

CONTENTS

4

APPLICATION OF TRANSFORM CONCEPTS TO ENGINEERING SYSTEMS 103

5

INTRODUCTION TO DYNAMIC SYSTEM SIMULATION 140

PREFACE

This text was written with the purpose of introducing the general concepts of system modeling and analysis to undergraduate engineers of the various disciplines. The authors feel strongly that there are basic concepts relative to engineering systems with which engineers of all disciplines should be familiar. The tools developed herein should also serve the practicing engineer who is confronted with predicting the performance of a dynamic system. The material in this text should set the stage for further study in the areas of control systems, mechanical vibration, network synthesis, process control, instrumentation, and digital signal processing. The prerequisite material for study of this text material is several semesters of calculus, statics, dynamics, and concurrent enrollment in a course in ordinary differential equations.

Chapter 1 lays the mathematical foundation for what follows by introducing the step function, the impulse function, and the solution of linear algebraic equations by Cramer's rule.

The Laplace transformation is introduced in Chapter 2 as a stand-alone tool for the solution of linear ordinary differential equations with constant coefficients. The authors feel strongly that system analysis today at this level should be done in the s domain, where the relationships between transient response and frequency response concepts are so much more evident. As a result, classical differential equation theory has been omitted deliberately as the authors want to encourage students to think about system dynamics in the s domain. This will be particularly useful for those students pursuing further studies in automatic control and network synthesis.

Chapter 3 introduces system dynamics by the vehicle of mechanical systems and electric circuits. Emphasis is on the formulation of governing equations by applications of a few first principles, namely, Newton's second law and Kirchhoff's two laws. Asking students to recall the materials from previous courses is simply good pedagogy and the authors feel that it enhances the students' education.

In Chapter 4 the previously developed mathematical tools are exploited to develop the idea of an input-output model of an engineering system. Applications in this chapter are restricted to mechanical and electrical systems. Applications to other types of engineering systems are given in the appropriate subsequent chapter.

The goal of Chapter 5 is to introduce students to system simulation concepts. The concept is introduced by the use of operational amplifiers to construct a simple linear analog computer. Although many engineers view analog simulation as antiquated, the concepts are useful for engineers doing digital simulation on large mainframe computers. The concepts of digital simulation are introduced and simulation languages are discussed using CSMP.

Fluid and thermal systems are examined in Chapter 6 by resorting to simple mass, energy, and momentum balances. The resulting nonlinear problems provide the vehicle for the discussion of linearization with respect to a steady operating point.

Electromechanical systems are examined in Chapter 7. Both inductive and capacitive systems are examined and the electromechanical energy conversion mechanisms are developed from first principles. It is seen that nonlinear models result in the capacitive case.

Since Fourier analysis serves as the basis for modern communication theory and digital signal processing, it is explored in Chapter 8 to the point where it will lead to the complex Fourier series. The effect of waveform symmetry on the evaluation of the integrals for the coefficients is discussed. The linear filtering of a periodic signal is given with examples and the approximation of the integrals by discrete sums is also considered and the discrete Fourier transform is introduced in a disguised fashion.

Any modern text on system analysis would be incomplete without a chapter on the state variable representation of linear time-invariant systems. This is the topic of Chapter 9. The utility of this representation is its use as a digital simulation tool.

A set of useful IBM-PC programs for system analysis is given in Appendix D.

The authors would like to thank the students who have used the manuscript for their constructive criticisms, which resulted in improvement of the manuscript. They also extend thanks to their colleagues who taught from the manuscript and freely gave their comments. In particular, the authors would like to thank Dr. James E. Lindsay of the University of Colorado at Denver who spent considerable effort in clarification of the concepts of Chapter 8.

The authors extend special thanks to the following reviewers whose contributions have enriched the text:

Professor Norman S. Eiss
Virginia Polytechnical University

Professor Richard Klein
University of Illinois at Urbana-Champaign

Dr. Frank Kreith
University of Colorado

John H. Painter
Altair Corporation

Dr. Struan R. Robertson
University of Lowell

Professor Charles A. Taft
University of New Hampshire

R. G. J.
F. M. L.

INTRODUCTION TO ENGINEERING SYSTEMS

1

INTRODUCTION AND MATHEMATICAL PRELIMINARIES

1.1 THE CONCEPT OF AN ENGINERING SYSTEM

An engineering system is a region of space inside a boundary or closed surface which is defined so as to isolate the elements of the system from the rest of the universe. A boundary will be defined for each system examined, and different sytems will often yield different boundaries. This definition of a system is consistent with the thermodynamic point of view.

Within this region of space are the system's physical elements which might be masses, springs, dampers, resistors, inductors, capacitors, hydraulic cylinders, pumps, fluid reservoirs, and any other elements which comprise common physical systems. The structure of the system is dependent not only on the elements contained in the system boundary but also their interconnection or topology. In other words, simply being given a set of elements does not provide sufficient information from which to deduce the nature of the system. What the engineer must know about the system is the physical arrangement of the elements that is the system topology. It is clear that the connection of two resistors in series in the system of Fig. 1.1(a) has different terminal properties from the two same resistors now connected in parallel, as in Fig. 1.1(b).

All engineering systems are characterized by variables which will be said to cross system boundaries.† Such variables may be divided into two types: (1) those which cause system action and are called stimuli or inputs or excitations, and (2) those which reflect the system action and are termed responses or outputs. This situation is illustrated in Fig. 1.2, where the stimuli or inputs are denoted by $y_1(t), \ldots, y_m(t)$ and responses or outputs are denoted by $x_1(t), \ldots, x_n(t)$. In general there may be one or more

†These variables crossing the boundary do not necessarily imply the flow of mass, energy, momentum flux, or electrical current across the boundary but perhaps only the flow of information across it.

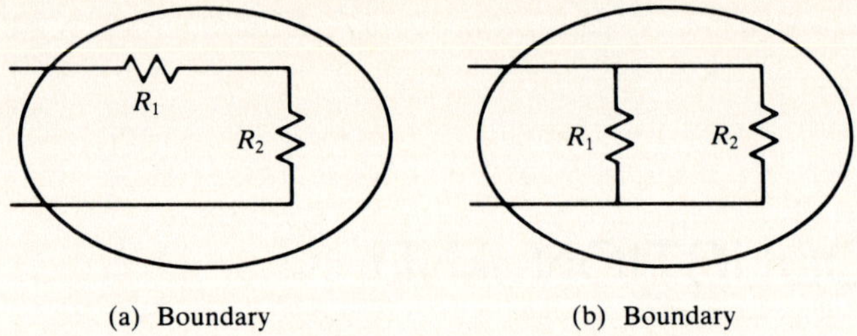

FIGURE 1.1 (a) Series resistors. (b) Parallel resistors.

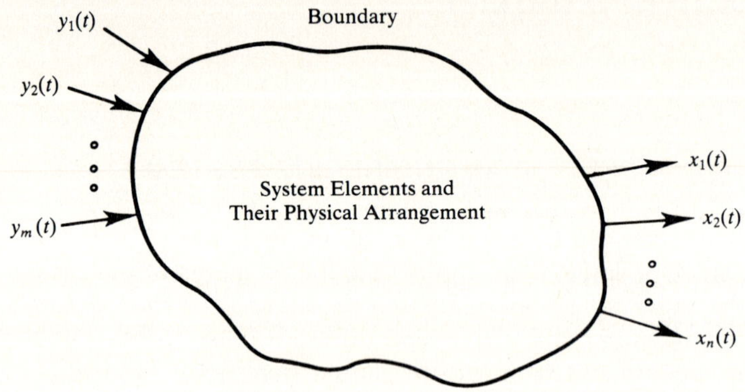

FIGURE 1.2 A typical multiple input–multiple output system.

stimuli and there may likewise be one or more outputs or responses and the number of responses and stimuli need not be equal. There may be other internal physical variables within the boundary which are not outputs but are necessary in the description of a particular problem.

One common type of system is one with only a single input $y(t)$ and a single output $x(t)$, as shown in Fig. 1.3.

The remainder of this text will be concerned with several mathematical and quantitative techniques for modeling and formulating system problems and with the solution of these mathematical models for predicting system performance to a particular set of stimuli.

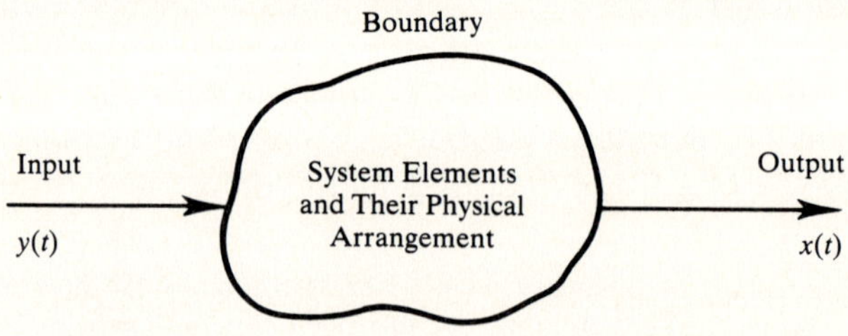

FIGURE 1.3 A single input–single output system.

In the remainder of this chapter we shall explore some mathematical ideas useful in the analysis of systems problems.

1.2 THE UNIT STEP FUNCTION

In a number of important physical problems we find that the excitations sometimes have a sudden beginning, as in the case of a closed switch in an electrical network or a suddenly applied load in a mechanical system. A useful tool in the description of such phenomenă is the unit step function $u(t)$, which is pictured in Fig. 1.4.

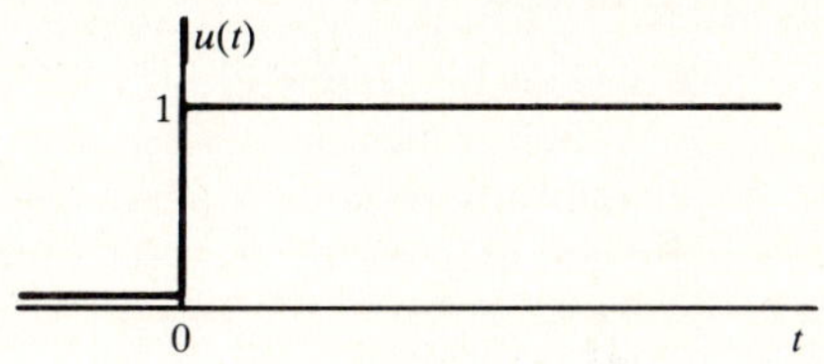

FIGURE 1.4 The unit step function.

The unit step function is defined by the relation†

$$u(t) = \begin{cases} 0 & t<0 \\ 1 & t>0 \end{cases} \quad \textbf{(1.2.1)}$$

It is also clear that, if the step is of some arbitrary height M, it could be represented as $Mu(t)$.

The unit step function is also useful as a switching function to "turn on" functions which need to start at $t=0$. Consider the function $A \sin \omega t$ illustrated in Fig. 1.5(a). The function $(A \sin \omega t)u(t)$ is illustrated in Fig.

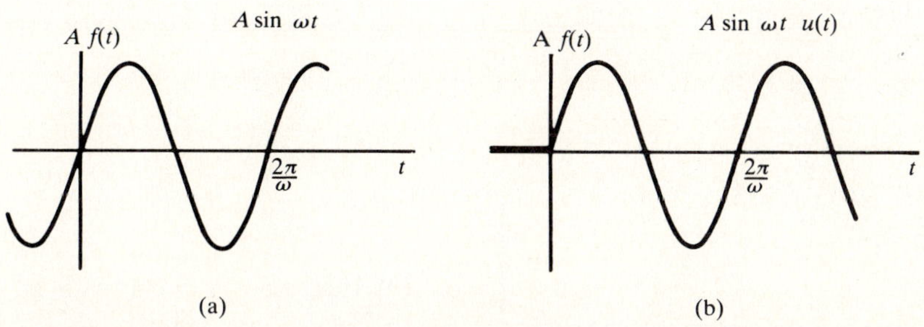

FIGURE 1.5 (a) A sinusoid existing for all time. (b) A sinusoid starting ut $(t=0)$ employing the unit step function.

---

†This definition is in the strict sense, not the mathematical definition of a function as there is no value of the dependent variable $u(t)$ associated with the independent variable t at $t=0$. We shall see that this, however, will be no problem in our work with engineering systems.

1.5(b) and it is seen that this function begins at $t = 0$. It might be representative of an ac electrical generator suddenly switched on at $t = 0$.

In this text the usual convention is followed of representing dc or constant values with capital letters and representing values which change with time by small letters with the independent variable in parentheses. For example, a battery voltage would be V (volts), and a sinusoidal time-varying voltage source would be

$$v(t) = V_{max} \sin \omega t \text{ volts}$$

Example 1.1

Consider the simple electrical circuit illustrated in Fig. 1.6. It has a battery and a switch which is closed at $t = 0$. We wish to express the voltage $v_{ab}(t)$ analytically to later evaluate the current in the network. The voltage $v_{ab}(t)$ is defined as zero for $t < 0$ so it may be expressed as a step function of height V or

$$v_{ab}(t) = Vu(t) \text{ volts, a constant for } t > 0$$

Now write the expression for the voltage $v_{cd}(t)$. From the terminal characteristics of a capacitor,

$$v_{cd}(t) = \frac{q(t)}{C} = \frac{1}{C}\int_0^t i(t)\,dt + \frac{q(0)}{C} \text{ volts}$$

a time-varying voltage. In fact, by Kirchhoff's voltage law,

$$Ri(t) + \frac{1}{C}\int_0^t i(t)\,dt + \frac{q(0)}{C} = V, \text{ a constant}$$

In real engineering problems the variables never change absolutely instantaneously. However, they often change rapidly enough that they can be represented by step functions to a high degree of accuracy.

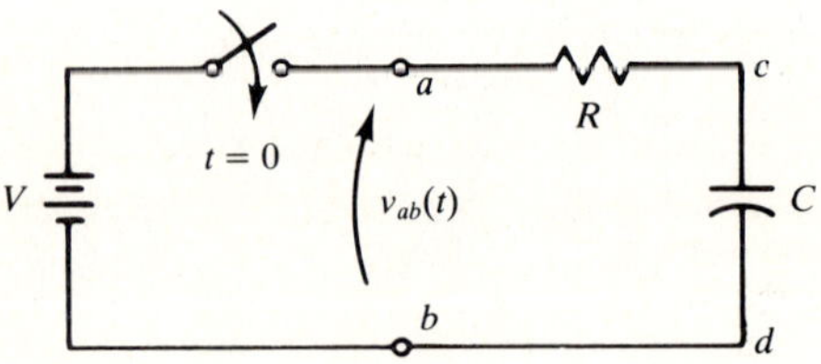

FIGURE 1.6 A circuit with a switched source voltage.

Often there is a need to employ a step function which begins at some time after $t = 0$. Consider the following definition of a delayed unit step function.

$$u(t-a)=\begin{cases}0 & t<a\\ 1 & t>a\end{cases} \tag{1.2.2}$$

Note that the jump in this step occurs when the argument inside the parentheses is zero, that is, when $t = a$. One can also note that this same property exists for the step function without delay, as defined in relation (1.2.1). Consider the application of the delayed unit step to the synthesis of an equation for an exponentially decaying pulse of finite duration.

Example 1.2

Write a single equation for the exponential pulse of Fig. 1.7(a). The pulse may be decomposed into two pieces $f_1(t)$ and $f_2(t)$, as illustrated in Fig. 1.7(b). The equations for $f_1(t)$ and $f_2(t)$ are, respectively,

$$f_1(t) = Ae^{-t}u(t)$$

and

$$f_2(t) = -Ae^{-t}u(t-1)$$

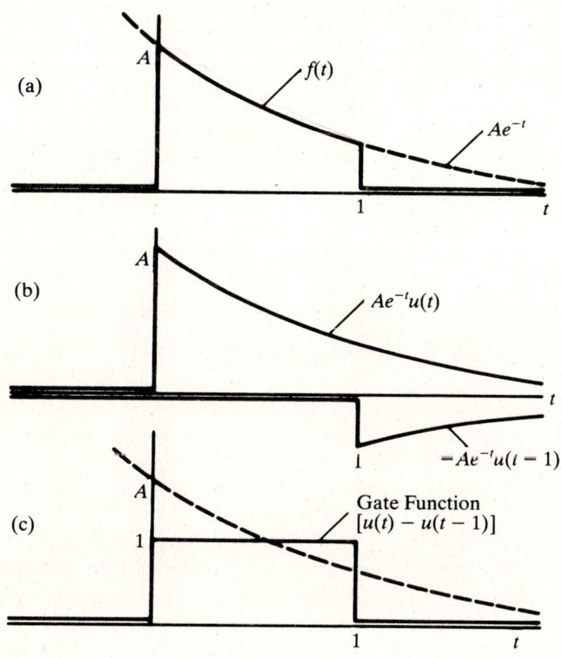

FIGURE 1.7 (a) Exponential pulse. (b) Component parts of the exponential pulse. (c) Exponential function and the "gate function."

The pulse is clearly the sum of the two components shown in Fig. 1.7(b). We write the equation for $f(t)$ as the sum of the two components or

$$f(t) = Ae^{-t}u(t) - Ae^{-t}u(t-1)$$

which may be simplified to

$$f(t) = Ae^{-t}[u(t) - u(t-1)]$$

This final expression is the product of the exponential function and the so-called rectangular pulse or "gate function," both of which are illustrated in Fig. 1.7(c). Had we known initially about the gate function, we might have written the function in the second form directly without all the initial arguments, but sometimes it is clearer to take a more fundamental approach. In this example we have illustrated use of step functions to both "turn on" and "turn off" a function which is nonzero for only a finite duration of time.

1.3 THE DIRAC DELTA OR UNIT IMPULSE FUNCTION

Let us now consider the pulse or gate function $d(t)$ illustrated in Fig. 1.8(a). This function $d(t)$ can be written as the sum of two step functions of height $1/\varepsilon$ using the gate function

$$d(t) = \frac{1}{\varepsilon}[u(t) - u(t-\varepsilon)] \qquad \textbf{(1.3.1)}$$

Note that the area of the pulse is unity regardless of the value of ε. We shall define the Dirac delta function as the limit of $d(t)$ as we shrink ε to zero or

$$\delta(t) = \lim_{\varepsilon \to 0} d(t) \qquad \textbf{(1.3.2)}$$

The properties of this newly defined function are that the width is zero and

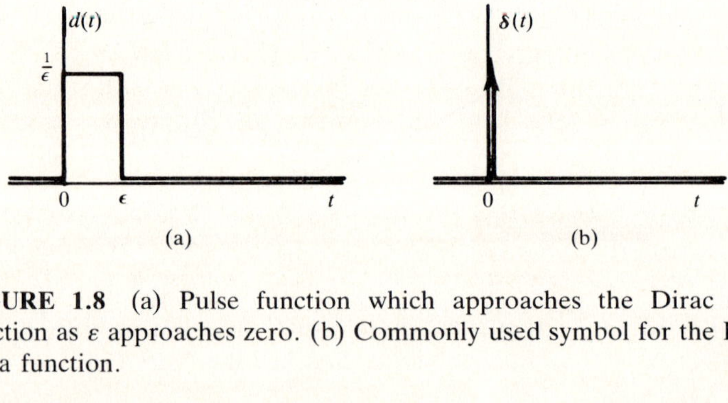

FIGURE 1.8 (a) Pulse function which approaches the Dirac delta function as ε approaches zero. (b) Commonly used symbol for the Dirac delta function.

the height is infinite in such a fashion that their product is unity. This unity area property is commonly written for some $c > 0$ as

$$\int_{-c}^{c} \delta(t)\, dt = 1 \tag{1.3.3}$$

and

$$\delta(t) = 0 \text{ for } t \neq 0 \tag{1.3.4}$$

No attempt is made to define $\delta(t)$ at $t = 0$ other than the integral of (1.3.3).† Expressions (1.3.3) and (1.3.4) are commonly used as definitions of the Dirac delta function.

Let us consider the delayed Dirac delta function which occurs at $t = a$. It can be written as $\delta(t - a)$. It is clear that definitions similar to Eqs. (1.3.3) and (1.3.4) can be written as

$$\int_{b}^{c} \delta(t - a)\, dt = \begin{cases} 1 & b < a < c \\ 0 & \text{Otherwise} \end{cases} \tag{1.3.5}$$

and

$$\delta(t - a) = 0 \quad t \neq a \tag{1.3.6}$$

Although it is impossible to have an exact Dirac function in engineering practice, pulses are sometimes narrow enough to be represented as such. If the pulse area is different from unity, say of area A, the pulse may be represented by a weighted Dirac delta function $A\delta(t - a)$.

Let us now consider another very interesting property of a Dirac delta function located at $t = a$. This property, commonly referred to as the sampling or sifting property, is that of

$$\int_{b}^{c} \delta(t - a) f(t)\, dt = \begin{cases} f(a) & b < a < c \\ 0 & \text{Otherwise} \end{cases} \tag{1.3.7}$$

Thus, the product of the Dirac delta function $\delta(t - a)$ with a continuous function $f(t)$, when integrated, gives the value of $f(t)$ at $t = a$. This can be shown, at least heuristically, by examination of the situation shown in Fig. 1.9(a), wherein the delta function is portrayed as a narrow gate function.

†It is clear that this definition of the Dirac function is in violation of the strict definition of a function which states that a function of t, say $f(t)$, is a rule by which we associate a value of $f(t)$ for every t. A completely acceptable mathematical definition of this function does not exist but requires the theory of distributions, which is beyond the mathematical level assumed in this text. However, the above "working" or operational definition is more than adequate for our needs.

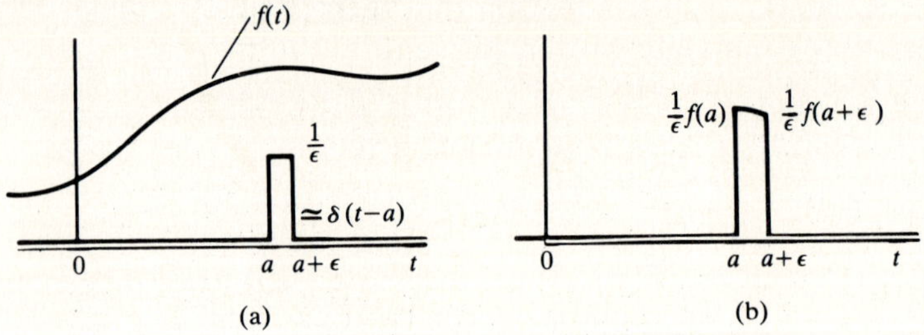

FIGURE 1.9 (a) Two elements of the integrand of Eq. (1.3.7). (b) The integrand of Eq. (1.3.7).

The integral can be evaluated at least approximately as the area of the trapezoid in Fig. 1.9(b) or

$$\int_b^c \delta(t-a)f(t)\,dt \simeq \left[\frac{f(a)/\varepsilon + f(a+\varepsilon)/\varepsilon}{2}\right][(a+\varepsilon)-a] \qquad \textbf{(1.3.8)}$$

If we now take the limit as ε becomes small and assume $f(t)$ to be a continuous function at $t = a$, then $f(a+\varepsilon)$ approaches $f(a)$ or

$$\int_b^c \delta(t-a)f(t)\,dt = \frac{1}{2}\,[2f(a)] = f(a) \qquad \textbf{(1.3.9)}$$

and expression (1.3.7) is proven. It is appropriate to note here that the term "sampling property" arises from the fact that the integration of $f(t)$ weighted with a delta function at $t = a$ samples or results in the value $f(t)$ at $t = a$ or $f(a)$. The term "sifting property" arises from the fact that the integration of $f(t)$ weighted with a Dirac function at $t = a$ "sifts" out the value $f(a)$. This sampling property is useful in digital computer control systems and in digital signal processing.

1.4 SOLUTION OF LINEAR EQUATIONS BY CRAMER'S RULE

In a wide variety of physical problems, one of the final steps in the solution is that of solving a set of n simultaneous linear equations for n unknowns, at least some of which are the desired numerical or functional solutions to the problem. We shall develop a method here which may be used to solve a general set of linear equations by hand computation, but remember that this method is not efficient enough to be implemented on a digital computer. The technique of Gaussian elimination is preferred if there are more than three equations and three unknowns.

Consider a general set of n linear equations in n unknowns x_1, $x_2, \ldots, x_n$ or

$$
\begin{aligned}
a_{11}x_1 + a_{12}x_2 + \cdots + a_{1n}x_n &= y_1 \\
a_{21}x_1 + a_{22}x_2 + \cdots + a_{2n}x_n &= y_2 \\
\vdots \qquad \vdots \qquad\quad\ \vdots \quad &\ \ \vdots \\
a_{n1}x_1 + a_{n2}x_2 + \cdots + a_{nn}x_n &= y_n
\end{aligned}
\tag{1.4.1}
$$

We shall assume in the following that the a_{ij} and the y_k are either constants or functions and are known. In matrix-vector notation the set of Eqs. (1.4.1) could be written simply as

$$\mathbf{Ax} = \mathbf{y} \tag{1.4.2}$$

where the matrix $\mathbf{A}$ is defined as the array of the coefficients of Eq. (1.4.1) as follows:

$$\mathbf{A} = \begin{bmatrix} a_{11}a_{12} \cdots a_{1n} \\ a_{21}a_{22} \cdots a_{2n} \\ \vdots \qquad\quad \vdots \\ a_{n1}a_{n2} \cdots a_{nn} \end{bmatrix} \tag{1.4.3}$$

while the vectors $\mathbf{x}$ and $\mathbf{y}$ are defined by

$$\mathbf{x} = \begin{bmatrix} x_1 \\ \vdots \\ x_n \end{bmatrix} \tag{1.4.4}$$

and

$$\mathbf{y} = \begin{bmatrix} y_1 \\ \vdots \\ y_n \end{bmatrix} \tag{1.4.5}$$

Cramer's rule states that the solution for the jth unknown x_j is given by

$$x_j = \frac{|\mathbf{A}_j|}{|\mathbf{A}|} \tag{1.4.6}$$

where $|\mathbf{A}|$ denotes the determinant of matrix $\mathbf{A}$ and $|\mathbf{A}_j|$ denotes the determinant of $\mathbf{A}$ with the jth column replaced by the column vector $\mathbf{y}$. (For a proof of this rule the reader is referred to any elementary text on linear algebra.)

If the solution for all n of the unknowns x_j is required, the numerator determinant must be evaluated n times. *Note that if* $|\mathbf{A}| = 0$, *there are no solutions to the problem because the original equations were not linearly independent*. This is perhaps an indication that some step in the formulation of a physical problem has been done incorrectly and the engineer should try a fresh approach to the equation formulation.

The method is best illustrated by several examples which follow.

Example 1.3

Solve the algebraic equations below for unknowns x_1 and x_2.

$$4x_1 + 3x_2 = 2$$

$$2x_1 + 5x_2 = 1$$

The **A** matrix is

$$\mathbf{A} = \begin{bmatrix} 4 & 3 \\ 2 & 5 \end{bmatrix}$$

The **x** vector is

$$\mathbf{x} = \begin{bmatrix} x_1 \\ x_2 \end{bmatrix}$$

and the **y** vector is

$$\mathbf{y} = \begin{bmatrix} 2 \\ 1 \end{bmatrix}$$

The solution for x_1 using relation (1.4.6) is

$$x_1 = \frac{\begin{vmatrix} 2 & 3 \\ 1 & 5 \end{vmatrix}}{\begin{vmatrix} 4 & 3 \\ 2 & 5 \end{vmatrix}} = \frac{10-3}{20-6} = \frac{7}{14} = \frac{1}{2}$$

and for x_2

$$x_2 = \frac{\begin{vmatrix} 4 & 2 \\ 2 & 1 \end{vmatrix}}{14} = \frac{0}{14} = 0$$

One can verify the correctness of the solutions by substituting to see that they satisfy *both* equations simultaneously. This example was nearly trivial and could have been solved by substitution or subtraction of the equations; however, the authors do not recommend this practice as there is considerable potential for numerical error.

Example 1.4

Consider the multiloop dc electrical circuit shown in Fig. 1.10. It is desired to solve for loop current I_3, the current through the 5 Ω resistor.

Solution: Writing Kirchhoff's voltage law around each closed loop we obtain

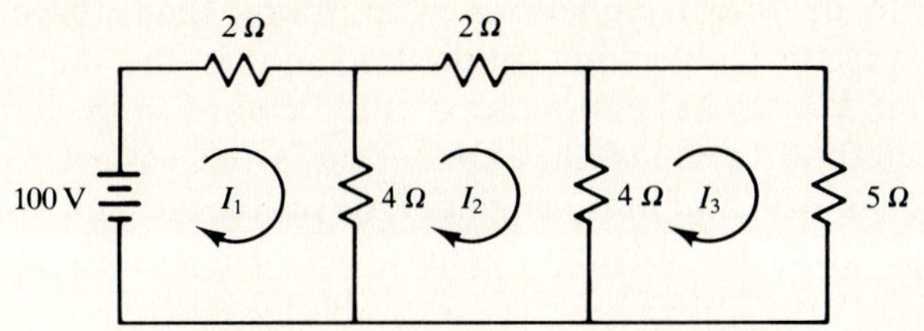

FIGURE 1.10 A two-loop circuit.

$$100 = 2I_1 + 4(I_1 - I_2)$$

$$0 = 2I_2 + 4(I_2 - I_3) + 4(I_2 - I_1)$$

$$0 = 5I_3 + 4(I_3 - I_2)$$

Rearranging yields

$$6I_1 - 4I_2 = 100$$

$$-4I_1 + 10I_2 - 4I_3 = 0$$

$$-4I_2 + 9I_3 = 0$$

The matrix equation for this example is

$$\mathbf{RI} = \mathbf{V}$$

where the matrices **R**, **I**, and **V** are

$$\mathbf{R} = \begin{bmatrix} 6 & -4 & 0 \\ -4 & 10 & -4 \\ 0 & -4 & 9 \end{bmatrix}$$

$$\mathbf{I} = \begin{bmatrix} I_1 \\ I_2 \\ I_3 \end{bmatrix} \qquad \mathbf{V} = \begin{bmatrix} V_1 \\ V_2 \\ V_3 \end{bmatrix} = \begin{bmatrix} 100 \\ 0 \\ 0 \end{bmatrix}$$

Application of Eq. (1.4.6) to solve for the loop current I_3 yields

$$I_3 = \frac{\begin{vmatrix} 6 & 4 & 100 \\ -4 & 10 & 0 \\ 0 & -4 & 0 \end{vmatrix}}{\begin{vmatrix} 6 & -4 & 0 \\ -4 & 10 & -4 \\ 0 & -4 & 9 \end{vmatrix}} = \frac{1600}{540 - (96 + 144)}$$

$$I_3 = \frac{1600}{300} = 5.333 \text{ amperes}$$

The solution for the other two loop currents would yield similar results.

For systems of linear equations with more than three unknowns, Cramer's rule is not a viable solution method, nor is any method involving hand calculation. The digital computer should be used in these higher order cases and the method of Gaussian elimination is the easiest to program in general and works well for most engineering problems with fewer than 10 unknowns. A program for solving a general set of linear equations is given in Appendix D.

1.5 THE SYSTEMS OF UNITS

When modeling and predicting the performance of physical systems, we need to define a consistent set of units. In this text we shall work in one of two possible sets of units, namely, the British Engineering System and SI (System International of Units). Although the United States is committed to a conversion to the SI system, the transition is occurring more slowly than had been anticipated by the proponents of the SI system. As a result, there is considerable amount of engineering which still uses the British Engineering System. The exception occurs in electrical problems where there is a

TABLE 1-1 Commonly Employed Units for Physical Quantities

Quantity	Common Symbol	British Engineering	SI
Time	t	second (s)	second (s)
Length	l	foot (ft)	meter (m)
Mass	m	slug	kilogram (kg)
Force	f	pound (lb)	newton (N)
Energy	w	British thermal unit (Btu)	joule (J)
Power	p	horsepower (hp)	watt (W)
Density	p	slug/ft^3	kg/m^3
Pressure	p	lb/ft^2	pascal (Pa)
Viscosity	μ	lb · s/ft^2	N · s/m^2
Volumetric flow rate	q	ft^3/s	m^3/s
Temperature	T	°R	K
Specific heat	C	Btu/slug · °R	J/kg · K
Current	i	—	ampere (A)
Voltage	v	—	volt (V)
Resistance	R	—	ohm (Ω)
Inductance	L	—	henry (H)
Capacitance	C	—	farad (F)
Magnetic flux	ϕ	—	weber (Wb)
Magnetic flux density	B	—	tesla (T)
Magnetic field intensity	H	—	A/m
Electric field intensity	E	—	V/m

long history of metric units although some of the former Rationalized MKS units have been replaced with new units to recognize major contributors to electrical science. The common units for many physical quantities are given in Table 1-1.

1.6 SUMMARY

In this chapter we have introduced the idea of an engineering system. We have defined the unit step function and examined its role as a switching function. The Dirac delta function was defined and some of its properties were explored. Because system problems often require the solution of a set of linear equations, Cramer's rule was introduced as a hand-computation method for the solution of linear equations with three or fewer unknowns.

PROBLEMS

1-1. Write equations for the waveforms given below using step functions where necessary.

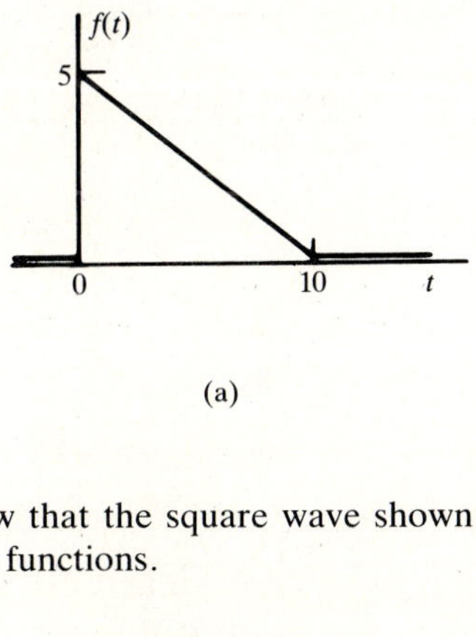

(a)

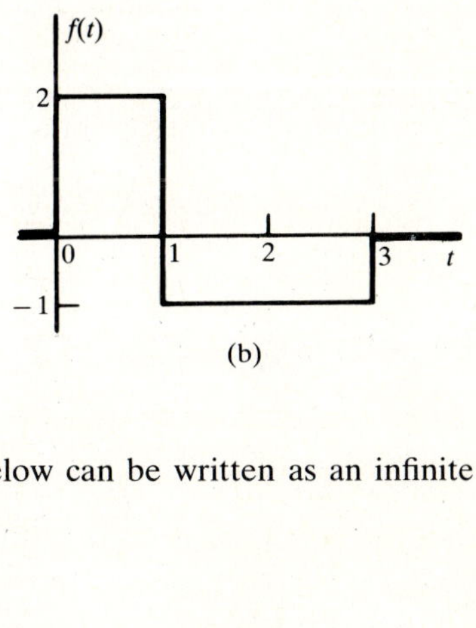

(b)

1-2. Show that the square wave shown below can be written as an infinite series of step functions.

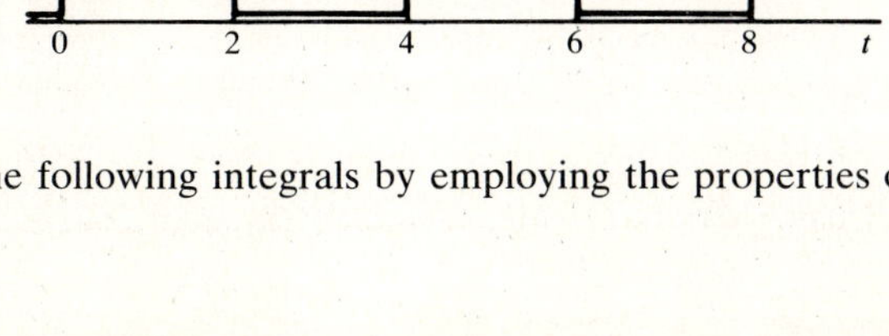

1-3. Evaluate the following integrals by employing the properties of the Dirac delta function:

(a) $I_a = \int_0^{\infty} \delta(t-1)e^{-t}\,dt$

(b) $I_b = \int_2^{5} \delta(t-1)e^{-t}\,dt$

(c) $I_c = \int_0^{10} \delta\left(t - \frac{\pi}{4}\right) \sin t\,dt$

1-4. Solve the following linear algebraic equations:

$$4x_1 + 3x_2 = 2$$

$$2x_1 + x_2 = 1$$

1-5. Solve the following equations for the unknown x_2:

$$x_1 + x_2 + x_3 = 1$$

$$x_1 + 2x_2 = 0$$

$$x_2 + x_3 = 0$$

1-6. **(a)** For the electrical network shown, show that equations resulting from application of Kirchhoff's voltage law for each loop are

$$100 = 2I_1 + 4(I_1 - I_2)$$

$$0 = 3I_2 + 5(I_2 - I_3) + 4(I_2 - I_1)$$

$$0 = 4I_3 + 5(I_3 - I_2)$$

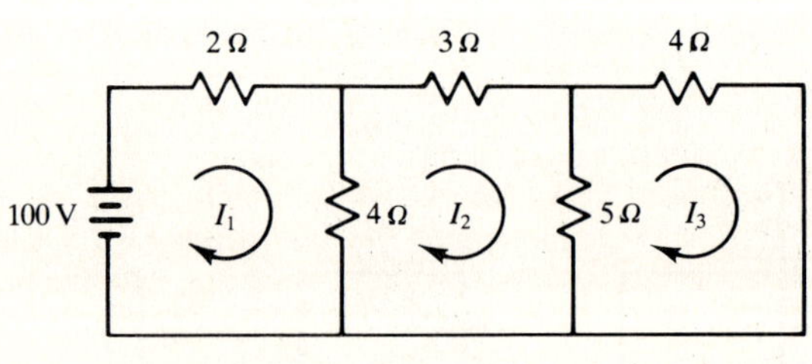

(b) Put these equations in matrix form:

$$\mathbf{Ax} = \mathbf{y}$$

(c) Solve for I_3 using Cramer's rule. (As a check, $|\mathbf{A}| = 354$.)

1-7. Consider the nonsymmetric truss shown loaded with 100 lb. Summation of vertical and horizontal forces at joint A results in two linear equations:

$$0.707F_1 + 0.6F_2 = 100$$

$$0.707F_1 - 0.8F_2 = 0$$

Solve for F_1 and F_2 by employing Cramer's rule. (As a check, $|\mathbf{A}| = -0.9898$.)

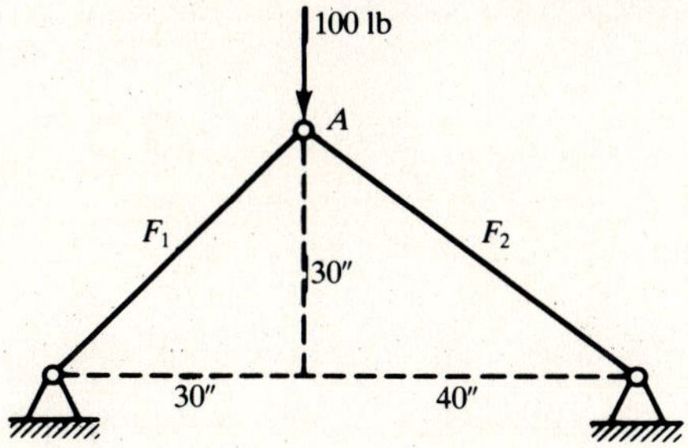

2

THE LAPLACE TRANSFORMATION

2.1 INTRODUCTION

In the following chapters we shall see how it is possible to formulate the governing equations for physical systems by applying the principles of physics. These equations are often dependent on assumptions made in applying physical laws. Frequently they must be simplified before they may be solved by the mathematical tools at the disposal of an engineer. In this chapter we shall introduce one very powerful tool in the study of engineering systems, the Laplace transformation.

2.2 AN ANALOGY WITH LOGARITHMS

To illustrate the role of the Laplace transformation, let us consider a simple analogous application which all readers have undoubtably encountered in their past education, namely the use of logarithms to calculate the product of two numbers X and Y. This process is tabulated in the sequence of operations given below:

1. Use the table of logarithms to find the respective logarithms of X and Y.
2. Add the two logarithms together to get the logarithm of the product: $(\log(XY) = \log(X) + \log(Y))$.
3. Use the log table in the reverse direction to take the antilog to yield the product XY.

If we examine this process we see that the more complex operation of multiplication was replaced by the simpler operation of addition but at the cost of the additional complication of using log tables.

We may similarly summarize the role of the Laplace transform in the solution of linear ordinary differential equations having constant coefficients. As for logarithms we also employ a table, in this case a table of transforms. The procedure is:

1. Use the transform tables to transform each term of the differential equation from the time domain to the transform domain.
2. Perform the algebraic operations to solve for the unknown solution.
3. Employ the transform tables to convert this solution back to the time domain.

These steps are summarized in Fig. 2.1. In this technique we see that the inherently more complicated operations of integration and differentiation have been replaced by the simpler operations of algebra and reliance on the existence of a table of transforms.

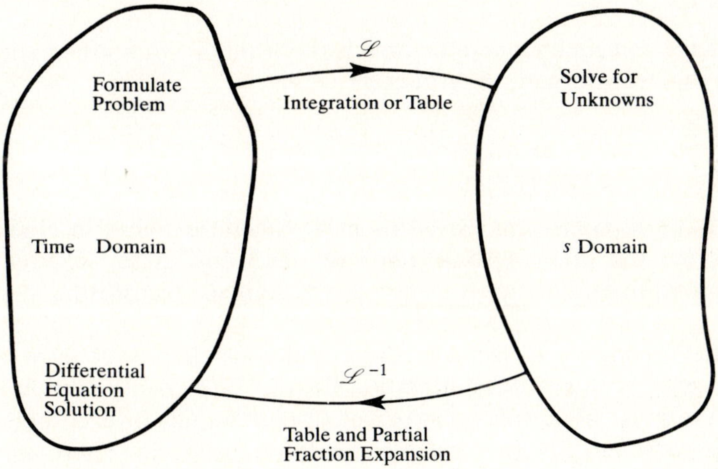

FIGURE 2.1 Summary of steps in solving a differential equation with the Laplace transformation.

2.3 DEFINITION AND EXISTENCE OF THE LAPLACE TRANSFORMATION

In many engineering applications we are interested in predicting ahead of time how a system will respond to a given stimulus which begins at some time which we shall refer to as $t = 0$. We shall consider here only functions which are zero for all negative time. Given some function of time $f(t)$ defined for all positive t, there is another function $F(s)$ associated with $f(t)$ for $t > 0$ such that $F(s)$ is given by the integral transformation†

†The lower limit is to be interpreted as being the right side of the origin, which we shall refer to as $t = 0+$ later in this text.

$$\mathscr{L}[f(t)] = \int_0^\infty f(t)e^{-st}\,dt \tag{2.3.1}$$

where the symbol $\mathscr{L}[\ \ ]$ is read as "the Laplace transform of." Note that this definite integral results in a function of s alone, which we shall call $F(s)$. This function $F(s)$ is called the Laplace transform of $f(t)$. For commonly defined engineering functions, there will be one and only one associated $F(s)$ for each $f(t)$ or it may be said that the transformation is unique. Hence, we might suspect that it is possible to invert the transform uniquely by constructing a table of transform pairs, and this is true. In general the transform variable s will be a complex variable with real part σ and imaginary part ω or

$$s = \sigma + j\omega \tag{2.3.2}$$

where j is the imaginary operator (given the symbol i in mathematics). We often see this transformation written as

$$F(s) = \mathscr{L}[f(t)] = \int_0^\infty f(t)e^{-st}\,dt \tag{2.3.3}$$

The Laplace transform exists for most functions found in engineering work but, for the sake of some rigor we shall now state, without proof, sufficient conditions for the existence of the Laplace transform.

1. The function $f(t)$ is piecewise continuous or has at most a finite number of finite discontinuities. In the case of periodic functions, there must be a finite number of discontinuities in each period.
2. The function $f(t)$ must be of exponential order, i.e., there must exist a σ such that

$$\int_0^\infty |f(t)|e^{-\sigma t}\,dt < \infty \tag{2.3.4}$$

For the class of problems considered in this text, the resulting Laplace transforms are ratios of polynomials in the complex variable s or

$$F(s) = \frac{b_m s^m + \cdots + b_1 s + b_0}{s^n + a_{n-1}s^{n-1} + \cdots + a_1 s + a_0} \qquad n \geq m \tag{2.3.5}$$

where the a's and b's are real numbers. Each of the polynomials can be written in factored form as

$$F(s) = \frac{b_m(s - z_1)(s - z_2)\cdots(s - z_m)}{(s - p_1)(s - p_2)\cdots(s - p_n)} \tag{2.3.6}$$

where the complex numbers $z_1, z_2, \ldots, z_m$ are values of the complex variable s which cause the numerator polynomial to vanish; hence, $F(s)$ vanishes. These values are referred to as the *zeros* of $F(s)$. The complex numbers $p_1, p_2, \ldots, p_n$ are values of s for which the denominator vanishes; hence the function $F(s)$ becomes infinite. As the complex variable s approaches one of the values p_i, the function $F(s)$ becomes large in magnitude and as a result the values p_i are called the *poles* of $F(s)$.

2.4 DEVELOPMENT OF A TABLE OF TRANSFORM PAIRS FOR SOME SIMPLE FUNCTIONS

Let us now apply the definition of Eq. (2.3.1) to get the transforms of some simple functions used in engineering applications.

(a) Unit Step Function: In Chapter 1 we discussed and defined the unit step function to be

$$f(t) = u(t) = \begin{cases} 0 & t<0 \\ 1 & t>0 \end{cases} \tag{2.4.1}$$

From the definition, the transform of the step function is

$$\mathscr{L}[u(t)] = \int_0^\infty 1e^{-st}\, dt \tag{2.4.2}$$

or

$$\mathscr{L}[u(t)] = \left.\frac{e^{-st}}{-s}\right|_0^\infty = \frac{1}{s} \tag{2.4.3}$$

The conclusion is that the Laplace transformation of the unit step function is $1/s$.

(b) Ramp Function: If

$$f(t) = tu(t) \tag{2.4.4}$$

then from the definition (Eq. (2.3.3))

$$F(s) = \int_0^\infty te^{-st}\, dt \tag{2.4.5}$$

Integration by parts yields

$$F(s) = -\left.\frac{te^{-st}}{s}\right|_0^\infty - \frac{1}{s}\int_0^\infty e^{-st}\, dt$$

$$F(s) = -\left.\left[\frac{t}{s} + \frac{1}{s^2}\right]e^{-st}\right|_0^\infty$$

Evaluating at the limits gives

$$\mathscr{L}[tu(t)] = \frac{1}{s^2} \tag{2.4.6}$$

(c) Many common engineering functions are of polynomial form, so we should investigate the transform of a function of order n or

$$f(t) = t^n u(t) \tag{2.4.7}$$

Using the example of the ramp function, we can show that, by integrating by parts n times, the result is

$$F(s) = \mathscr{L}[t^n u(t)] = \frac{n!}{s^{n+1}} \tag{2.4.8}$$

(d) A number of important engineering problems have solutions which have exponential character or

$$f(t) = e^{-at} u(t) \tag{2.4.9}$$

where a is either positive or negative. From the definition

$$\mathscr{L}[e^{-at} u(t)] = \int_0^\infty e^{-at} e^{-st}\, dt \tag{2.4.10}$$

or

$$\mathscr{L}[e^{-at} u(t)] = \int_0^\infty e^{-(s+a)t}\, dt \tag{2.4.11}$$

Upon integration

$$\mathscr{L}[e^{-at} u(t)] = \left. \frac{e^{-(s+a)t}}{-(s+a)} \right|_0^\infty \tag{2.4.12}$$

and evaluation at the limits yields

$$\mathscr{L}[e^{-at} u(t)] = \frac{1}{s+a} \tag{2.4.13}$$

(e) Another important function from an engineering point of view is the sine wave or

$$f(t) = \sin \omega t\, u(t) \tag{2.4.14}$$

We know from the Euler identities that this function may be written as

$$f(t) = \left[\frac{e^{j\omega t} - e^{-j\omega t}}{2j}\right] u(t) \tag{2.4.15}$$

or

$$\sin \omega t \, u(t) = \frac{1}{2j}\,[e^{j\omega t} - e^{-j\omega t}]u(t) \tag{2.4.16}$$

These functions are simply exponentials which may be transformed using the work of part **(d)** of this section.

$$\mathcal{L}[\sin \omega t \, u(t)] = \frac{1}{2j}\left[\frac{1}{s - j\omega} - \frac{1}{s + j\omega}\right] \tag{2.4.17}$$

$$= \frac{1}{2j}\,\frac{\not{s} + j\omega - \not{s} + j\omega}{s^2 + \omega^2} \tag{2.4.18}$$

The result is then

$$\mathcal{L}[\sin \omega t \, u(t)] = \frac{\omega}{s^2 + \omega^2} \tag{2.4.19}$$

The cosine function will be used as an exercise at the end of this chapter.

(f) Damped Sine Function

$$f(t) = e^{-at} \sin \omega t \, u(t) \tag{2.4.20}$$

Using the Euler identities, the function may be written as

$$f(t) = \frac{e^{-at}}{2j}\,[e^{j\omega t} - e^{-j\omega t}] \tag{2.4.21}$$

This is transformed to be

$$F(s) = \frac{1}{2j}\left[\frac{1}{s + a - j\omega} - \frac{1}{s + a + j\omega}\right] \tag{2.4.22}$$

Finding a common denominator yields

$$F(s) = \frac{1}{2j}\,\frac{s + a + j\omega - s - a + j\omega}{(s + a)^2 + \omega^2} \tag{2.4.23}$$

The result is, after simplification,

$$\mathcal{L}[e^{-at} \sin \omega t \, u(t)] = \frac{\omega}{(s + a)^2 + \omega^2} \tag{2.4.24}$$

Similarly, we can show that the transform of the damped cosine function is

$$\mathscr{L}[e^{-at}\cos\omega t\, u(t)] = \frac{(s+a)}{(s+a)^2+\omega^2} \tag{2.4.25}$$

In the previous work we developed the Laplace transformations of some simple functions. In the engineering literature it is common to tabulate these functions with the time-domain function in one column and the associated s-domain function in the adjacent column. From this obvious pairing we arrive at the term Laplace transform pair. The common transforms just developed are given in Table 2-1 and a larger table is given in Table A-2 of Appendix A.

TABLE 2-1 Table of Common Laplace Transform Pairs

$f(t)\ t \geq 0$	$F(s) = \mathscr{L}[f(t)]$
1. $u(t)$	$\frac{1}{s}$
2. t	$\frac{1}{s^2}$
3. t^2	$\frac{2}{s^3}$
4. t^n	$\frac{n!}{s^{n+1}}$
5. e^{-at}	$\frac{1}{s+a}$
6. $\sin\omega t$	$\frac{\omega}{s^2+\omega^2}$
7. $\cos\omega t$	$\frac{s}{s^2+\omega^2}$
8. $e^{-at}\sin\omega t$	$\frac{\omega}{(s+a)^2+\omega^2}$
9. $e^{-at}\cos\omega t$	$\frac{(s+a)}{(s+a)^2+\omega^2}$

2.5 EFFECT OF DENOMINATOR ON INVERSE TRANSFORM FORM

In general, the s-domain functions for which we are seeking the inverse transform will be the ratio of two polynomials in the complex variable s or

$$F(s) = \frac{N(s)}{s^n + a_{n-1}s^{n-1} + \cdots + a_1 s + a_0} \tag{2.5.1}$$

The denominator polynomial is of degree n and will have n roots. Its roots are real or complex numbers which, when substituted for s, make the denominator polynomial vanish. The roots of the denominator are called the *poles* of $F(s)$ as they cause the function $F(s)$ to become infinite. We shall designate these poles as $p_1, p_2, \ldots, p_n$ and as such we can write the function $F(s)$ with a factored denominator or Eq. (2.5.1) can be written as

$$F(s) = \frac{N(s)}{(s - p_1)(s - p_2)\cdots(s - p_n)} \tag{2.5.2}$$

It is also common to examine the roots of the numerator polynomial $N(s)$, which are commonly referred to as the *zeros* of $F(s)$.

It should be noted that, if they are complex, poles must occur in conjugate pairs because of the real values of a_i in the denominator polynomial. If the value of p_i is real, this will contribute to a term in the inverse transform of the form $e^{p_i t}$. If the root is complex ($p_i = -\alpha_i + j\beta_i$), there will be a complex conjugate root $(-\alpha_i - j\beta_i)$. These two roots yield complex exponential solutions which can be combined by the Euler identities to yield inverse terms of the form $e^{-\alpha_i t}\cos\beta_i t$ and $e^{-\alpha_i t}\sin\beta_i t$, as indicated by entries 8 and 9 in Table 2-1. Repeated roots of $p_i = 0$ yield polynomial terms in the temporal variable t. As a particular matter, if a pair of complex or imaginary roots occurs, it is better to leave this quadratic term in unfactored form and to complete the square in the quadratic to yield a term of the form of the denominators of entries 8 and 9 in Table 2-1.

2.6 TRANSFORMS OF SOME COMMON MATHEMATICAL OPERATIONS

In order for the Laplace transform to be a useful tool in the solution of differential equations which arise in a variety of physical problems, it will be necessary to develop transforms for mathematical operations on as yet unknown functions of time. The most common operations encountered are those of differentiation and integration.

2.6.1 Transform of the Derivative

If we denote the transform of $f(t)$ as $F(s)$, it is not beyond the scope of this discussion to inquire as to what might be the transform of the rate of change of $f(t)$ or df/dt. From the definition of the transform (Eq. (2.3.1)), the transform of the derivative can be written conceptually as

$$\mathscr{L}\left[\frac{df}{dt}\right] = \int_0^\infty \frac{df}{dt}\, e^{-st}\, dt \tag{2.6.1}$$

If we now integrate by parts by selecting

$$u = e^{-st} \qquad dv = \frac{df}{dt}\,dt = df$$

and (2.6.2)

$$du = -se^{-st}\,dt \qquad v = f$$

the result is

$$\mathcal{L}\left[\frac{df}{dt}\right] = f(t)e^{-st}\Big|_0^\infty - \int_0^\infty -sf(t)e^{-st}\,dt \tag{2.6.3}$$

When the limits of the first term are evaluated, the result is

$$\mathcal{L}\left[\frac{df}{dt}\right] = f(\infty)e^{-\infty} - f(0) + s\int_0^\infty f(t)e^{-st}\,dt \tag{2.6.4}$$

where we have again interpreted the lower limit as $t = 0+$.

If $f(t)$ has the property of being of exponential order, the exponential function in the first term will be stronger than $f(t)$ at $t = \infty$ and this term will vanish, leaving

$$\mathcal{L}\left[\frac{df}{dt}\right] = s\int_0^\infty f(t)e^{-st}\,dt - f(0) \tag{2.6.5}$$

Now it is easy to recognize the remaining integral as the transform of $f(t)$ or $F(s)$, thus yielding

$$\mathcal{L}\left[\frac{df}{dt}\right] = sF(s) - f(0) \tag{2.6.6}$$

To derive the expression for the transform of the second derivative, one must integrate twice by parts and use the property that the first derivative of $f(t)$ is also of exponential order. The result of such a derivation is

$$\mathcal{L}\left[\frac{d^2f}{df^2}\right] = s^2F(s) - sf(0) - \frac{df(0)}{dt} \tag{2.6.7}$$

where $F(s)$ is the transform of $f(t)$. The result for the general nth derivative is tabulated in Table A-1 of Appendix A.

2.6.2 Transform of the Definite Integral

In a number of electric circuit problems the formulated equations are in the form of integro-differential equations which have terms involving integrals of the dependent variable. Let us consider the transform of the definite integral

$$\mathscr{L}\left[\int_0^t f(t')\,dt'\right] = \int_0^\infty \left[\int_0^t f(t')\,dt'\right] e^{-st}\,dt \tag{2.6.8}$$

where the prime denotes the dummy variable in the definite integral. Now let us integrate by parts by selecting

$$u = \int_0^t f(t')\,dt' \qquad dv = e^{-st}\,dt$$

and (2.6.9)

$$du = f(t)\,dt \qquad v = -\frac{1}{s}\,e^{-st}$$

Carrying out the integration

$$\mathscr{L}\left[\int_0^t f(t')\,dt'\right] = -\int_0^t f(t')\,dt'\,\frac{1}{s}\,e^{-st}\bigg|_0^\infty + \frac{1}{s}\int_0^\infty f(t)e^{-st}\,dt \tag{2.6.10}$$

If the function is of exponential order, the first term vanishes and the remaining integral is recognized as the transform of $f(t)$, so

$$\mathscr{L}\left[\int_0^t f(t')\,dt'\right] = \frac{F(s)}{s} \tag{2.6.11}$$

By similar arguments the transform of an indefinite integral can be shown to be

$$\mathscr{L}\left[\int f(t')\,dt'\right] = \frac{F(s)}{s} + \frac{1}{s}\int_{-\infty}^0 f(t')\,dt' \tag{2.6.12}$$

Example 2.1

Let us now use all of our previously developed tools to solve a differential equation, namely

$$\frac{dx}{dt} + 4x = u(t), \qquad x(0) = 2$$

Let us take the Laplace transform

$$\mathscr{L}\left[\frac{dx}{dt} + 4x\right] = \mathscr{L}[u(t)]$$

Noting that the transform is linear, it may be applied term by term to the left side to yield

$$\mathcal{L}\left[\frac{dx}{dt}\right] + 4\mathcal{L}[x] = \mathcal{L}[u(t)]$$

Now use the transform of the derivative and of the unit step to give

$$sX(s) - x(0) + 4X(s) = \frac{1}{s}$$

Noting that $x(0)$ is specified in the problem, we get after substituting it in

$$(s+4)X(s) = \frac{1}{s} + 2$$

Solving for $X(s)$ yields

$$X(s) = \frac{1}{s(s+4)} + \frac{2}{s+4}$$

Employing transform entries from Table A-2 gives the inverse transform

$$x(t) = \left[\frac{1}{4}(1 - e^{-4t}) + 2e^{-4t}\right]u(t)$$

2.7 INVERSION OF TRANSFORMS—THE PARTIAL FRACTION EXPANSION

It is possible that a problem will be large enough that the tables given here or elsewhere will not be sufficient. As a method of extending the tables to larger problems, we shall examine, in some detail, the method of *partial fraction expansion*.

2.7.1 Distinct Poles of $F(s)$—The Brute Force Method

If all the p_i's of expression (2.5.2) are different, it is possible to write the function $F(s)$ as a sum, as in relation (2.7.1):

$$\frac{N(s)}{(s-p_1)(s-p_2)\ldots(s-p_n)} = \frac{A_1}{s-p_1} + \frac{A_2}{s-p_2} + \cdots + \frac{A_n}{s-p_n} \tag{2.7.1}$$

The equality of relation (2.7.1) must hold for all s. If a common denominator is found on the right side, then the numerators of both sides must be equal and hence the coefficients of like powers in the numerators must be equal. The result of this is a set of n linear equations with the A_i as the unknowns. An example will serve to illustrate the technique.

Example 2.2

Invert the following s-domain function by the brute force technique.

$$F(s) = \frac{s+1}{s(s+2)(s+3)} = \frac{A_1}{s} + \frac{A_2}{s+2} + \frac{A_3}{s+3}$$

The result of finding a common denominator on the right side is

$$\frac{s+1}{s(s+2)(s+3)} = \frac{A_1(s+2)(s+3) + A_2 s(s+3) + A_3 s(s+2)}{s(s+2)(s+3)}$$

Equating coefficients of like powers of s, the result is a set of linear equations in the unknown coefficients:

$$s^2: \quad 0 = A_1 + A_2 + A_3$$

$$s^1: \quad 1 = 5A_1 + 3A_2 + 2A_3$$

$$s^0: \quad 1 = 6A_1$$

Thus, $A_1 = \frac{1}{6}$ and the remaining two equations can be solved simultaneously to give $A_2 = \frac{1}{2}$ and $A_3 = -\frac{2}{3}$. The resulting inverse function is given by using Table 2-1 on the partial fraction expansion.

$$f(t) = \frac{1}{6}\,u(t) + \frac{1}{2}\,e^{-2t} - \frac{2}{3}\,e^{-3t}$$

There is, however, some economy to be had if the solution of the simultaneous equations can be eliminated, as we shall indicate in the following section.

2.7.2 The Heaviside Expansion Method

If we multiply both sides of Eq. (2.7.1) by $(s - p_1)$ we get

$$\frac{(s-p_1)N(s)}{(s-p_1)(s-p_2)\ldots(s-p_n)} = A_1 + \frac{A_2(s-p_1)}{s-p_2} + \cdots + \frac{A_n(s-p_1)}{s-p_n} \tag{2.7.2}$$

For equality, this relation must hold for all s so we may evaluate both sides at $s = p_1$, which will make all but the first term of the right side vanish. So

$$A_1 = \left.\frac{N(s)}{(s-p_2)(s-p_3)\ldots(s-p_n)}\right|_{s=p_1} \tag{2.7.3}$$

which easily gives us A_1. We can similarly evaluate the rest of the coefficient and in general we can get the ith coefficient from

$$A_i = \lim_{s \to p_i} (s - p_i)F(s) \tag{2.7.4}$$

In the case where the pole p_i is complex, A_i will also be a complex number. The technique is illustrated by the following example.

Example 2.3

Invert the following s-domain function by applying the Heaviside method.

$$F(s) = \frac{s+1}{s(s^2 + 5s + 6)} = \frac{s+1}{s(s+2)(s+3)}$$

The form of the partial fraction expansion will be

$$F(s) = \frac{A_1}{s} + \frac{A_2}{s+2} + \frac{A_3}{s+3}$$

Coefficient A_1 is given by Eq. (2.7.4) to be

$$A_1 = \left. \frac{s(s+1)}{s(s+2)(s+3)} \right|_{s=0} = \frac{1}{(2)(3)} = \frac{1}{6}$$

Similarly

$$A_2 = \left. \frac{\cancel{(s+2)}(s+1)}{s\cancel{(s+2)}(s+3)} \right|_{s=-2} = \frac{-2+1}{(-2)(-2+3)} = \frac{1}{2}$$

and

$$A_3 = \left. \frac{\cancel{(s+3)}(s+1)}{s(s+2)\cancel{(s+3)}} \right|_{s=-3} = \frac{-3+1}{(-3)(-3+2)} = \frac{-2}{3}$$

Thus the inverse function is given as

$$f(t) = \left[\frac{1}{6} u(t) + \frac{1}{2} e^{-2t} - \frac{2}{3} e^{-3t} \right] u(t)$$

which is the same result given by the brute force technique (as it must be).

Another example will serve to illustrate a computational problem associated with the Heaviside expansion method when dealing with complex denominator roots.

Example 2.4

Invert the following transform function by the Heaviside expansion method.

$$F(s) = \frac{s+3}{(s+1)(s^2 + 4s + 13)}$$

If we solve for the roots of the quadratic, we find that they are at $s = -2 \pm j3$. The denominator may then be written in factored form to give

$$F(s) = \frac{s+3}{(s+1)(s+2+j3)(s+2-j3)}$$

The expansion is of the form

$$F(s) = \frac{A_1}{s+1} + \frac{A_2}{s+2+j3} + \frac{A_3}{s+2-j3}$$

We can get A_1 in the usual manner to give

$$A_1 = \left. \frac{s+3}{s^2+4s+13} \right|_{s=-1} = \frac{2}{10}$$

and we get A_2 as

$$A_2 = \left. \frac{s+3}{(s+1)(s+2-j3)} \right|_{s=-2-j3} = \frac{-2-j3+3}{(-2-j3+1)(-j6)}$$

$$= \frac{1-j3}{-18+j6} = \frac{6-j8}{-60}$$

Similarly

$$A_3 = \left. \frac{s+3}{(s+1)(s+2+j3)} \right|_{s=-2+j3} = \frac{6+j8}{-60}$$

Then the inverse transform is

$$f(t) = \left[\frac{2}{10} e^{-t} + \frac{6-j8}{-60} e^{(-2-j3)t} + \frac{6+j8}{-60} e^{(-2+j3)t} \right] u(t)$$

or

$$f(t) = \left[\frac{2}{10} e^{-t} - \frac{6}{60} e^{-2t}(e^{j3t} + e^{-j3t}) + \frac{j8}{60} e^{-2t}(e^{-j3t} - e^{j3t}) \right] u(t)$$

Now employing the Euler identities the last two terms are damped cosines and sines, respectively, or

$$f(t) = \left[\frac{2}{10} e^{-t} - \frac{2}{10} e^{-2t} \cos 3t + \frac{8}{30} e^{-2t} \sin 3t \right] u(t)$$

The reader is urged to work out the algebraic details of this example to appreciate the potential for errors in the complex algebra. When completed, the answer should be a real function of time.

In the next section we shall consider a technique to eliminate the complex algebra of the preceding example.

2.7.3 Complex Roots—The Easy Way

In our previous work we have seen that complex roots led, in general, to both damped sine and damped cosine terms in the inverse transform. For this reason it seems that, instead of dealing with all the complex numbers involved we could make the original expansion in terms of the transforms of damped sines and cosines. This means that we need not factor the quadratic but rather, due to the form of the damped sine and cosine transforms, we should complete the square. Let us briefly review how to complete the square, which is done by forming a squared factor of s plus one-half the coefficient of the linear term in s. Then the constant added from this squaring operation must be subtracted, as follows:

$$s^2 + bs + c = \left(s + \frac{b}{2}\right)^2 + c - \frac{b^2}{4} \tag{2.7.5}$$

$$= (s + \alpha)^2 + \beta^2$$

where $-\alpha$ is the real part of the characteristic root and β is the imaginary part. For a numerical example consider

$$s^2 + 4s + 13 = (s + 2)^2 + 13 - 4 = (s + 2)^2 + 3^2$$

Example 2.5

Let us invert the function of Example 2.4

$$F(s) = \frac{s + 3}{(s + 1)(s^2 + 4s + 13)}$$

If we solve for the roots of the quadratic we find that they are $s_{1,2} = -2 \pm j3$. These complex roots tell us that we should complete the square to give

$$F(s) = \frac{s + 3}{(s + 1)[(s + 2)^2 + 3^2]}$$

Since the quadratic has complex roots, damped sines and cosines are indicated in the inverse function. The transform pairs of damped sines and cosines are included in Table 2-1. The form of the expansion is

$$F(s) = \frac{A}{s + 1} + \frac{B(s + 2)}{(s + 2)^2 + 3^2} + \frac{C3}{(s + 2)^2 + 3^2}$$

We get A by the usual Heaviside method or

$$A = \frac{-1 + 3}{1 - 4 + 13} = \frac{2}{10}$$

Now we must find B and C by the brute force method by finding a common denominator and equating the numerators to give

$$s+3 = A(s^2+4s+13) + B(s+2)(s+1) + C3(s+1)$$

Equating the coefficients of like powers of s, the resulting set of linear equations is

$$s^2\text{:}\quad 0 = A + B$$

$$s^1\text{:}\quad 1 = 4A + 3B + 3C$$

$$s^0\text{:}\quad 3 = 13A + 2B + 3C$$

Since we have already solved for A, this set of linear equations is not so formidable as it initially appears, so

$$B = -A = -\frac{2}{10}$$

and

$$3C = 1 - 4A - 3B$$

or

$$C = \frac{1}{3}\left[1 - \frac{8}{10} + \frac{6}{10}\right] = \frac{8}{30}$$

The inverse transform is given directly without further algebra as

$$f(t) = \left[\frac{2}{10}e^{-t} - \frac{2}{10}e^{-2t}\cos 3t + \frac{8}{30}e^{-2t}\sin 3t\right]u(t)$$

Note that this solution was accomplished without any complex algebra, alleviating a highly probable source of error when the partial fraction coefficients are calculated by hand.

2.7.4 Repeated Real Factors

When the denominator of $F(s)$ has repeated factors, either real or complex, special precautions are required to make the partial fraction expansion equal to the originally given function $F(s)$.

Let us consider a function which has only one repeated root, which is repeated k times. This $F(s)$ has the form

$$F(s) = \frac{N(s)}{(s-p_1)^k(s-p_2)\ldots(s-p_n)} \qquad \textbf{(2.7.6)}$$

The only form of partial fraction expansion which can be made to equal Eq. (2.7.6) is that which contains denominators with factors of $(s - p_1)$ to various powers up to and including $(s - p_1)^k$ or of the form

$$F(s) = \sum_{j=1}^{k} \frac{B_j}{(s - p_1)^j} + \sum_{i=2}^{n} \frac{A_i}{s - p_i} \tag{2.7.7}$$

where the second sum represents the terms contributed by the remaining denominator terms. Any other form of the first term of Eq. (2.7.7) will make equality between Eqs. (2.7.6) and (2.7.7) impossible. The coefficients A_i in Eq. (2.7.7) can be evaluated by the Heaviside method, as demonstrated earlier, and one B_j coefficient, B_k, can also be evaluated by the Heaviside method or

$$B_k = (s - p_1)^k F(s)\big|_{s=p_1} \tag{2.7.8}$$

The difficulty here is that the remaining B_j cannot be found in a similar fashion as all multipliers other than $(s - p_1)^k$ result in indeterminate solutions. Although there are more sophisticated methods for evaluation of the B_j, they are primarily of academic value and the authors have found that the brute force technique is preferred in real problems. In general, evaluate B_k and all the A_i first by the Heaviside method. With these values known, the linear equations generated by the brute force method are easily solved for $B_1, B_2, \ldots, B_{k-1}$. The example which follows illustrates the technique.

Example 2.6

Evaluate the inverse transform of the function $F(s)$ given below:

$$F(s) = \frac{s + 3}{(s + 1)^3 (s + 2)} \tag{*}$$

The correct form of the partial fraction expansion is

$$F(s) = \frac{B_1}{s + 1} + \frac{B_2}{(s + 1)^2} + \frac{B_3}{(s + 1)^3} + \frac{A}{s + 2} \tag{**}$$

First, evaluate A to get

$$A = \frac{(s + 2)(s + 3)}{(s + 1)^3 (s + 2)}\bigg|_{s=-2} = \frac{-2 + 3}{(-2 + 1)^3} = -1$$

and similarly B_3 to get

$$B_3 = \frac{(s + 1)^3 (s + 3)}{(s + 1)^3 (s + 2)}\bigg|_{s=-1} = \frac{-1 + 3}{-1 + 2} = 2$$

Now B_1 and B_2 can be evaluated by the brute force technique. Finding a common denominator in Eq. (**) and equating numerators of Eqs. (*) and (**) yields

$$s + 3 = B_1(s + 1)^2(s + 2) + B_2(s + 1)(s + 2) + 2(s + 2) - (s + 1)^3$$

Equating coefficients of like powers of s gives

$$s^3: \quad 0 = B_1 - 1$$

$$s^2: \quad 0 = 4B_1 + B_2 - 3$$

Therefore

$$B_1 = 1$$

and

$$B_2 = 3 - 4B_1 = 3 - 4 = -1$$

and the resulting expansion is

$$F(s) = \frac{1}{s+1} + \frac{-1}{(s+1)^2} + \frac{2}{(s+1)^3} - \frac{1}{s+2}$$

Now use Table A-2 to find the inverse transform

$$f(t) = \left[e^{-t} - te^{-t} + \frac{2}{2}\, t^2e^{-t} - e^{-2t}\right]u(t)$$

If there are other sets of multiple real roots, the technique is then applied repeatedly.

2.8 ADDITIONAL TRANSFORM THEOREMS

Earlier in this chapter we developed several theorems associated with the transform of derivatives and integrals but there are some other theorems which further extend Table 2-1. We shall also present other theorems as needed throughout the text. The final result of each theorem will be enclosed in a box and is also given in Table A-1.

2.8.1 Multiplication by *t*

Consider now the transform of $tf(t)$, which can be investigated by considering the derivative of $F(s)$ with respect to s. If we consider the definition of the transform, this can be written as

$$\frac{d}{ds}\,[F(s)] = \frac{d}{ds}\int_0^\infty f(t)e^{-st}\,dt \tag{2.8.1}$$

Since the limits of integration are not functions of the variable s, the differentiation process can be moved inside the integral to operate on the integrand or

$$\frac{dF(s)}{ds} = \int_0^\infty f(t)(-t)e^{-st}\,dt \tag{2.8.2}$$

Multiplying both sides by -1 yields

$$\int_0^\infty t\,f(t)e^{-st}\,dt = -\frac{dF(s)}{ds} \tag{2.8.3}$$

We can immediately recognize the integral as the transform of $tf(t)$ or

$$\boxed{\mathcal{L}[tf(t)] = -\frac{dF(s)}{ds}} \tag{2.8.4}$$

This theorem can be applied repeatedly to yield the transform of a function multiplied by t^2, t^3, and so on. We may generalize by continued differentiation with respect to s to give

$$\mathcal{L}[t^n f(t)] = (-1)^n \frac{d^n F(s)}{ds^n} \tag{2.8.5}$$

Example 2.8

Employ relation (2.8.4) to find the transform of te^{-at}. The transform of e^{-at} is $1/(s+a)$ so, employing relation (2.8.4), the result is

$$\mathcal{L}[te^{-at}] = -\frac{d}{ds}\left[\frac{1}{s+a}\right]$$

and, upon differentiating the quotient, the result is

$$\mathcal{L}[te^{-at}] = \frac{1}{(s+a)^2}$$

2.8.2 Multiplication by e^{-at} (Modulation Theorem)

Let us now examine the transform of a function $f(t)$ when multiplied by the exponential function which, from the definition of the transform, is

$$\mathcal{L}[e^{-at}f(t)] = \int_0^\infty f(t)e^{-at}e^{-st}\,dt \tag{2.8.6}$$

Combination of the exponential terms yields

$$\mathscr{L}[e^{-at}f(t)] = \int_0^\infty f(t)e^{-(s+a)t}\,dt \qquad \textbf{(2.8.7)}$$

The integral may be recognized as being the same as the definition of the transform with the variable s replaced by $s + a$ or

$$\boxed{\mathscr{L}[e^{-at}f(t)] = F(s + a)} \qquad \textbf{(2.8.8)}$$

Example 2.9

Apply the result of Eq. (2.8.8) to find the transform of $e^{-at}\sin\omega t$. The transform of the sine function is $\omega/(s^2 + \omega^2)$ so, from relation (2.8.8), the desired transform is

$$\mathscr{L}[e^{-at}\sin\omega t] = \frac{\omega}{(s+a)^2 + \omega^2}$$

2.8.3 Delay Theorem

Let us consider the function $f(t)$ which exists for all t, as illustrated in Fig. 2.2(a). We shall now consider the function when multiplied by various step functions and/or when delayed by a seconds. The function $f(t)u(t)$, which exists only for $t > 0$, is illustrated in Fig. 2.2(b). The function $f(t - a)$ is illustrated in Fig. 2.2(c), while Fig. 2.2(d) illustrates the undelayed function switched at $t = a$ given by $f(t)u(t - a)$. The delayed function turned on at $t = 0$ is given by $f(t - a)u(t)$ and illustrated in Fig. 2.2(e). The delayed function turned on at $t = a$ seconds given by $f(t - a)u(t - a)$ is shown in Fig. 2.2(f).

We shall concern ourselves here with the type of function illustrated in Fig. 2.2(f). We shall consider the transform of the function $f(t - a)u(t - a)$. If we consider the definition of the transform

$$\mathscr{L}[f(t-a)u(t-a) = \int_a^\infty f(t-a)e^{-st}\,dt \qquad \textbf{(2.8.9)}$$

where the lower limit has been modified to reflect the delayed step function which makes the integrand zero for $0 < t < a$. Let us make a change of variable, letting $t - a = \tau$ in the integral or

$$\mathscr{L}[f(t-a)u(t-a)] = \int_0^\infty f(\tau)e^{-s(\tau+a)}\,d\tau \qquad \textbf{(2.8.10)}$$

We may factor e^{-sa} out of the integral to give

$$\mathscr{L}[f(t-a)u(t-a)] = e^{-sa}\int_0^\infty f(\tau)e^{-s\tau}\,d\tau \qquad \textbf{(2.8.11)}$$

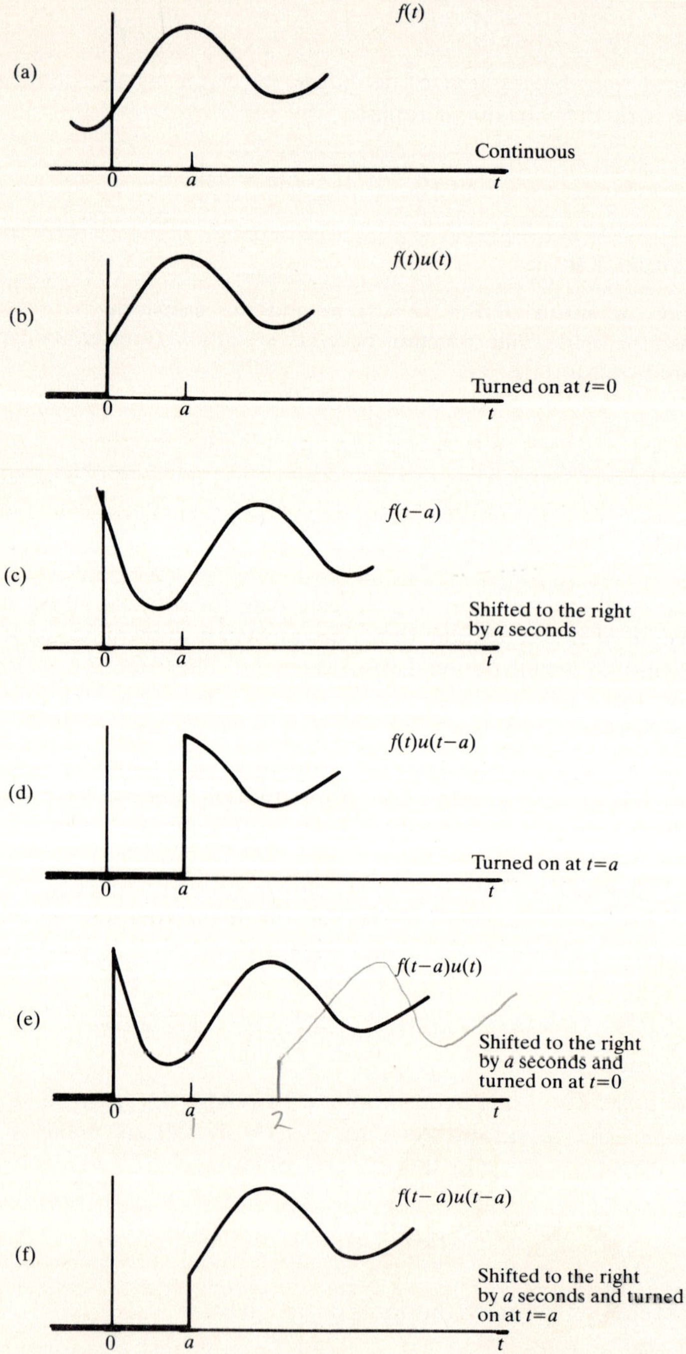

FIGURE 2.2 Various delayed and switched functions.

and now we recognize the remaining integral as $F(s)$, the transform of the undelayed function $f(t)$. Thus

$$\boxed{\mathscr{L}[f(t-a)u(t-a)] = e^{-as}F(s)} \qquad \textbf{(2.8.12)}$$

Example 2.10

Employing the delay theorem of Eq. (2.8.12), find the transform of the function $(t-3)u(t-3)$. The associated undelayed function $f(t)$ is t. The transform of t is $1/s^2$, so the transform of $(t-3)u(t-3)$ is

$$\mathscr{L}[(t-3)u(t-3)] = e^{-3s}\frac{1}{s^2}$$

The converse of the theorem is frequently employed when inverting transforms and will be given here with an application to illustrate how it is used. If we take the inverse Laplace transform of expression (2.8.12), the result is

$$\boxed{f(t-a)u(t-a) = \mathscr{L}^{-1}[e^{-as}F(s)]} \qquad \textbf{(2.8.13)}$$

The following example will serve to illustrate the usefulness of this form of the theorem.

Example 2.11

Invert the function $G(s)$ given below.

$$G(s) = e^{-2s}\frac{s}{s^2+4^2}$$

It is clear from the table of transforms (Table 2-1) that the inverse transform of the polynomial ratio part of this transform is $\cos 4t$ but employing relation (2.8.13)

$$g(t) = \mathscr{L}^{-1}[G(s)] = [\cos 4(t-2)]u(t-2)$$

We shall employ this theorem extensively later when working with physical systems.

2.8.4 Final-Value Theorem

Sometimes it is useful to know the constant final value that a function $f(t)$ (with associated transform $F(s)$) reaches when given only $F(s)$. If a function $F(s)$ has an inverse transform $f(t)$ which has a constant part, it must contain a single factor s in the denominator or must be of the form

$$F(s) = \frac{N(s)}{s(s - p_1) \ldots (s - p_n)} \tag{2.8.14}$$

If we make a partial fraction expansion, then

$$F(s) = \frac{A}{s} + \sum_{i=1}^{n} \frac{B_i}{s - p_i} \tag{2.8.15}$$

and the inverse function is

$$f(t) = Au(t) + \sum_{i=1}^{n} B_i e^{p_i t} \tag{2.8.16}$$

For the function $f(t)$ to reach a constant value for large t, the exponential terms due to the $(s - p_i)$ denominator factors must vanish for large t. This will be true only if the real parts of all the p_i are negative. In other words, all the poles of $F(s)$ must lie in the left half of the complex s plane, except for the single pole at the origin. From relation (2.8.16) we can see then that, if the real parts of all the p_i are negative, then

$$\lim_{t \to \infty} f(t) = A \tag{2.8.17}$$

It is also of interest to note that A is a coefficient in the partial fraction expansion of Eq. (2.8.15) and can be given by the Heaviside method to be

$$A = \lim_{s \to 0} sF(s) \tag{2.8.18}$$

Equating Eqs. (2.8.17) and (2.8.18) gives us the desired result, known as the *final-value theorem* or

$$\boxed{\lim_{t \to \infty} f(t) = \lim_{s \to 0} sF(s)} \tag{2.8.19}$$

Example 2.12

For the function $F(s)$, find the final value of $f(t)$ without performing the complete transform inversion

$$F(s) = \frac{s + 3}{s(s^2 + 4s + 13)}$$

Employing the theorem,

$$\lim_{t \to \infty} f(t) = \lim_{s \to 0} \cancel{s}\left[\frac{s + 3}{\cancel{s}(s^2 + 4s + 13)}\right]$$

Upon evaluation the final value is

$$f(\infty) = \frac{3}{13}$$

which is much easier than accomplishing the complete transform inversion.

2.8.5 Initial-Value Theorem

Once in a while we might be interested in the value of some function $f(t)$ at $t = 0$ when only the transform function $F(s)$ is given. Consider the transform of the first derivative given in Section 2.6:

$$\int_0^\infty \left(\frac{df}{dt}\right) e^{-st}\, dt = sF(s) - f(0) \tag{2.8.20}$$

If we now let s approach infinity,

$$\lim_{s \to \infty} \int_0^\infty \frac{df}{dt} e^{-st}\, dt = \lim_{s \to \infty} [sF(s) - f(0)] \tag{2.8.21}$$

If df/dt is transformable, the integral on the left side of Eq. (2.8.21) exists and is zero and since $f(0)$ is independent of s, then

$$\boxed{0 = \lim_{s \to \infty} sF(s) - f(0)} \tag{2.8.22}$$

and the theorem is proven to yield

$$\boxed{f(0) = \lim_{s \to \infty} sF(s)} \tag{2.8.23}$$

Example 2.13

Find the initial value of the inverse transform of the following function:

$$F(s) = \frac{5s^2 + 3s + 2}{s(s^2 + 4s + 13)}$$

Then

$$sF(s) = \frac{5s^2 + 3s + 2}{s^2 + 4s + 13}$$

and if we take the limit $s \to \infty$, we get an indeterminate form. If we divide the numerator and denominator by s^2 to yield

$$sF(s) = \frac{5 + 3/s + 2/s^2}{1 + 4/s + 13/s^2}$$

the initial value of $f(t)$ is found to be

$$f(0) = 5$$

2.8.6 Transform of Periodic Functions

Consider a periodic function with period T which goes to zero for $t < 0$, as illustrated in Fig. 2.3(a). Let us define the first cycle function $f_1(t)$ as illustrated in Fig. 2.3(b) and defined by

$$f_1(t) = \begin{cases} f(t) & 0 < t \leq T \\ 0 & \text{Otherwise} \end{cases} \tag{2.8.24}$$

We may write the function $f(t)$ as the sum of an infinite number of delayed version of $f_1(t)$ or

$$f(t) = \sum_{n=0}^{\infty} f_1(t - nT) \tag{2.8.25}$$

Let us define the transform of $f_1(t)$ to be $F_1(s)$. We may then employ the delay theorem to calculate the transform of $f(t)$ to be

$$F(s) = \sum_{n=0}^{\infty} e^{-nTs} F_1(s) \tag{2.8.26}$$

or

$$F(s) = F_1(s)[1 + e^{-Ts} + e^{-2Ts} + \cdots] \tag{2.8.27}$$

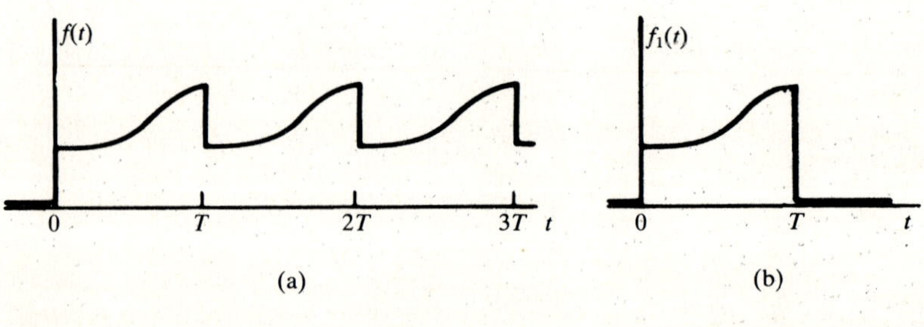

FIGURE 2.3 (a) A function that is periodic for $t > 0$. (b) The first cycle function $f_1(t)$.

Although it is not obvious, it can be shown that the infinite series in Eq. (2.8.27) has a closed form sum or

$$\boxed{F(s) = F_1(s)\left[\frac{1}{1 - e^{-Ts}}\right]} \quad \textbf{(2.8.28)}$$

For the skeptical reader this result may be verified by performing long division on the quotient of relation (2.8.28) to yield relation (2.8.27). A simple example should clarify the application of this theorem.

Example 2.14

Find the Laplace transform of the pulse waveform starting at $t = 0$ illustrated in Fig. 2.4. The first cycle function $f_1(t)$ is given by

$$f_1(t) = 5[u(t) - u(t-2)]$$

and the transform of this function is

$$F_1(s) = 5\left[\frac{1}{s} - \frac{e^{-2s}}{s}\right] = \frac{5}{s}\,[1 - e^{-2s}]$$

Now, if we note that the period T is 3 seconds, then we may employ relation (2.8.27) to give the transform of the function.

$$F(s) = \frac{5}{s}\,[1 - e^{-2s}]\left[\frac{1}{1 - e^{-3s}}\right]$$

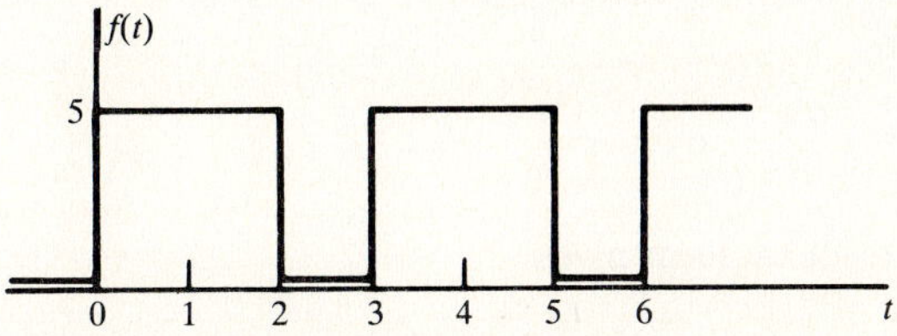

FIGURE 2.4 Square-wave periodic function.

All of the theorems derived thus far are given in tabular form in Table A-1 of Appendix A.

2.9 SOLVING LINEAR ORDINARY DIFFERENTIAL EQUATIONS WITH CONSTANT COEFFICIENTS

The full utility of the Laplace transform method is best illustrated by some detailed examples of the solution of linear ordinary integro-differential equations with constant coefficients. To accomplish this objective we make extensive use of the linear property of the Laplace transform. For instance, if we are transforming a linear combination of terms of the form

$$\mathscr{L}\left[a_1 \frac{dx}{dt} + a_2 x(t) + a_3 \int_0^t x(t)\,dt\right] = \mathscr{L}[f(t)] \tag{2.9.1}$$

the transform is the linear combination of the transforms or

$$a_1 \mathscr{L}\left[\frac{dx}{dt}\right] + a_2 \mathscr{L}[x(t)] + a_3 \mathscr{L}\left[\int_0^t x(t)\,dt\right] = F(s) \tag{2.9.2}$$

Using the transforms of operations from Table A-1 the result is

$$a_1[sX(s) - x(0)] + a_2 X(s) + a_3 \frac{X(s)}{s} = F(s) \tag{2.9.3}$$

We should note also that the transform of the product of two time domain functions is *not* the product of the transforms of the functions and likewise the inverse transform of the product of two s-domain functions is *not* the product of associated time-domain functions. We shall discuss this topic further in Chapter 4, where we shall derive the convolution theorem. Let us now consider several examples which should serve to tie together the ideas of this chapter.

Example 2.15

Solve the integro-differential equation given below subject to the initial condition given.

$$2\frac{dy}{dt} + 16y + 50\int_0^t y\,dt = 3t \cdot u(t), \qquad y(0) = 2$$

Transforming term by term gives

$$2[sY(s) - y(0)] + 16Y(s) + \frac{50}{s}\,Y(s) = \frac{3}{s^2}$$

Transporting the initial condition term to the right gives

$$\left[2s + 16 + \frac{50}{s}\right]Y(s) = \frac{3}{s^2} + 4 = \frac{4s^2 + 3}{s^2}$$

and solving for the unknown $Y(s)$ yields

$$Y(s) = \frac{4s^2 + 3}{2s(s^2 + 8s + 25)}$$

Examination of the denominator tells us that there are poles at $s = 0$ and $s = -4 \pm j3$ so it is appropriate to complete the square in the quadratic term of the denominator

$$Y(s) = \frac{4s^2 + 3}{2s[(s+4)^2 + 3^2]}$$

These pole locations indicate that the response $y(t)$ should be composed of a step function and damped sines and cosines, so we make the partial fraction expansion of the form, having extracted the two from the denominator.

$$Y(s) = \frac{1}{2}\left[\frac{A}{s} + \frac{B(s+4)}{(s+4)^2 + 3^2} + \frac{C3}{(s+4)^2 + 3^2}\right]$$

We can get A by the Heaviside method or

$$A = \frac{3}{25}$$

If we now find a common denominator and equate the numerators, the result is

$$4s^2 + 3 = A(s^2 + 8s + 25) + Bs(s+4) + C3s$$

Equating the coefficients of like powers of s gives

$$s^2: \quad 4 = A + B$$

$$s^1: \quad 0 = 8A + 4B + 3C$$

$$B = 4 - A = 4 - \frac{3}{25} = \frac{97}{25}$$

and

$$C = \frac{1}{3}\left[-\frac{8(3)}{25} - \frac{4(97)}{25}\right] = \frac{-412}{75}$$

The inverse transform is

$$y(t) = \frac{1}{2}\left[\frac{3}{25}\,u(t) + \frac{97}{25}\,e^{-4t}\cos 3t - \frac{412}{75}\,e^{-4t}\sin 3t\right]u(t)$$

And as a check we can easily evaluate the initial condition

$$y(0) = \frac{1}{2}\left[\frac{3}{25} + \frac{97}{25}\right] = \frac{4}{2} = 2$$

which indicates that we may not have made an error.

Example 2.16

Let us now consider an example of simultaneous differential equations for which we want the solution $x(t)$.

$$\frac{dx}{dt} + 2x - y = 2t$$

$$\frac{dy}{dt} + y - x = 0$$

$$x(0) = y(0) = 0$$

If we transform the differential equations,

$$(s+2)X(s) - Y(s) = \frac{2}{s^2}$$

$$-X(s) + (s+1)Y(s) = 0$$

Solving the equations with Cramer's rule, the result is

$$X(s) = \frac{\begin{vmatrix} \frac{2}{s^2} & -1 \\ 0 & s+1 \end{vmatrix}}{\begin{vmatrix} s+2 & -1 \\ -1 & s+2 \end{vmatrix}} = \frac{2(s+1)}{s^2(s^2+3s+1)}$$

The roots of the denominator quadratic are

$$s_{1,2} = -\frac{3}{2} \pm \sqrt{\frac{9}{4} - 1} = -1.5 \pm 1.118$$

$$= -2.618, -0.382$$

The solution with a factored denominator is

$$X(s) = \frac{2(s+1)}{s^2(s+2.618)(s+0.382)}$$

The indicated form for the partial fraction expansion is

$$X(s) = 2\left[\frac{A}{s} + \frac{B}{s^2} + \frac{C}{s+2.618} + \frac{D}{s+0.382}\right]$$

We can obtain B, C, and D by the Heaviside method or

$$B = \frac{1}{1} = 1$$

and

$$C = \frac{(-2.618+1)}{(-2.618)^2(-2.618+0.382)} = \frac{1.618}{-15.325} = 0.1056$$

and

$$D = \frac{(-0.382+1)}{(-0.382)^2(-0.382+2.618)} = \frac{0.618}{0.3263} = 1.8940$$

Finding a common denominator and equating numerators gives

$$s + 1 = As(s^2 + 3s + 1) + B(s^2 + 3s + 1)$$

$$+ Cs^2(s + 0.382) + Ds^2(s + 2.618)$$

$$s^3: \quad 0 = A + C + D$$

Then

$$A = -C - D = -0.1056 - 1.8940 = -2$$

The inverse transform is then

$$x(t) = 2[-2u(t) - t + 0.1056e^{-2.618t} + 1.894e^{-0.382t}]u(t)$$

and if we check the initial condition the result is

$$x(0) = 2[-2 + 0.1056 + 1.844] = 0$$

In a similar fashion we could evaluate the $y(t)$ solution. To this point, in this section, we have not needed to employ the delay theorem to get the solution to a differential equation problem. Let us now pose a problem where the delay theorem will be needed twice. This is typical of problems arising in mechanical shock and electrical pulse circuits.

Example 2.17

Let us consider the following differential equation:

$$\frac{dy}{dt} + y = u(t) - u(t-2)$$

$$y(0) = 0$$

where $u(t)$ and $u(t-2)$ are unit step functions at $t = 0$ and $t = 2$ seconds, respectively. The forcing function on the right side of the equation is illustrated in Fig. 2.5. If we take the transform of the differential equation employing the delay theorem the result is

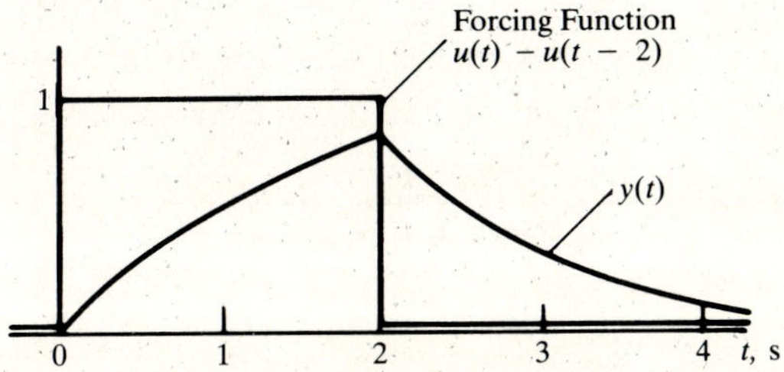

FIGURE 2.5 Pulse forcing function and differential equation solution.

$$(s+1)Y(s) = \frac{1}{s} - \frac{e^{-2s}}{s}$$

and solving for the unknown results in

$$Y(s) = \frac{1}{s(s+1)} - \frac{e^{-2s}}{s(s+1)}$$

The first term may be inverted in a straightforward manner but the second term is of the form $F(s)e^{-as}$, which is covered in entry six in Table A-1 of Appendix A. The solution to the differential equation is

$$y(t) = [1 - e^{-t}]u(t) - [1 - e^{-(t-2)}]u(t-2)$$

This solution is also plotted in Fig. 2.5.

2.10 AN EASILY PROGRAMMED PARTIAL FRACTION EXPANSION ALGORITHM

Let us consider the inversion of a transform function which has a denominator polynomial of degree r. We shall assume that the denominator polynomial roots have been found and that the only repeated roots are at the origin ($s = 0$). This is not a terribly restrictive assumption since in real physical problems there are seldom real repeated roots elsewhere. The function we wish to invert is of the form

$$F(s) = \frac{h_{r-1}s^{r-1} + h_{r-2}s^{r-2} + \cdots + h_0}{s^{n_0} \prod_{i=1}^{n_1} (s+\beta_i) \prod_{j=1}^{n_2} (s^2 + 2\alpha_j s + \omega_j^2)} \tag{2.10.1}$$

In this expression, the β_i are real and the complex roots are left in quadratic form. There are n_0 origin roots, n_1 distinct real roots, and $2n_2$ distinct complex roots. The parameter r is given by $r = n_0 + n_1 + 2n_2$. The repeated origin roots contribute polynomial terms in t in the inverse function, while the real poles contribute exponential functions and the quadratic terms contribute damped sines and cosines. These facts dictate that the partial fraction expansion associated with Eq. (2.10.1) should be

$$F(s) = \sum_{i=1}^{n_0} \frac{A_i}{s^i} + \sum_{i=1}^{n_1} \frac{B_i}{s+\beta_i} + \sum_{j=1}^{n_2} \frac{C_j\sqrt{\omega_j^2 - \alpha_j^2}}{s^2 + 2\alpha_j s + \omega_j^2} + \frac{D_j(s+\alpha_j)}{s^2 + 2\alpha_j s + \omega_j^2} \tag{2.10.2}$$

The problem here is that of finding the A_i, B_i, C_j, and D_j. If we equate the right sides of Eqs. (2.10.1) and (2.10.2) for r particular real values of s, say $s = p_1, p_2, \ldots, p_r$, a set of real linear equations in the r unknown coefficients results. These linear equations can be solved by a programmable hand calculator or digital computer for the unknown coefficients. When the coefficients are known, the inverse transform $f(t)$ may be simply written down as

$$f(t) = \sum_{i=1}^{n_0} \frac{A_i t^{i-1}}{(i-1)!} + \sum_{i=1}^{n_1} B_i e^{-\beta_i t} + \sum_{j=1}^{n_2} C_j e^{-\alpha_j t} \sin \Omega_j t + D_j e^{-\alpha_j t} \cos \Omega_j t \tag{2.10.3}$$

where the frequency parameter Ω_j is given by

$$\Omega_j^2 = \omega_j^2 - \alpha_j^2 \tag{2.10.4}$$

It should be noted that, in choosing the values p_i to evaluate the right sides of Eqs. (2.10.1) and (2.10.2), there are unacceptable choices for p_i. In particular, none of the p_i should be chosen coincident with the poles of $F(s)$. In other words, the p_i should not be selected at $s = 0$ or at $s = -\beta_i$. The authors have also found that the relative size of the p_i should be selected such that they are about the same order of magnitude as the distance of the poles of $F(s)$ from the origin of the s plane. A microcomputer program which performs this process is given in Appendix D. Inputs to this program are n_0, n_1, n_2; the β_i, α_j, and ω_j; and the numerator coefficients h_i. The user is also asked to input the seed numbers $p_1, p_2, \ldots, p_r$.

2.11 SUMMARY

In this and the previous chapter we have developed some of the analytical tools necessary for the analysis of engineering systems. This chapter has been devoted to exploration of the Laplace transformation as a tool for the solution of linear ordinary integro-differential equations with constant coefficients. It has been shown that this method has considerable merit because initial conditions are incorporated into the problem at an early stage and both complementary and particular solutions to the differential equation are generated simultaneously. The unit step function has proved helpful for handling forcing functions which are discontinuous or piecewise continuous.

In subsequent chapters we shall use all of these tools in the solution of physical problems. The value of these tools is their applicability to a wide class of physical problems.

REFERENCES

1. Savant, C. J. *Fundamentals of the Laplace Transform*. McGraw-Hill, New York, 1962.
2. Cheng, D. K. *Analysis of Linear Systems*. Addison-Wesley, Reading, MA, 1959.
3. Fredrick, D. K., and Carlson, A. B. *Linear Systems in Communication and Control*. John Wiley & Sons, New York, 1971.
4. Ogata, K. *System Dynamics*. Prentice-Hall, Englewood Cliffs, NJ, 1978.
5. Sturm, R. D., and Ward, J. R. *Laplace Transform Solution of Differential Equations*. Prentice-Hall, Englewood Cliffs, NJ, 1968.

PROBLEMS

2-1. Using the definition of the Laplace transform, show that it is a linear transformation. To be a linear transformation it must satisfy relations (1) and (2) below.

$$1.\ \mathscr{L}[Af(t)] = AF(s)$$

$$2.\ \mathscr{L}[Af(t) + Bg(t)] = AF(s) + BG(s)$$

where A and B are scalar constants.

2-2. Show that the Laplace transform of a cosine function $\cos \omega t$ is

$$\mathscr{L}[\cos \omega t\, u(t)] = \frac{s}{s^2 + \omega^2}$$

2-3. Given the identity for the hyperbolic sine function

$$\sinh \beta t = \frac{1}{2}[e^{\beta t} - e^{-\beta t}]$$

find the Laplace transform of $\sinh \beta t$.

2-4. Find the Laplace transform of the following functions, relying on the definition of the transform.
(a) $t^2 u(t)$ (Hint: integrate by parts)
(b) $\cosh \beta t$ (Hint: resolve the hyperbolic cosine into exponential form)

2-5. Using the Laplace transform tables, find the Laplace transforms of the following functions for $t > 0$.

(a) $f(t) = 3 \sin 5t + 2 \cos 3t + 6t^3$

(b) $f(t) = 3\, e^{-2t} \sin 5t + 2\, e^{-3t} + 5t^2$

2-6. Find the Laplace transform of the following functions.

(a) $f(t) = \sin^2 \omega t$ **(b)** $f(t) = t \sin \omega t$ **(c)** $f(t) = te^{-at}$

2-7. Find the Laplace transform of the following functions by employing the appropriate theorems.

(a) $f(t) = t^2 e^{-mt} \quad t \geq 0$

(b) $f(t) = (t + 2)^2 u(t - 1)$

where $u(t - 1)$ is a delayed unit step function.

(c) $f(t) = e^{-5t} \cosh(3t)$

2-8. For the functions shown, write equations for them using the delayed unit step function and find the associated Laplace transform of each signal using the delay theorem where necessary.

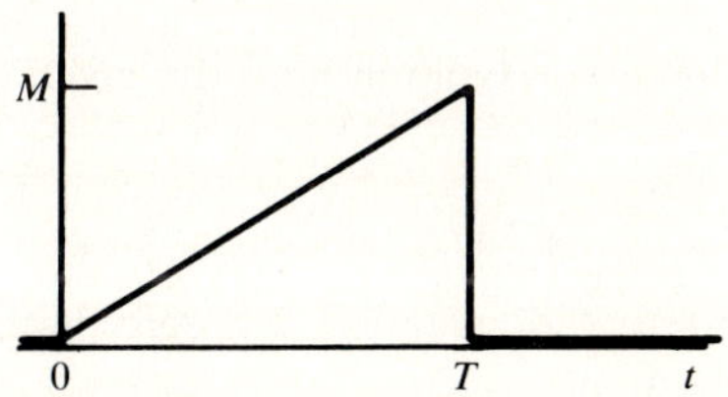

(a)

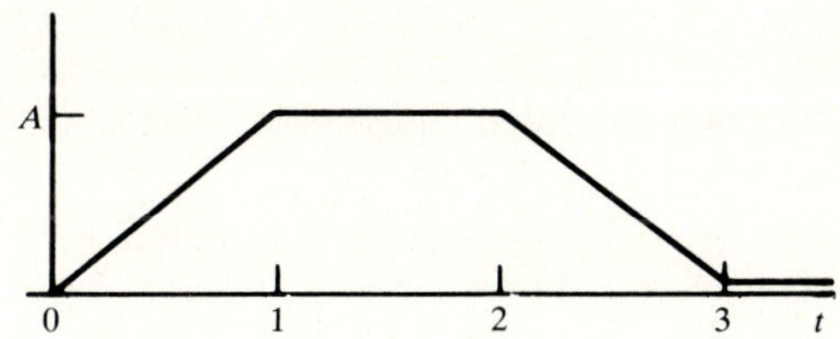

(b)

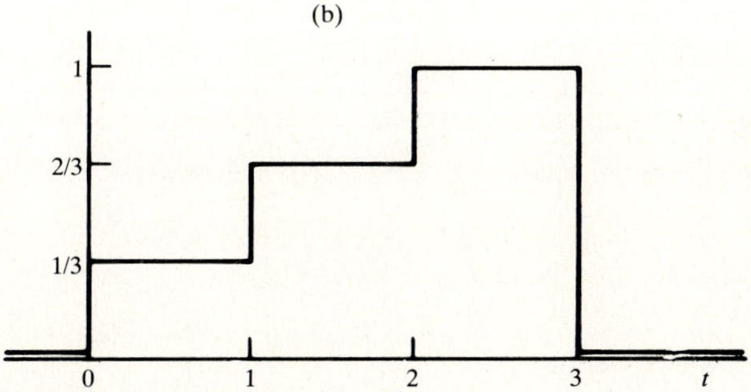

(c)

2-9. Show that the Laplace transform of a delayed step function $u(t - a)$ is

$$\mathcal{L}[u(t - a)] = \frac{e^{-as}}{s}$$

2-10. Find, by the method of partial fraction expansion, the inverse transform of the following s-domain functions.

(a) $\dfrac{s}{(s+1)(s+2)(s+3)}$ **(b)** $\dfrac{s+1}{s^2(s+2)(s+3)}$

(c) $\dfrac{s+2}{(s^2+3^2)(s+1)}$ **(d)** $\dfrac{s+2}{(s+1)^2(s^2+4s+13)}$

(e) $\dfrac{s+2}{s^2+6s+10}$ **(f)** $\dfrac{s+4}{s(s^2+11s+30)}$

(g) $\dfrac{s+2}{(s+1)^2(s^2+2^2)}$ **(h)** $\dfrac{s+3}{(s+2)(s^2+8s+25)}$

(i) $\dfrac{6}{s^3(s+3)}$ **(j)** $\dfrac{e^{-2s}}{s(s+2)(s+3)}$

2-11. Solve the following differential equation with the given initial condition:

$$\dot{y} + 4y = \begin{cases} 0 & t<0 \\ \sin t & t>0 \end{cases}, \qquad y(0) = 3$$

2-12. Solve the following differential equation for zero initial conditions by the Laplace transform method:

$$\ddot{x} + 2\dot{x} + 2x = 5u(t)$$

2-13. Solve the following differential equation for $x(t)$.

$$\frac{d^2x}{dt^2} + 4\frac{dx}{dt} + 8x = e^{-2t}$$

$$x(0) = \frac{dx(0)}{dt} = 0$$

2-14. Solve the following integro-differential equation system for $y(t)$:

$$2\frac{dx}{dt} + \int_0^t x\,dt + 3y = u(t)$$

$$\frac{dy}{dt} + 3x = 0$$

$$y(0) = x(0) = 0$$

2-15. Solve the following set of coupled ordinary differential equations for $x(0) = 1$ and $y(0) = 0$:

$$\dot{x} + 2(x - y) = 0$$

$$\dot{y} + 4y - 3x = tu(t)$$

2-16. Solve the following differential equations with the initial conditions given.

(a) $\ddot{x} + 4\dot{x} = u(t), \quad x(0) = \dot{x}(0) = 0$

(b) $\ddot{x} + 6\dot{x} + 9x = 4u(t), \quad x(0) = \dot{x}(0) = 0$

2-17. Find the Laplace transform of the periodic functions illustrated below.

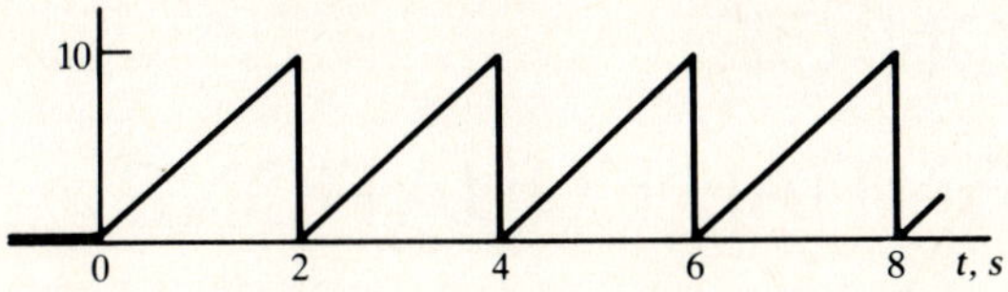

(a)

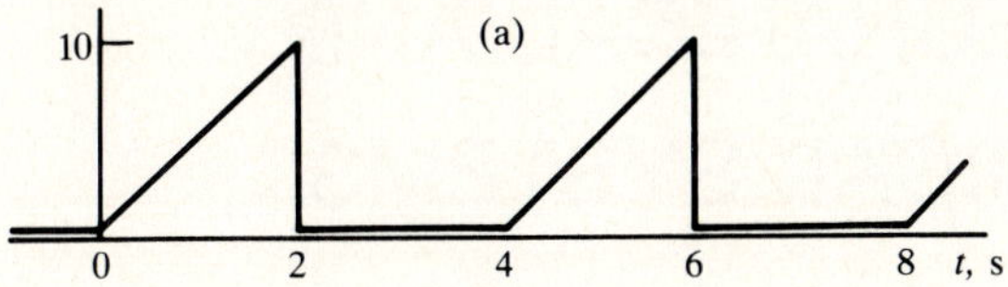

(b)

3

FORMULATION AND SOLUTION OF DIFFERENTIAL EQUATIONS FOR MECHANICAL AND ELECTRICAL SYSTEMS

3.1 INTRODUCTION

In Chapter 2 we explored the Laplace transform method for the solution of linear ordinary differential equations with constant coefficients. Differential equations of that type can be used to model a large number of real physical systems. In this chapter we shall examine how to formulate these equations correctly, for if so much as one sign is incorrect so will be the predicted system response.

There are many other methodologies for formulation of system equations (see, for example Ref. 1). We shall rely on rigorous statements of a few physical laws and carefully defined sign conventions for our development of these formulations. It is the authors' strong opinion that a firm physical understanding of these problems is far preferable to any nonphysical arguments which might yield the correct system equations.

3.2 BASIC IDEAS OF SYSTEM MODELING

A question which commonly arises among those who have not done system modeling is "What is a system model?" We shall define a system model as a set of equations which adequately predicts the behavior of a system to a set of known inputs (driving functions and initial conditions). The value of any model lies in its capability to predict real-system behavior accurately for different sets of inputs, so that the real physical system itself need not be subjected to these inputs. With an acceptable system model we need not have a physical system prototype to test responses to various inputs and disturbances. The idea of having a mathematical model can often save costly construction and modification of a prototype because the model can be easily and inexpensively modified if, from its predictions, it is deemed inadequate. New predictions will then tell us if the modified system is adequate for our purposes.

The process of establishing a mathematical system model is iterative in that the first model formulated is usually very simple and often does not adequately describe the system performance to a known set of inputs. It will be replaced by a revision which, in turn, will be replaced by yet another revision, etc., as better information is added. The process of modeling is given in block diagram form in Fig. 3.1.

Initially, we formulate in our minds a conceptual model of the system based on our experiences in the physical world and initial investigations of the system behavior. Based on physical laws and the mathematical statements of these laws, along with assumptions about unknown portions of the system, we formulate the mathematical model, which is a group of equations which we hypothesize describes the system behavior. When solved, either analytically or by digital or analog computer, model responses are generated for a set of inputs. To do the complete modeling task, these responses should be compared to actual system responses under identical conditions. If these two sets of responses are not sufficiently alike, we should revise the

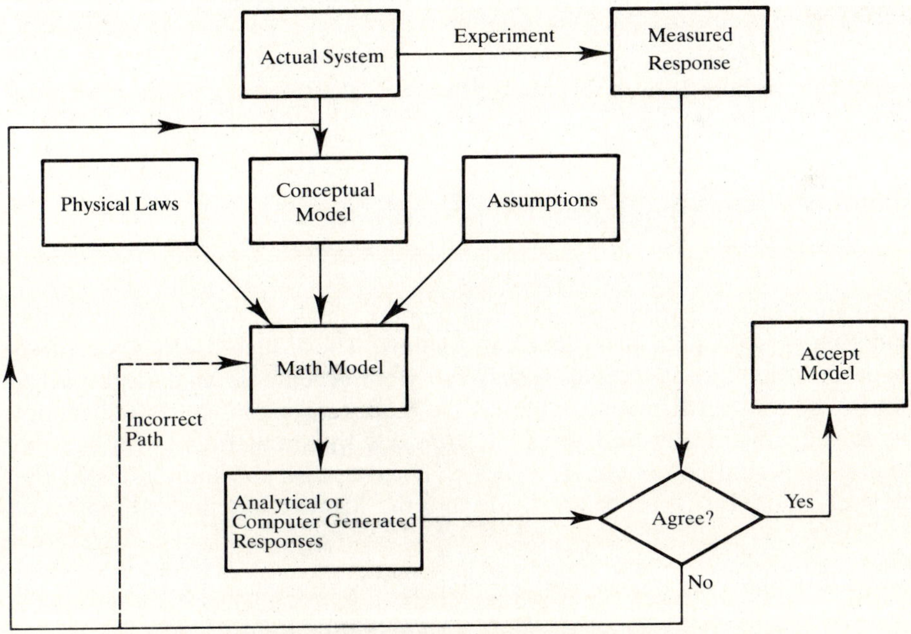

FIGURE 3.1 Block diagram of the system modeling process.

model. The revision should occur beginning with the conceptual model, not just at the level of the math model.

It is not uncommon for system modelers to revise the math model without attention to the physical implications of such modifications. This step can often lead to violations of physical principles. When model response characteristics do not agree with real-system response, the correct iterative path is to return first to the conceptual model and again work through the modeling process, as illustrated in Fig. 3.1, ensuring that physical laws are not violated by the model.

3.2.1 Linear Systems

A system is linear if it is described by a system of linear integro-differential equations. Consider a system where a stimulus $y_1(t)$ produces a response $x_1(t)$ and stimulus $y_2(t)$ produces a response $x_2(t)$. The system is said to be linear if the following two conditions hold:

1. A stimulus $Ay_1(t)$ produces a response $Ax_1(t)$ and $Ay_2(t)$ produces $Ax_2(t)$.
2. A stimulus $Ay_1(t) + By_2(t)$ produces response $Ax_1(t) + Bx_2(t)$.

These two conditions allow us to state an important principle associated with linear systems, i.e., the superposition principle. A simple statement of the principle of superposition is that if a system has two simultaneous stimuli, the response to each may be calculated separately and the response to the sum of both is then the sum of the two calculated single responses. Of course, the principle may be extended to any number of stimuli.

3.3 REAL AND IDEAL MODELS IN MECHANICAL SYSTEMS

We commonly deal with mathematical models which are idealizations of the real elements present in mechanical systems. These idealizations are often made in order to give models which will yield acceptable analytical predictions of system performance, whereas the more realistic models often cannot be solved analytically and must be attacked by simulation on either the analog or digital computer. In order that we realize the limitations of the idealized model it is important to discuss the idealized elements employed.

3.3.1 Springs

If we consider a typical force deflection curve for a coil spring, as shown in Fig. 3.2(a) when a positive force is applied and the spring is in tension

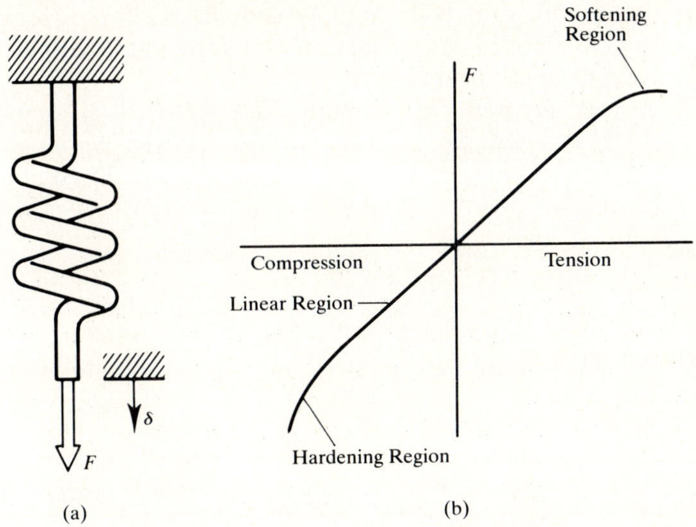

FIGURE 3.2 Coil spring which softens in one direction and hardens in the other direction.

(stretched from its equilibrium position), the force-displacement relationship will be reasonably linear until the material of the spring is strained past the proportional limit of the material (plastic deformation), at which point the deflection per unit force increases, as illustrated in Fig. 3.2(b) or the spring is said to soften. If, on the other hand, the spring is put in compression, it will behave linearly until the spring coils "bottom out" against each other and the spring is said to harden.

A second spring type is the leaf spring illustrated in Fig. 3.3(a), which exhibits the force deflection curve shown in Fig. 3.3(b). The stiffness is due to the bending of the beam and will thus soften under load with the portion

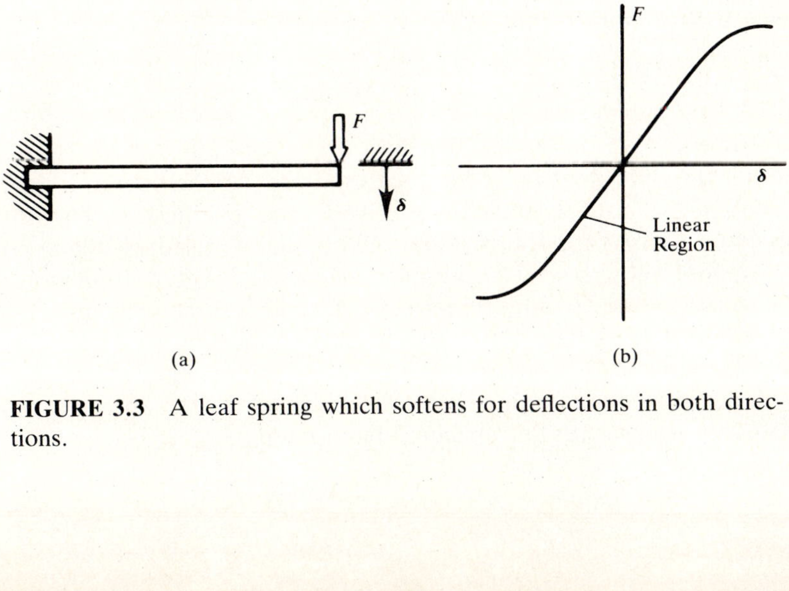

FIGURE 3.3 A leaf spring which softens for deflections in both directions.

of the beam in tension (stretched) being forced into the plastic region. Note that this deflection curve is symmetrical in that softening occurs with forces (and deflections) in opposite directions.

We have just qualitatively examined the stiffness character of two springs commonly encountered in mechanical assemblies. We found them to have a common linear character for small deflections and different nonlinear character for large deflections. The linear character can be described by

$$F = k\delta \tag{3.3.1}$$

where the parameter k is referred to as the *spring constant*, *stiffness*, or *spring modulus*. These terms will be used interchangeably in the remainder of the text.

3.3.2 Energy Dissipation Models in Mechanical Systems

When modeling mechanical systems, one is often confronted with the task of modeling the energy dissipation in the system. Before the modeling task can be addressed, it is advisable first to investigate the phenomena by which energy is actually dissipated. Dissipation is important because it limits what might otherwise be infinitely large responses. We shall now discuss commonly encountered energy dissipation models.

Aerodynamic Drag Let us consider the steady motion of a body through a fluid at some velocity u which is high enough that the fluid flow about the body is turbulent. There is considerable experimental evidence—in the form of wind tunnel tests under the conditions assumed above—that the drag force is proportional to the square of the velocity u of the body and opposite in direction to that of the velocity u or

$$F = -\frac{1}{2}\rho A C_D u|u| \tag{3.3.2}$$

where ρ is the fluid density, A the projected area of the body normal to the flow, and C_D the experimentally derived drag coefficient. The form of expression (3.3.2) with the absolute value sign is used so the force will always oppose velocity u. The drag coefficient C_D is often a function of the fluid viscosity, velocity, and object size. A typical force-velocity relation is illustrated in Fig. 3.4(a). Although it is not rigorous to extrapolate the steady flow concepts just discussed to transient or oscillatory flow cases, it is often done (with reasonable success) for lack of better knowledge of the mechanisms of transient aerodynamic phenomena.

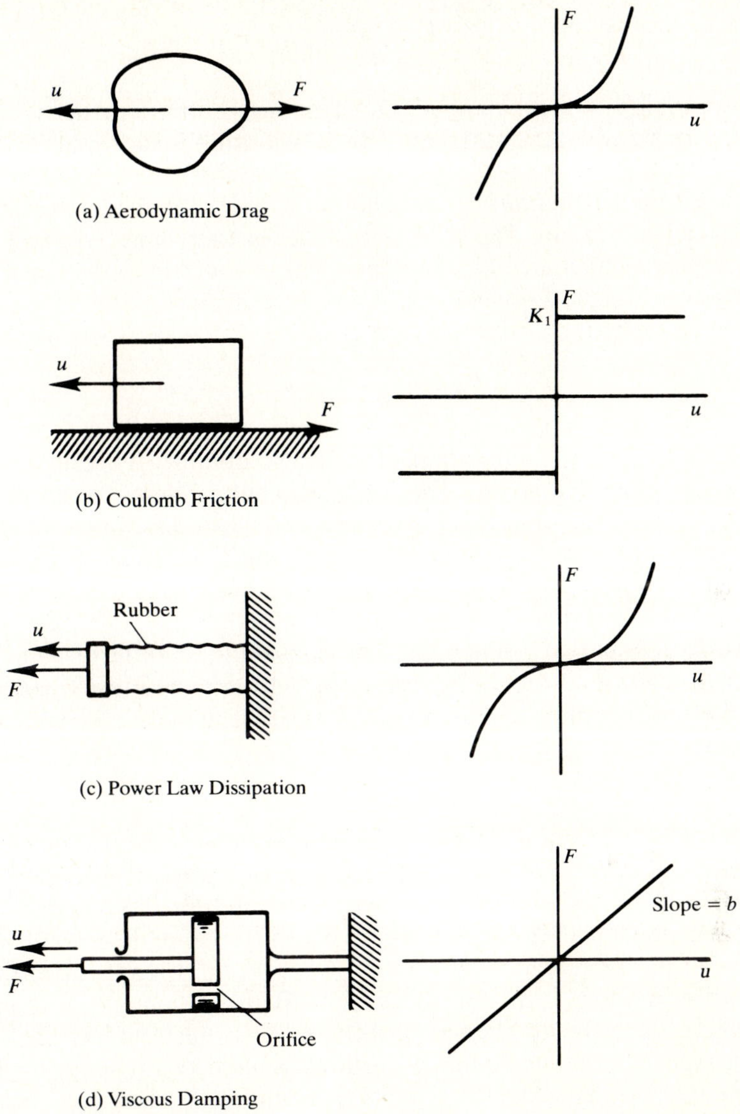

FIGURE 3.4 Models for dissipation in mechanical systems.

Coulomb or Sliding Friction When two dry or slightly lubricated surfaces slide on one another, as is usually the case with shafts supported by bushings, the frictional force is often dependent only on the normal force between the surfaces (nearly), is *independent* of velocity, and is in a direction opposite to the relative velocity, u, between the surfaces. It is described by

$$F = -K_1 \frac{u}{|u|} \tag{3.3.3}$$

The constant K_1 is a function of the dynamic friction coefficient and the normal force between the surfaces. This relationship is shown in Fig. 3.4(b).

Power-Law Dissipation It is common to use rubber to solve problems of unwanted vibration. The reason for this is that most types of rubber generate forces which are velocity-dependent in addition to the usual elastic- or spring-type forces. For such a material in a given physical configuration, the velocity-*dependent* forces may be given by

$$F = -K_2 u|u^{n-1}| \tag{3.3.4}$$

where the constant K_2 and exponent n can be obtained experimentally for the configuration of interest. This power-law dependence is illustrated in Fig. 3.4(c). This type of velocity dependence is also descriptive of the drag forces created in non-Newtonian fluids for which the shear forces are not proportional to the spatial rate of change of velocity.

Viscous Dissipation When the fluid flow about an object is laminar, we find experimentally that the drag forces present depend linearly on velocity and hence the dissipative forces are of the form

$$F = bu \tag{3.3.5}$$

(such flow is called Stokes' flow), which is an extremely simple linear model. The coefficient b is commonly called the viscous damping coefficient or the damping constant. All forces previously discussed have had some type of nonlinear dependence on the velocity u. This viscous model often is used to characterize dashpots (shock absorbers) wherein fluid is forced through an orifice when motion is in one direction and the other way through that orifice when the velocity changes direction, as is illustrated in Fig. 3.4(d).

This type of dynamic damping is often assumed as a first approximation when the damping mechanism is unknown, and also when analytical solutions to the system's differential equations are desired. Other types of damping are often approximated by viscous damping for ease of mathematical solution. The other types of damping discussed earlier in this section do not lend themselves to analytical differential equation solutions and hence either analog or digital computer solutions must be sought.

3.4 RECTILINEAR MECHANICAL SYSTEMS

We shall now consider mechanical systems in which the masses translate in one direction only, with multiple translations and rotations to be considered

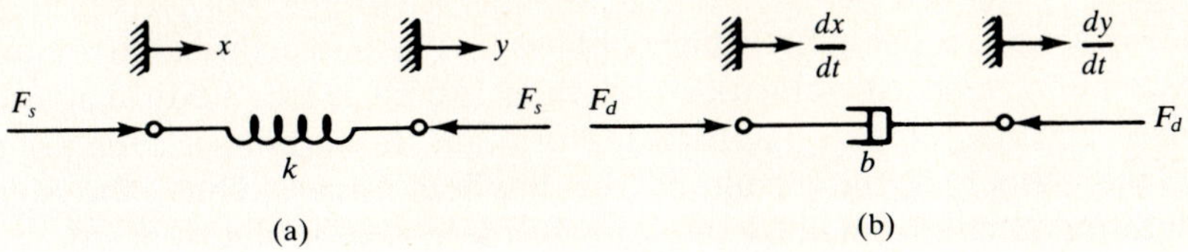

FIGURE 3.5 Linear elements in translational mechanical systems. (a) Linear spring and (b) viscous dashpot.

in a later section of this chapter. The masses involved will be connected by linear springs and viscous dashpots (often called dampers), elements which are illustrated in Fig. 3.5(a) and (b).

Recall that the spring force F_s is proportional to the relative displacement between the ends of the spring. This relative displacement in terms of the notation of Fig. 3.5(a) is $(x - y)$ and thus the spring force is

$$F_s = k(x - y) \tag{3.4.1}$$

From this relation it is clear that, when the displacements x and y are equal, the force F_s will be zero. In a similar fashion, the force developed in the dashpot or damper is proportional to the difference in the velocities of the end points of the dashpot (the relative velocity) or for the sign convention established in Fig. 3.5(b):

$$F_d = b\left(\frac{dx}{dt} - \frac{dy}{dt}\right) \tag{3.4.2}$$

The appropriate tool for formulation of system equations of motion is Newton's second law, which states that rate of change of linear momentum of a mass is equal to the sum of the external forces. Written analytically,

$$\frac{d}{dt}(mv) = \sum F_{\text{ext}} \tag{3.4.3}$$

Expanding the derivative of the product yields

$$m\frac{dv}{dt} + v\frac{dm}{dt} = \sum F_{\text{ext}} \tag{3.4.4}$$

Although the second term is important in problems where the mass is changing with time (e.g., rockets and jet airplanes), we shall treat only systems where the mass remains constant, so that the second term in Eq. (3.4.4) is zero or

$$m\frac{dv}{dt} = m\frac{d^2x}{dt^2} = \sum F_{\text{ext}} \tag{3.4.5}$$

where the velocity v is the rate of change of displacement x. One must first establish a sign convention which defines the positive direction of forces, displacements, and/or velocities and then sketch a free-body diagram for the mass showing all external forces. Note that the term $m(d^2x/dt^2)$ is *not* a force but rather is the result of the applied forces. Thus, the sign on acceleration is determined by the signs of the terms on the right side, which are the externally applied forces. This technique is best illustrated by the thoroughly documented examples which follow.

Example 3.1

Find the differential equation of motion for mass m for the system illustrated in Fig. 3.6(a) if the forcing function $f(t)$ is known. Assume the rollers to be frictionless. If we assume that $x > 0$ and $dx/dt > 0$ (although we could have made the other assumption with no change in the result), the free-body diagram is as shown in Fig. 3.6(b). In the original system diagram, Fig. 3.6(a), the positive direction for x is indicated, so application of Eq. (3.4.5) yields

$$m \frac{d^2x}{dt^2} = -kx - b\frac{dx}{dt} + f(t) \tag{3.4.6}$$

This equation could be solved for $x(t)$ by the Laplace transform method if $f(t)$ and the initial values $x(0)$ and $dx(0)/dt$ are known.

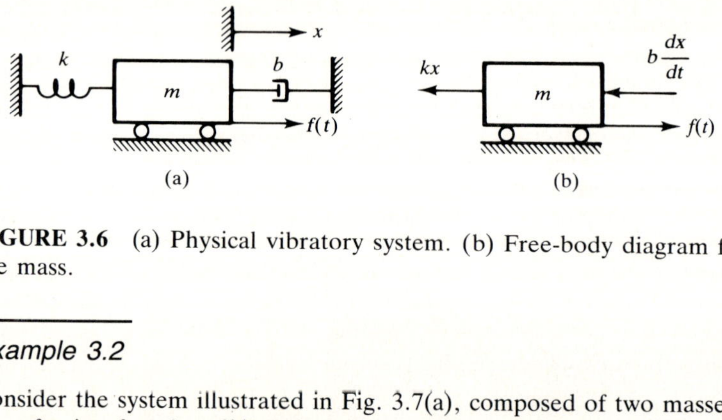

FIGURE 3.6 (a) Physical vibratory system. (b) Free-body diagram for the mass.

Example 3.2

Consider the system illustrated in Fig. 3.7(a), composed of two masses driven by a forcing function $f(t)$ acting on m_2. If we assume $x_1 > x_2$ and $\dot{x}_1 > \dot{x}_2$† (these are arbitrary independent assumptions), the free-body diagrams are as shown in Fig. 3.7(b). Noting that positive directions are defined to the right, the equations of motion for the masses are, respectively:

For mass m_1

$$m_1\ddot{x}_1 = -b_1\dot{x}_1 - b_2(\dot{x}_1 - \dot{x}_2) - k_1(x_1 - x_2)$$

† In the remainder of this text overdots will be employed to denote differentiation with respect to time.

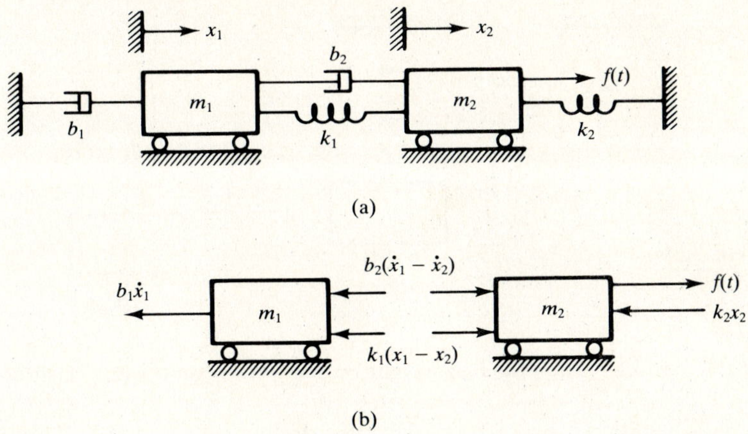

FIGURE 3.7 (a) Two-mass system. (b) Free-body diagrams for the masses.

For mass m_2

$$m_1\ddot{x}_2 = f(t) + k_1(x_1 - x_2) + b_2(\dot{x}_1 - \dot{x}_2) - k_2x_2$$

These equations, along with initial conditions $x_1(0)$, $x_2(0)$, $\dot{x}_1(0)$, and $\dot{x}_2(0)$ and a known forcing function $f(t)$, are sufficient to describe the motion of the two masses. Note that the two equations are *not* independent and must be solved simultaneously for $x_1(t)$ and $x_2(t)$. Also, note that any other combination of assumptions about the relative sizes of x_1 and x_2 and $\dot{x}_1$ and $\dot{x}_2$ will yield different free-body diagrams but that any assumptions will yield the same equations of motion.

In this discussion we have assumed that the elastic elements (springs) are massless when, in fact, they are not. One way to handle spring mass is to lump one-half of the spring mass at each of the points of attachment of the spring, i.e., on each mass attached to the spring. For spring two, this implies that one-half the mass is added to m_2 but, since the other end is not moving, there is no mass effect there.

3.4.1 The Effect of Gravity

So far in this chapter we have considered only rectilinear systems in which the motion is horizontal, while the gravity forces (weights) act vertically and hence do not influence the motion. We shall, in this section, consider a single mass hanging on a spring, as illustrated in Fig. 3.8.

When the weight is attached to the end of the spring, the spring deforms by a distance δ. The deflection of the spring from the unstretched

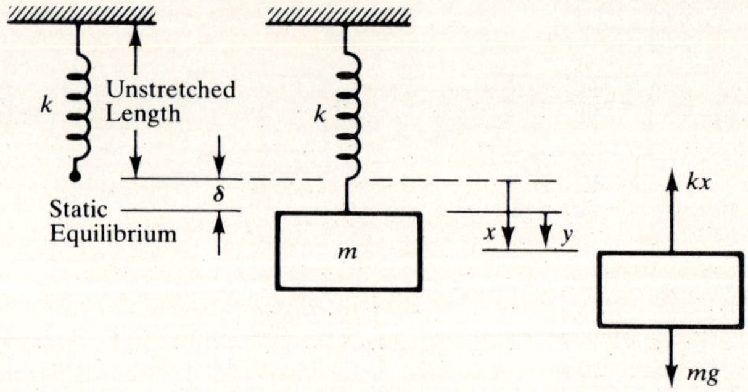

FIGURE 3.8 The effect of weight on the static equilibrium position.

spring length will be termed x, while the deflection from the static equilibrium position will be termed y and the two are related by

$$x = y + \delta \tag{3.4.7}$$

The static equilibrium force balance gives

$$k\delta = mg \tag{3.4.8}$$

The free-body diagram for the mass is shown in Fig. 3.8 and the equation of motion is

$$m\ddot{x} = -kx + mg \tag{3.4.9}$$

If we differentiate relation (3.4.7) twice, then

$$\ddot{x} = \ddot{y} \tag{3.4.10}$$

We may substitute Eqs. (3.4.7) and (3.4.8) into Eq. (3.4.9) to give

$$m\ddot{y} = -k(y + \delta) + mg \tag{3.4.11}$$

Embedded in this relation is the static equilibrium relation (Eq. (3.4.8)) and hence the static force terms mg and $k\delta$ drop out to give

$$m\ddot{y} = -ky \tag{3.4.12}$$

This is the equation which would result if the coordinate were originally defined from the static equilibrium position, as y was, and if the only force considered on the free-body diagram were ky.

In general, we may say that gravity forces may be omitted from free-body diagrams if only spring forces proportional to the deflection from static equilibrium are considered. The only exception to this rule is the case where restoring torques due to gravity are also functions of the coordinate of rotation, as in the problem of a pendulum.

3.5 TORSIONAL MECHANICAL SYSTEMS

In the last section we considered systems composed of masses which were free to translate in one direction only. Here we shall treat systems which are free to rotate about one axis only. Problems of combined translation and rotation will be treated later in this chapter.

For bodies which are free to rotate only, the appropriate physical law for analysis is Newton's law for rotation or, for the case of constant mass moment of inertia J,

$$J \frac{d^2\theta}{dt^2} = \sum T \tag{3.5.1}$$

where J is the mass moment of inertia about the axis about which the torques are being summed. It may be either a fixed pivot or an instantaneous center. The mass moment of inertia is defined by

$$J = \int_M r^2 \, dM \tag{3.5.2}$$

Commonly used mass moments of inertia are given in Table 3-1. Other elements may be present in the system and these elements are illustrated in Figs. 3.9(a) and (b). The torsional spring of Fig. 3.9(a) is assumed to be elastic such that the torque developed, T_s, is proportional to the net twist of the shaft:

$$T_s = k(\theta_1 - \theta_2) \tag{3.5.3}$$

Note that, if θ_1 and θ_2 are equal, the torque developed is zero and that if one rotation is zero, the torque is proportional to the remaining nonzero rotation.

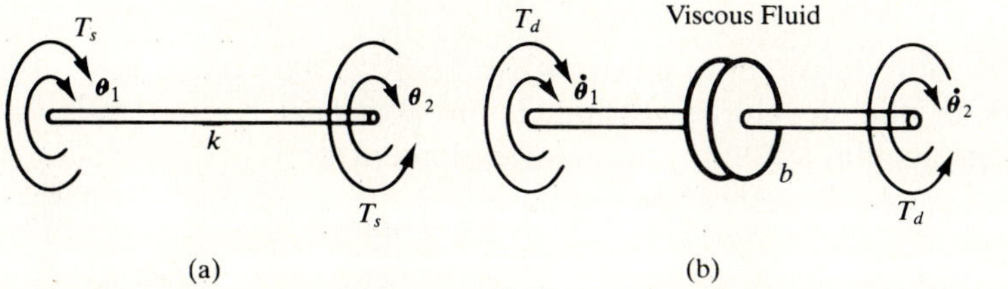

FIGURE 3.9 (a) Torsional spring or torsion bar. (b) Torsional damper.

TABLE 3-1 Mass Moments of Inertia for Common Engineering Shapes

Thin Disk

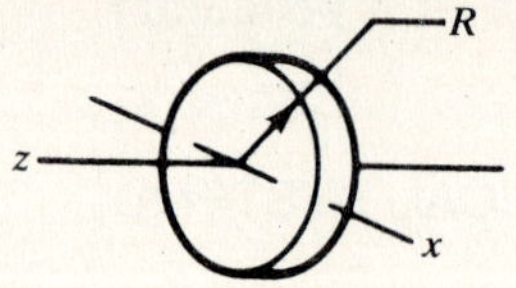

$J_z = \frac{1}{2}MR^2$

$J_x = \frac{1}{4}MR^2$

Thin Ring

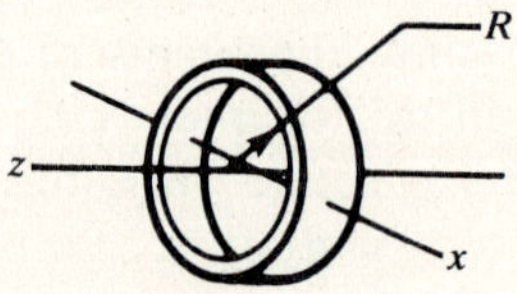

$J_z = MR^2$

Solid Cylinder

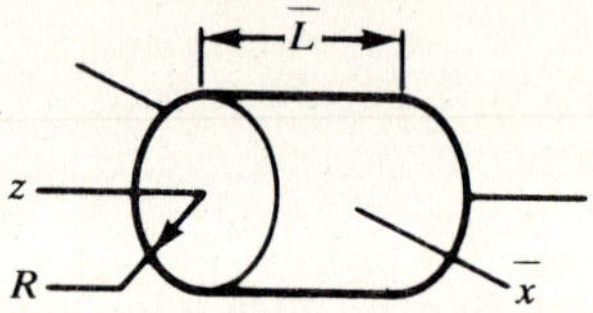

$J_z = \frac{1}{2}MR^2$

$J_x = \frac{1}{12}M(3R^2 + L^2)$

Solid Sphere

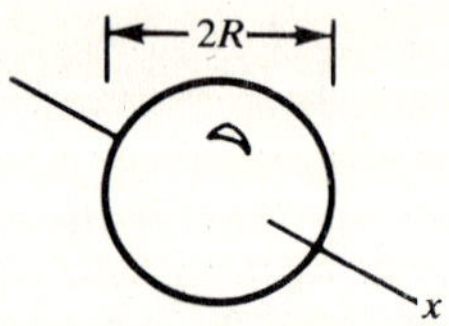

$J_x = \frac{2}{5}MR^2$

Slender Rod

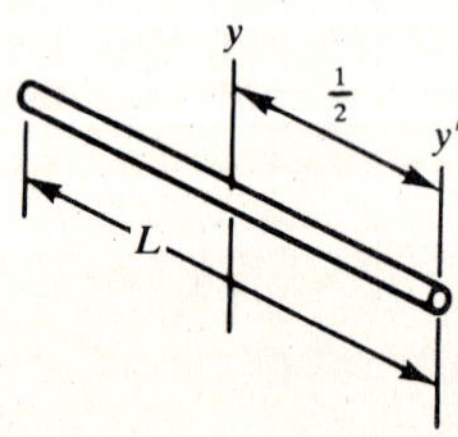

$J_y = \frac{1}{12}ML^2$

$J_{y'} = \frac{1}{3}ML^2$

The other element to be considered here is the viscous torsional damper for which the developed torque is proportional to the relative rate of rotation (angular velocity) across the element or

$$T_d = b\left(\frac{d\theta_1}{dt} - \frac{d\theta_2}{dt}\right) \qquad \textbf{(3.5.4)}$$

If both ends of this element turn at the same rate there is no torque developed. Likewise, if one end is grounded or has zero angular velocity, then the torque developed is proportional to the remaining nonzero rate of rotation. This is often a simple model used to describe friction in the bearings of a piece of rotating machinery.

Example 3.3

Consider the mechanical system illustrated in Fig. 3.10(a). The free-body diagrams for the two inertial disks are shown in Fig. 3.10(b). In this case, it was assumed that $\theta_1 > \theta_2$. Based on this assumption, the torques are assigned to the free-body diagram; this makes the task of writing the equations of motion simple. Application of Eq. (3.5.1) to each of the masses while observing the positively assigned sign convention yields

$$J_1\ddot{\theta}_1 = -k_1\theta_1 - k_2(\theta_1 - \theta_2)$$

and

$$J_2\ddot{\theta}_2 = -k_3\theta_2 + k_2(\theta_1 - \theta_2) + T(t)$$

If the torque $T(t)$ is known, along with the initial conditions $\theta_1(0)$, $\theta_2(0)$, $d\theta_1(0)/dt$, and $d\theta_2(0)/dt$, the problem can be solved for the rotations and/or the angular velocities.

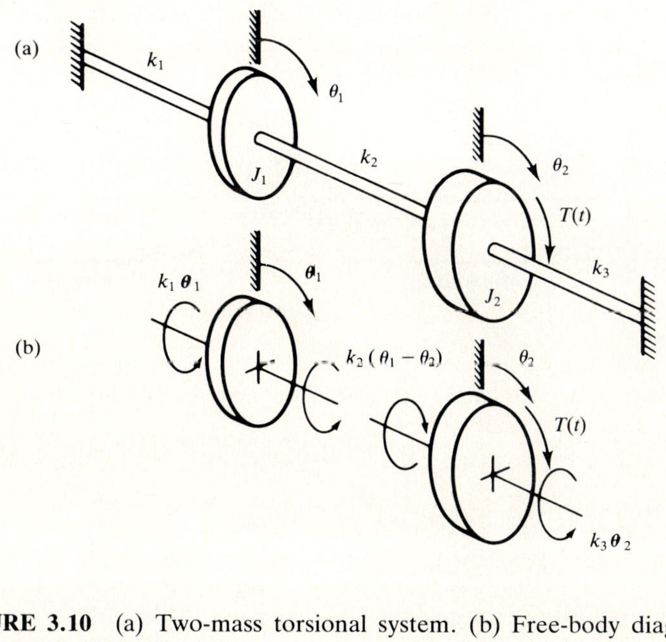

FIGURE 3.10 (a) Two-mass torsional system. (b) Free-body diagram for the disks assuming $\theta_1 > \theta_2$.

3.6 DISPLACEMENT-DRIVEN SYSTEMS

To this point in this chapter we have considered mechanical systems which are driven by external forces, but often systems are excited by a known motion of a boundary point. Consider as an example the system of Fig. 3.11(a), in which the displacement $y(t)$ is a known function of time and the resultant displacement $x(t)$ is to be found. The free-body diagrams for the mass m and the massless left-hand support are given in Fig. 3.11(b). If we write the equation of motion of the mass observing the positive sign convention we get

$$m\ddot{x} = k(y - x) + b(\dot{y} - \dot{x}) \tag{3.6.1}$$

If $y(t)$ is known, and thus its derivative, then expression (3.6.1) can be solved for the motion $x(t)$. Suppose we wish to examine a force balance for the left support, assuming it to be massless. We get

$$0 = f(t) - k(y - x) - b(\dot{y} - \dot{x}) \tag{3.6.2}$$

where we have included a reaction force $f(t)$ of the surroundings on the support in order to create $y(t)$. If in fact the displacement $y(t)$ is known, then Eq. (3.6.1) is a sufficient description of the system and Eq. (3.6.2) can be used to evaluate the unknown force $f(t)$. If, on the other hand, the force $f(t)$ is known and the displacement $y(t)$ is unknown, both expressions (3.6.1) and (3.6.2) must be solved simultaneously for $x(t)$ and $y(t)$. There are problems at the end of the chapter which illustrate both force and displacement excitation.

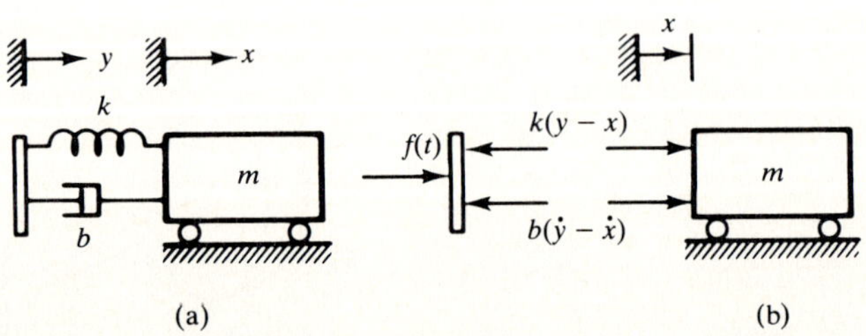

FIGURE 3.11 (a) Displacement-excited system. (b) Free-body diagrams of the mass and support.

3.7 COMPOSITE MECHANICAL SYSTEMS

Many mechanical systems are composed of elements some of which undergo translation and others which undergo rotation. Often these elements are linked by pins which, in the first approximation, we shall assume to be frictionless. Where gears are present we shall represent them as cylinders which roll on one another without slipping. This formulation technique is illustrated in the several examples which follow.

Example 3.4

Consider the system composed of a mass and rigid bar and two springs, illustrated in its static equilibrium position in Fig. 3.12(a). The mass moment of inertia of the rod about pivot A is J_{rod}. If we assumed that the displacements are positive and small (such that vertical displacement may be approximated linearly by $a\theta$), the free-body diagram of Fig. 3.12(b) results. Now employing the free-body diagrams we can write the two equations of motion. For the pivoting rod the summation of torques yields

$$J_{rod}\ddot{\theta} = -k_1(a\theta - x)a - k_2(a+b)^2\theta$$

and for the mass

$$m\ddot{x} = f(t) + k_1(a\theta - x)$$

If the forcing function $f(t)$ is known, and the initial values of displacements and velocities are known, these two expressions can be solved for the motions $x(t)$ and $\theta(t)$. For illustration, let us assign the following numerical values to the problem:

$$a = b = 1\text{ m} \qquad m = 10\text{ kg}$$

$$k_1 = k_2 = 400\text{ N/m} \qquad J_{rod} = 16\text{ kg}\cdot\text{m}^2/\text{rad}$$

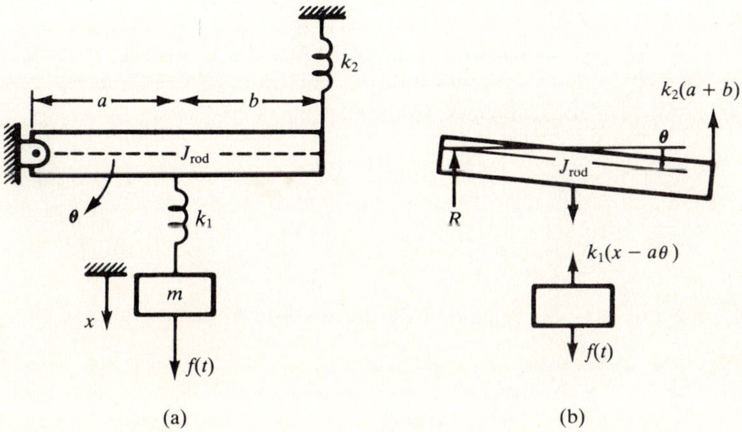

FIGURE 3.12 (a) Composite mechanical system. (b) Free-body diagram for $x > a\theta$.

Let the forcing function be a suddenly applied (at $t = 0$) constant force of 10 N. The differential equations become, for $t > 0$,

$$16\ddot{\theta} + 2000\theta - 400x = 0$$

$$10\ddot{x} + 400x - 400\theta = 10u(t)$$

If we transform these equations, assuming that initial displacements and velocities are zero, the resulting algebraic equations are

$$(16s^2 + 2000)\theta(s) - 400X(s) = 0$$

$$-400\theta(s) + (10s^2 + 400)X(s) = \frac{10}{s}$$

If we employ Cramer's rule, the solution for $\theta(s)$ is

$$\theta(s) = \frac{\begin{vmatrix} 0 & -400 \\ \frac{10}{s} & 10s^2 + 400 \end{vmatrix}}{\begin{vmatrix} 16s^2 + 2000 & -400 \\ -400 & 10s^2 + 400 \end{vmatrix}}$$

After evaluation of the determinants, the result is

$$\theta(s) = \frac{25}{s[s^4 + 165s^2 + 4000]}$$

If we factor the denominator, we find that there are two pairs of imaginary roots, as one would expect in such a system in the absence of damping.

$$\theta(s) = 25\left[\frac{1}{s(s^2 + 11.639^2)(s^2 + 5.43^2)}\right]$$

The form of this denominator dictates the form of the partial fraction expansion which must be composed of a step and two sines and two cosines of $5.43r/s$ and $11.639r/s$, respectively, or

$$\theta(s) - 25\left[\frac{A}{s} + \frac{B11.639}{s^2 + 11.639^2} + \frac{Cs}{s^2 + 11.639^2} + \frac{D5.43}{s^2 + 5.43^2} + \frac{Es}{s^2 + 5.43^2}\right]$$

The constants may be evaluated by the method given in Chapter 2:

$$A = 2.5 \times 10^{-4}$$

$$B = D = 0$$

$$C = 0.696 \times 10^{-4}$$

$$E = -3.20 \times 10^{-4}$$

The inverse transform is

$$\theta(t) = 25[2.5 \times 10^{-4}u(t) + 0.696 \times 10^{-4}\cos(11.639t)$$

$$- 3.20 \times 10^{-4}\cos(5.43t)] \text{ radians}$$

This function is plotted in Fig. 3.13. We should perhaps explain the strange shape of this response function. It is not purely sinusoidal because it is the sum of (1) a constant and (2) two cosine functions of different frequency. (Sums of sinusoids of different frequencies are *not* sinusoids.)

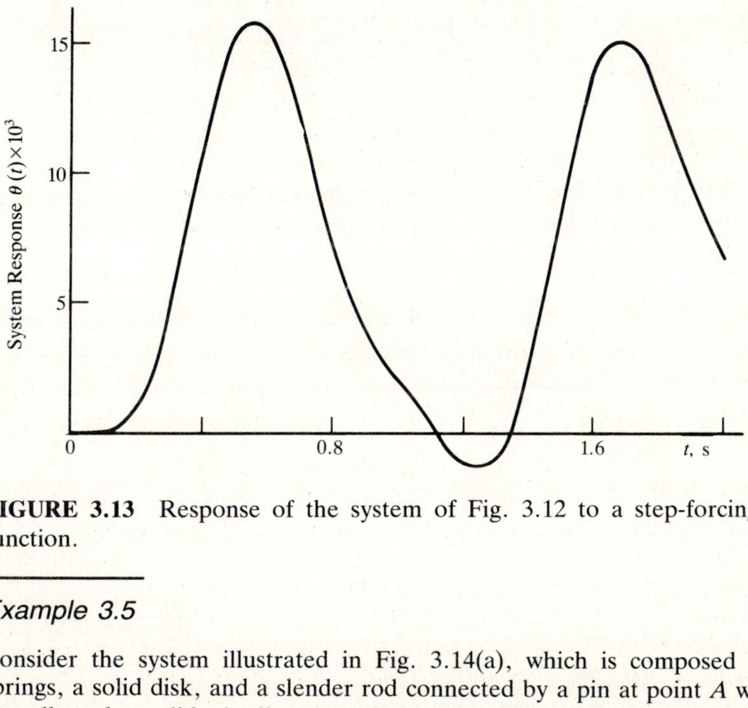

FIGURE 3.13 Response of the system of Fig. 3.12 to a step-forcing function.

Example 3.5

Consider the system illustrated in Fig. 3.14(a), which is composed of two springs, a solid disk, and a slender rod connected by a pin at point A with the pin allowed to slide in the slot of the rod. The respective small angular displacements are θ and ϕ and referenced as shown in Fig. 3.14(a). Excitation $p(t)$ is applied to the rod at the midpoint. Assume that, at pin A, the pin transmits an unknown force $f(t)$ from the rod to the disk which is represented in arbitrary but opposite directions on the two system elements. Using these free-body diagrams, we can write the equations of motion. For the disk we get

$$J_1\ddot{\theta} = f(t)R - k_1R^2\theta \qquad \textbf{(a)}$$

and for the rod

$$J_{\text{rod}}\ddot{\phi} = -f(t)(a) + p(t)\left(\frac{a}{2}\right) - k_2\left[\frac{a}{2}\right]^2\phi \qquad \textbf{(b)}$$

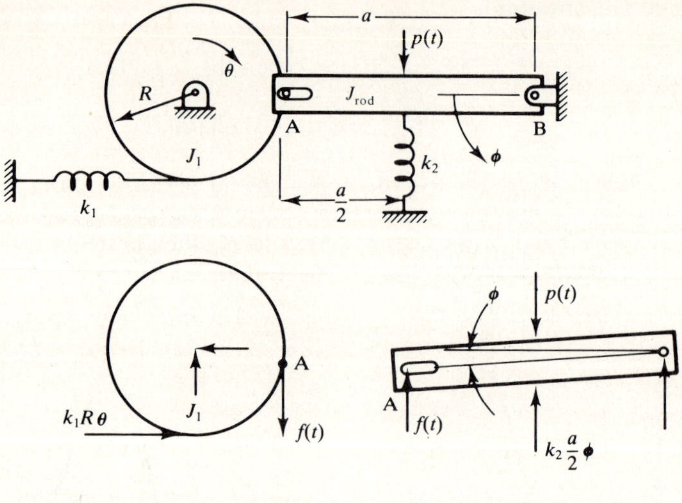

FIGURE 3.14 (a) A composite mechanical system. (b) A free-body diagram of the elements of the system assuming positive displacement.

where J_{rod} is the moment of inertia of the rod about pin B. In these equations we have three unknowns, namely $f(t)$, $\phi(t)$, and $\theta(t)$. We can also note that the vertical displacements of the rod and disk at point A are the same, or, for small angles, assuming a changes insignificantly with ϕ (it does change slightly),

$$R\theta = a\phi \qquad (*)$$

This is the third kinematic relation which, although it is algebraic, is sufficient to allow us to solve for $\phi(t)$ or $\theta(t)$. A relation such as Eq. **(*)** is often referred to as a constraint among the coordinates. Note that

$$\theta = \frac{a}{R}\phi$$

and substitute this into Eq. **(a)** to give

$$J_1\frac{a}{R^2}\ddot{\phi} + k_1R\frac{u}{R}\phi = f(t)$$

when solved for $f(t)$. Now let us substitute this into relation **(b)** to give

$$J_{rod}\ddot{\phi} = -\left[J_1\frac{a}{R^2}\ddot{\phi} + k_1a\phi\right]a + p(t)\frac{a}{2} - k_2\frac{a^2}{4}\phi$$

or, when cleaned up, the result is

$$\left(J_{rod} + J_1\frac{a^2}{R^2}\right)\ddot{\phi} + \left(k_1a^2 + k_2\frac{a^2}{4}\right)\phi = p(t)\frac{a}{2}$$

Examination of the terms of Eq. **(a)** yields some important physical insight. The term J_1a^2/R^2 in the first parentheses is the effective or "reflected" inertia of the cylinder at the pivot of the rod. Similarly, the term k_1a^2 is the equivalent torsional stiffness of spring k_1 at the pivot of the rod.

Another system which fits into the category of problems we are currently examining is a system which includes meshing gears, which we shall treat as cylinders with diameters of the pitch diameters of the gears. It will also be assumed that the peripheral velocities of the cylinders are equal or that they roll on one another without slipping. The technique for formulating the equations of motion is best illustrated by an example.

Example 3.6

An important problem in the dynamics of machinery is the case where gear-train inertia and shaft flexibility of a drive mechanism may no longer be neglected. Consider the system given in Fig. 3.15(a). Let us assume that the gears J_2 and J_3 can be treated as cylinders which roll continuously on one another without slipping. As a result of this contact there is a tangential force $F(t)$ transmitted from J_2 to J_3 and vice versa. It is this force $F(t)$ which reflects the torque transmitted by a gear train. We shall treat $F(t)$ as an unknown. The free-body diagrams for the system elements are illustrated in Fig. 3.15(b) with the unknown force $F(t)$ included. Note that in the free-body diagram the force transmitted by the gears is included.

If we write the equation of motion for J_1 we get

$$J_1\ddot{\theta}_1 = T(t) - k(\theta_1 - \theta_2) \qquad \textbf{(a)}$$

and for J_2 the equation is

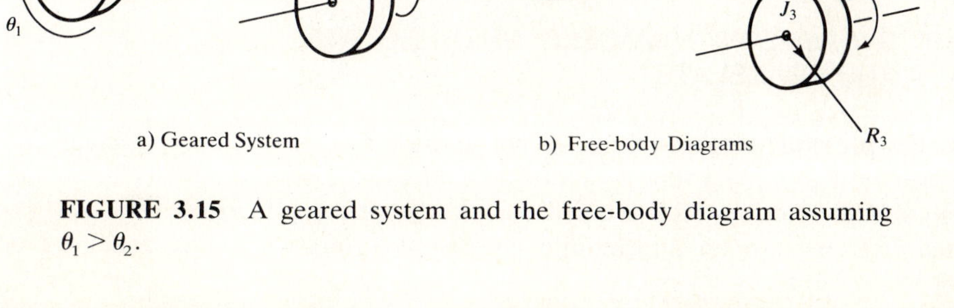

FIGURE 3.15 A geared system and the free-body diagram assuming $\theta_1 > \theta_2$.

$$J_2\ddot{\theta}_2 = k(\theta_1 - \theta_2) + F(t)R_2 \tag{b}$$

and for J_3

$$J_3\ddot{\theta}_3 = -F(t)R_3 \tag{c}$$

Note here that we have three differential equations and four unknowns, namely θ_1, θ_2, θ_3, and $F(t)$. There is, however, some information we have not yet employed and that is the fact that gears two and three must roll without slipping or

$$R_2\theta_2 = R_3\theta_3 \tag{d}$$

Hence we have the needed fourth equation. We could solve these four simultaneously. However, we might reduce the system order by substituting the second derivative of Eq. **(d)** into Eq. **(c)** for θ_3 to give

$$J_3 \frac{R_2}{R_3} \ddot{\theta}_2 = -F(t)R_3$$

We could now solve the remaining set of three equations, but let us continue to reduce the order by substitution. To do this, solve for $F(t)$ to give

$$F(t) = -J_3 \frac{R_2}{R_3^2} \ddot{\theta}_2 \tag{e}$$

Now substitute Eq. **(e)** for $F(t)$ in Eq. **(b)**, which gives

$$\left\{J_2 + J_3\left[\frac{R_2}{R_3}\right]^2\right\}\ddot{\theta}_2 + k(\theta_2 - \theta_1) = 0 \tag{f}$$

Note that the inertia J_3 is reflected back to gear two by the square of the gear ratio. Also, we can solve expressions **(f)** and **(a)** for $\theta_1(t)$ and $\theta_2(t)$ and then expression **(d)** can be used to give $\theta_3(t)$.

There are other nonphysical methodologies for the formulation of equations of motion of geared systems but the technique presented here is physically satisfying and reasonably easy to apply.

3.8 SYSTEMS WITH MASSES WHICH ROTATE AND TRANSLATE

In the previous sections we have considered masses which experience only translation or only rotation about a fixed axis. There are many important physical problems which are not in either of these categories. Examples of such systems are an automobile chassis or a large machine mounted on vibration isolators.

Equations for these mixed systems may be formulated in many ways but the most reliable technique using Newtonian dynamics is to define the motion as a combination of translation of the center of gravity and rotation about the center of gravity. Although this technique is not limited to the formulation of planar problems, we shall illustrate it with a planar example.

Example 3.7

Formulate the governing equation of motion for the machine which is suspended on elastic springs, as illustrated in Fig. 3.16(a). Assume that the springs have sufficient lateral stiffness that the mass may translate in the vertical direction only. We shall assume that the angles of rotation are small, such that $\theta = \sin\theta = \tan\theta$. With this assumption, the free-body diagram of Fig. 3.16(b) can be drawn. Now we may write the vertical equilibrium equation as

$$m\ddot{z} = -k_1(z - a\theta) - k_2(z + b\theta) + f(t)$$

while the rotational equilibrium equation is

$$J_A\ddot{\theta} = k_1(z - a\theta)a - k_2(z + b\theta)b + f(t)c$$

Note that these two equations are coupled and thus may be solved simultaneously only if the four initial conditions and the forcing function $f(t)$ are known.

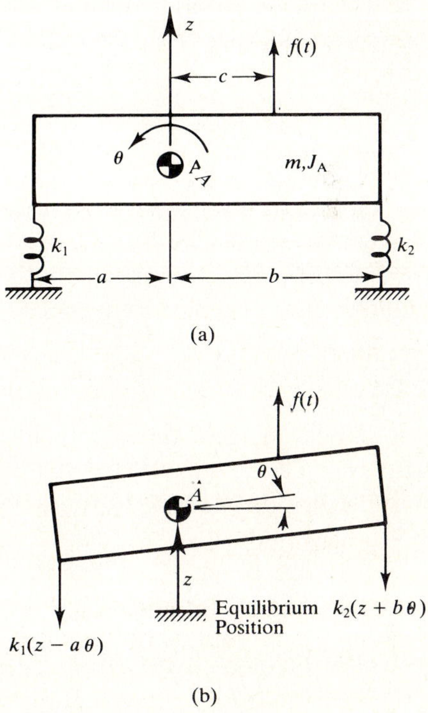

FIGURE 3.16 (a) A machine mounted on elastic springs. (b) Free-body diagrams for the machine with arbitrary displacements.

3.9 MODELING OF ELECTRICAL NETWORKS

To this point in this chapter we have considered the modeling and solution of problems involving dynamic mechanical systems. Now we shall examine the modeling and solution of electrical network problems by employing rigorous first principles just as we have done in the study of mechanical systems.

3.9.1 Electrical System Definitions and Units

In this section we shall make some fundamental definitions and discuss the conventional units to be used in the discussion of electrical networks. When discussing electrical phenomena, another fundamental physical quantity is needed in addition to the usual mechanical system measurements of length, mass, and time. This added unit is the measurement of electrical charge. The unit of measurement in MKS units is the coulomb (C) and is given the symbol $q(t)$. The charge on one electron, the smallest known physical unit of charge, is negative and is

$$q_e = -1.602 \times 10^{-19} \text{ C} \tag{3.9.1}$$

Indeed, many electrons are required to make up a charge of one coulomb.

Electrical current is defined as the rate of charge transport in coulombs/second, called amperes, through a surface (such as the cross section of a wire) and is given the symbol $i(t)$. Thus,

$$i(t) = \frac{dq}{dt} \text{ C/s or A} \tag{3.9.2}$$

Benjamin Franklin initially assumed that the charge transported in conductors was positive. However, with greater understanding of the atomic structure of matter, effectively both positive and negative charges have been found to exist. The flow of negative charges (electrons) results in a current in the opposite direction of the flow in the charge, whereas the flow of positive charge results in a current in the direction of charge flow. There are no *terminal* electrical experiments that can be performed to establish whether the mobile charges, those actually "carrying the current," are positive or negative. The engineering assumption, as one must be made, is that the positive direction for current is the direction of the transport of positive charge.

We shall define the potential, more commonly referred to as voltage, as a measure of the electrical forces available to cause mobile charge to be transported in a conductor. The unit of voltage is the volt, defined as the potential increase given to a charge of one coulomb when its energy is increased by one joule.

One note should be made here that the additional electrical unit used by the SI unit system is the unit of current or the ampere (A). Of course the unit of charge can then be defined as the ampere-second (A-s). This is the only point of departure of the SI system from the RMKS system.

3.9.2 Independent Electrical Sources

When examining the response of electrical networks, the response is due either to initial stored energy in capacitors or inductors or to that caused by energy sources such as batteries, electrical generators, or laboratory signal generators. Two types of ideal sources are shown in Fig. 3.17. They are, respectively, the ideal current source $i_s(t)$ and the ideal voltage source $v_s(t)$. The ideal current source is one which will deliver the specified current $i_s(t)$ amperes regardless of what load is attached at the terminals a − a′. Similarly, the ideal voltage source is one which delivers $v_s(t)$ volts regardless of the load attached at terminals b − b′.

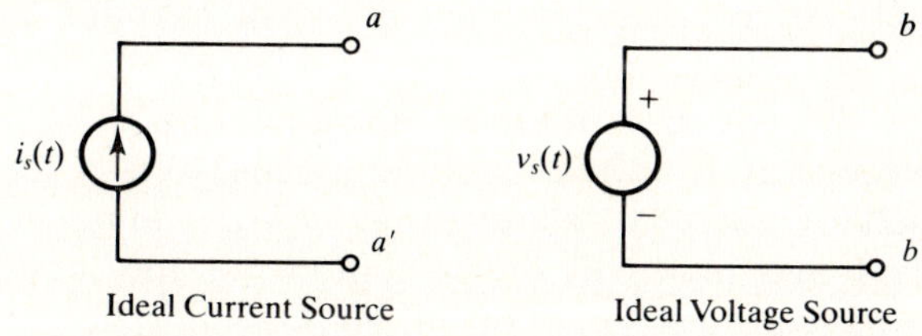

FIGURE 3.17 Ideal electrical sources.

We know that real electrical sources often do not deliver current or voltage which is independent of the load. For example, if we attach a small resistance to a flashlight battery the current will be high but the terminal battery voltage will drop well below the rated voltage listed on the battery case. The battery will maintain the rated terminal voltage only when the load resistance is relatively large. We shall discuss the modeling of real sources in the presence of real loads later in this chapter.

3.9.3 Passive Circuit Elements

Passive circuits are composed of constant elements called resistors, capacitors, and inductors. To these circuits we can add sources (excitations) and study the responses (typically voltages and/or currents). Passive circuits therefore contain no nonlinear or "active" elements such as diodes or transistors.

The most common element in electrical networks is the resistor. Its circuit symbol is shown in Fig. 3.18. This is an element for which the applied

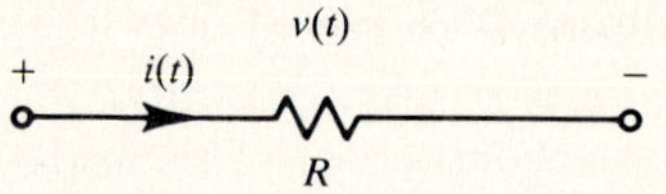

FIGURE 3.18 Resistor showing terminal voltage and current relationships.

voltage and resulting current are proportional, as long as the current (power dissipation) remains low enough that the material from which the resistor is composed is not damaged. The governing relation for the resistor is Ohm's law, which is

$$v(t) = Ri(t) \tag{3.9.3}$$

for $v(t)$ and $i(t)$, the terminal voltage and current, with polarity and direction shown in Fig. 3.18, such that the voltage drop occurs in the direction of the current. The proportionality factor R is referred to as the resistance in ohms if the voltage is specified in volts and the current is specified in amperes. This element stores no electrical energy but rather is an energy dissipator.

The second passive circuit element we shall examine is the inductor shown in Fig. 3.19, which stores energy in a magnetic field. It does not dissipate energy. Its circuit symbol is shown in Fig. 3.19. Although any current has energy stored in the associated magnetic field, an inductance is usually the result of adjacent windings of a coil, adding magnetic fields. Faraday observed that in a coil the magnitude of the induced voltage is proportional to the rate of change of linking magnetic flux or

$$|v(t)| = \frac{d(N\phi)}{dt} \tag{3.9.4}$$

where ϕ is the flux generated by one turn of the current and N is the number of turns in the coil. If the flux ϕ is the result of the current in the coil, we can write Eq. (3.9.4) for a fixed number of turns N as, dropping the magnitude sign for convenience,

$$v(t) = N\,\frac{d\phi}{di}\,\frac{di}{dt} \tag{3.9.5}$$

If, as is true in many cases, we may assume that ϕ is directly proportional to the current such that $d\phi/di$ is independent of i, and is therefore a constant, then we can write Eq. (3.9.5) as

$$v(t) = L\,\frac{di}{dt} \tag{3.9.6}$$

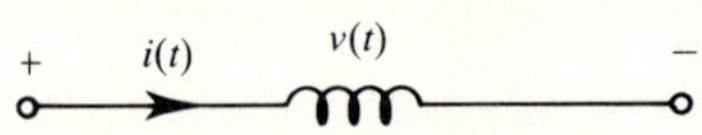

FIGURE 3.19 Inductor showing terminal voltage and current relationships.

where L is the inductance in henries and is defined from Eq. (3.9.5) as

$$L = N\frac{d\phi}{di} \tag{3.9.7}$$

Expression (3.9.6) represents the working terminal volt-ampere relationship for an inductor, including the correct voltage polarity for $v(t)$, a voltage drop in the direction of the current. If we separate the variables in Eq. (3.9.6) and integrate to find $i(t)$, we get

$$i(t) = i(0) + \frac{1}{L}\int_0^t v(t)\,dt \tag{3.9.8}$$

which gives terminal current $i(t)$ as a function of $v(t)$ with polarities and directions as shown in Fig. 3.19.

Now let us consider a final circuit element, the capacitor, which stores electrical energy in an electric field between charged conductors. The capacitor, like the inductor, does not dissipate energy; it only stores it. It is shown symbolically in Fig. 3.20. The terminal voltage is proportional to the charge difference between the conductors or

$$v(t) = \frac{1}{C}q(t) \tag{3.9.9}$$

where the proportionality factor C is referred to as the capacitance C in farads (or, more practically, usually in microfarads, μF).

The capacitance of the special geometry having parallel conducting plates of area A and spacing d, separated by a dielectric (insulator) with a permittivity or "dielectric constant" of ε, is given by

$$C = \frac{\varepsilon A}{d} \tag{3.9.10}$$

We know that current is defined as the rate of charge transport or

$$i(t) = \frac{dq}{dt} \tag{3.9.11}$$

If we differentiate Eq. (3.9.9) and substitute Eq. (3.9.11) for dq/dt then the result is the terminal voltage-current relationship

$$i(t) = C\frac{dv}{dt} \tag{3.9.12}$$

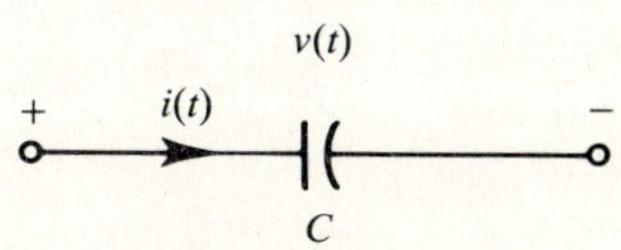

FIGURE 3.20 The capacitor showing terminal current and voltage relationships.

TABLE 3-2 Circuit Elements and Their Working Relations

Circuit Element	Resistor	Inductor	Capacitor
Symbol	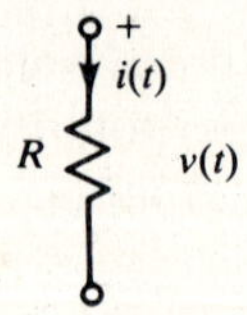	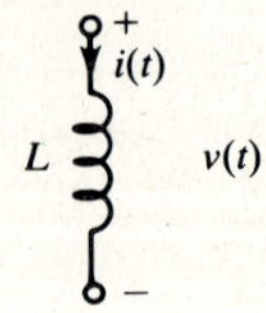	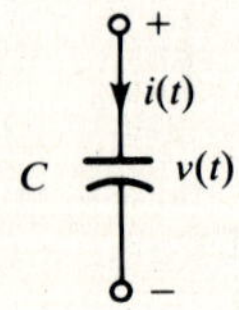
Terminal Voltage	$v(t) = Ri(t)$	$v(t) = L\dfrac{di}{dt}$	$v(t) = \frac{1}{C}\int_0^t i(t)\,dt + v(0)$
Terminal Current	$i(t) = \dfrac{v(t)}{R}$	$i(t) = \dfrac{1}{L}\int_0^t v(t)\,dt + i(0)$	$i(t) = C\dfrac{dv}{dt}$
Energy Stored or Power Dissipated	$p(t) = \dfrac{v^2(t)}{R} = i^2(t)R$	$W(t) = \frac{1}{2}Li^2(t)$	$W(t) = \frac{1}{2}Cv^2(t)$

Again, the voltage drop occurs in the direction of the current. If we separate the variables in this expression and integrate, the result is

$$v(t) = v(0) + \frac{1}{C}\int_0^t i(t)\,dt \tag{3.9.13}$$

All the working relations for electric circuit elements are summarized in Table 3-2.

3.10 KIRCHHOFF'S CIRCUIT LAWS

In the previous section of this chapter we developed the working relations for the elements which commonly comprise electric circuits. These relations are summarized in Table 3-2.

Although the German physicist Kirchhoff explored many problems, including problems of mechanics, he is perhaps best known for his fundamental work in the field of electricity. We shall now state two observations made by Kirchhoff, relative to his experimental investigations involving electric circuits, and then shall develop a methodology for applying these laws to the formulation of the linear equations describing electric circuits.

3.10.1 Kirchhoff's Voltage Law

Kirchhoff observed, after detailed measurements on a variety of electric circuits, that, in a complete traverse of a closed path (a loop) at a fixed time in an electrical network, the net change in electrical potential is zero, or that the sum of potential drops is equal to the sum of the potential rises around any closed loop. However, if we observe that potential rises are the negative of potential drops, then we may write Kirchhoff's voltage law for any closed loop of N segments:

$$\sum_{n=1}^{N} v_n(t) = 0 \text{ for any closed loop} \qquad \textbf{(3.10.1)}$$

If we consider a simple problem, as illustrated in Fig. 3.21, and consider a clockwise traverse of the loop beginning and ending at point P, then, for each segment with polarity change as indicated,

$$v_a + v_b - v_c + v_d - v_e - v_f - v_g = 0 \qquad \textbf{(3.10.2)}$$

where we have assumed that potential drops in the clockwise direction are positive.

The direction of traverse is arbitrary, as is what one calls positive. Once one decides what the positive quantity is, then this decision must be observed in all subsequent equation formulation.

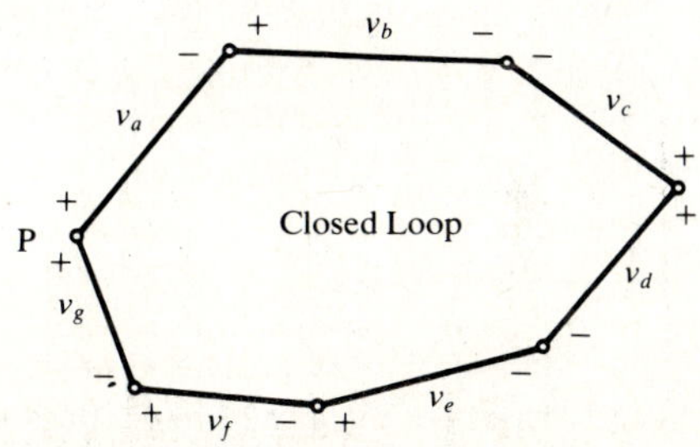

FIGURE 3.21 A closed path with potentials shown.

3.10.2 Kirchhoff's Current Law

The other observation Kirchhoff made based on his measurements is that, at any time, the sum of currents leaving a closed boundary is zero. Of course, this boundary could simply enclose only a single junction (a point where three or more currents combine) of circuit conductors or could comprise a whole group of circuit elements inside a closed surface. For a closed surface with N conductors crossing the surface

$$\sum_{n=1}^{N} i_n(t) = 0 \text{ for any closed surface} \qquad \textbf{(3.10.3)}$$

For the situation illustrated in Fig. 3.22 with current in the directions indicated, the statement of Kirchhoff's current law is

$$-i_1 + i_2 + i_3 - i_4 - i_5 = 0 \tag{3.10.4}$$

where current leaving the surface was arbitrarily assumed to be positive.

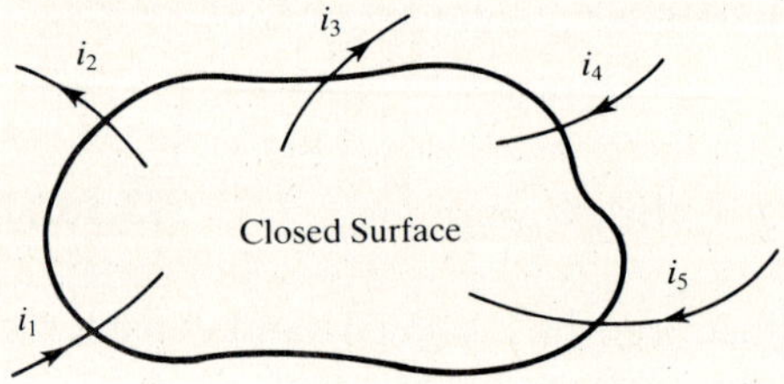

FIGURE 3.22 Closed surface for current summation.

3.10.3 The Method of Loop Currents

A popular method, based on Kirchhoff's voltage law for analysis of planar (can be drawn in a plane with no lines crossing) electrical networks, is the method of loop currents. In this method each interior circuit loop (or "window") is assigned a loop current with a preassumed positive direction. Once this is done, the branch currents in each circuit branch (a circuit segment connecting two junctions) can be given as the sum or difference of the loop currents in adjacent loops. Once these branch currents are known, the relation for the voltage rises and drops around each loop may be written in terms of the element working equations defined earlier, which now involve the branch currents. Several examples will help us understand this technique.

Example 3.8

Formulate the governing equation for the single-loop circuit shown in Fig. 3.23. The positive loop current has been assigned the clockwise direction and it is clear that, in this one-loop circuit, all branch currents are identically the

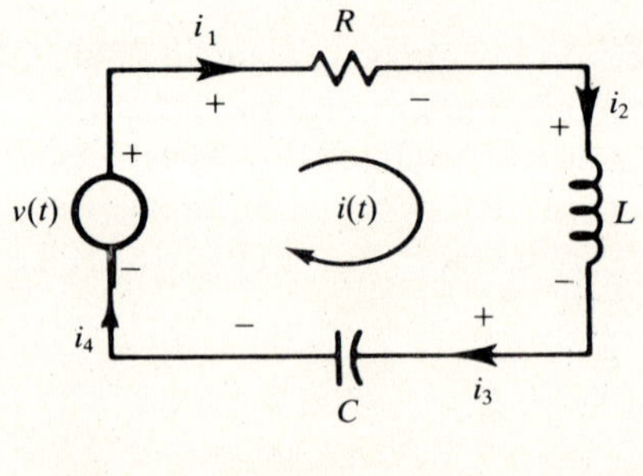

FIGURE 3.23 Single-loop R-L-C network.

same as the loop current $i(t)$. It is now possible to assign the potential polarity to the circuit elements using the information of Table 3-2. Then it is a simple matter to write down the single-loop equation by summing potential drops to get, starting from and returning to point P in a clockwise direction,

$$-v(t) + Ri(t) + L\frac{di}{dt} + \frac{1}{C}\int_0^t i\,dt + v_c(0) = 0$$

This is the equation describing the circuit and it has only one unknown, $i(t)$. If $v(t)$ and $v_c(0)$ were known, then the Laplace transformation could be employed to solve for $I(s)$ and thus $i(t)$.

Example 3.9

Consider the circuit of the preceding Example 3.8 for which the following circuit parameters are given:

$$L = 0.5 \times 10^{-3}\ \text{H}$$

$$R = 20\ \Omega$$

$$C = 1\ \mu\text{F}$$

For a suddenly applied battery of 10 volts at $t = 0$,

$$v(t) = 10u(t)\ \text{volts}$$

Assume initial conditions

$$v_c(0) = 0$$

$$i(0) = 0$$

Find the current $i(t)$ by the Laplace transform method. Substitution of the circuit parameters into the loop equation gives

$$20i(t) + 0.5 \times 10^{-3}\frac{di}{dt} + 10^6\int_0^t i\,dt = 10u(t)$$

Transforming this equation gives

$$\left(20 + 0.5 \times 10^{-3}s + \frac{10^6}{s}\right)I(s) = \frac{10}{s}$$

If we solve for the unknown $I(s)$, the result is

$$I(s) = \frac{10}{0.5 \times 10^{-3}s^2 + 20s + 10^6}$$

If we factor out 0.5×10^{-3} from the denominator, the result is

$$I(s) = 2 \times 10^4 \left[\frac{1}{s^2 + 4 \times 10^4 s + 2 \times 10^9} \right]$$

If we complete the square in the denominator we get

$$I(s) = 2 \times 10^4 \left[\frac{1}{(s + 2 \times 10^4)^2 + (4 \times 10^4)^2} \right]$$

Using transform pair 12 from Table A-2, the inverse transform is

$$i(t) = 0.5e^{-2\times 10^4 t} \sin 4 \times 10^4 t \text{ amperes}$$

The current waveform is illustrated in Fig. 3.24.

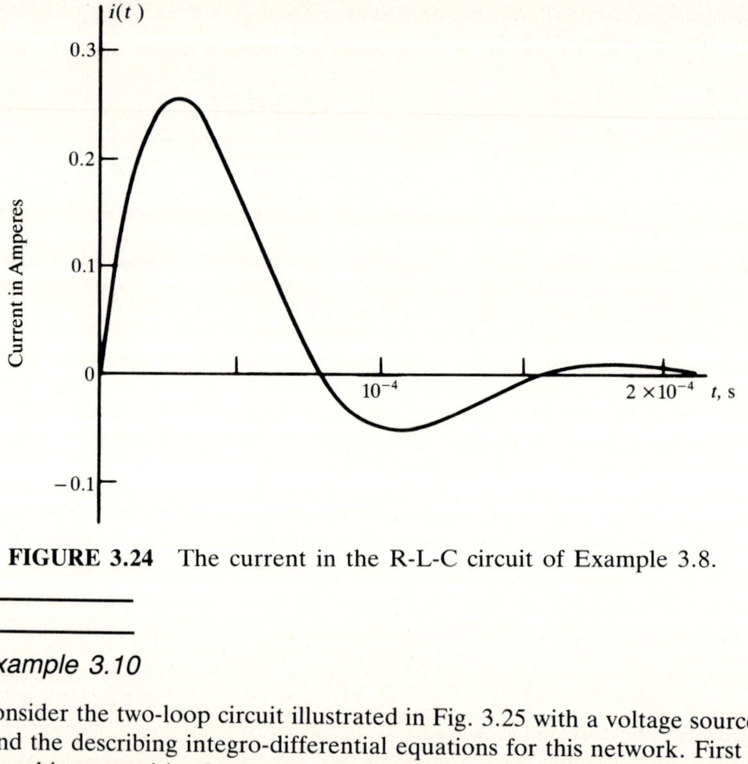

FIGURE 3.24 The current in the R-L-C circuit of Example 3.8.

Example 3.10

Consider the two-loop circuit illustrated in Fig. 3.25 with a voltage source $v(t)$. Find the describing integro-differential equations for this network. First assign the arbitrary positive loop currents as shown (in this case both were chosen in the clockwise direction). The branch currents are then evaluated and the directions are as indicated in Fig. 3.25. The current in the middle branch could have been chosen to be $i_2 - i_1$, in which case it would be shown in the upward direction. With the currents shown, the loop equation for loop 1 is, starting at point x and proceeding clockwise, summing voltage drops,

$$-v(t) + R_1 i_1(t) + L \frac{d}{dt}(i_1 - i_2) = 0 \tag{a}$$

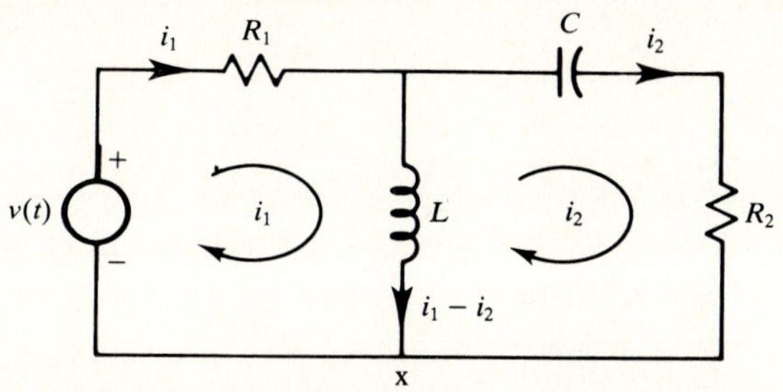

FIGURE 3.25 A two-loop electrical network.

For loop 2, starting at point x and proceeding in the clockwise direction, the equation is

$$L\frac{d}{dt}(i_2 - i_1) + \frac{1}{C}\int_0^t i_2\,dt + v_c(0) + R_2 i_2 = 0 \qquad \textbf{(b)}$$

Equations **(a)** and **(b)** represent a system of two linear integro-differential equations in the unknowns $i_1(t)$ and $i_2(t)$. If $v(t)$, $v_c(0)$, and $i_L(0)$ were known, these equations could be solved by the Laplace transform method for $i_1(t)$ and $i_2(t)$.

3.10.4 The Method of Node Voltages

A second method of circuit analysis based on Kirchhoff's current law is known as the node voltage method. We shall define a node as the junction of two or more circuit elements. We shall also choose one node in the network as the reference node or ground and give it the symbol ⏚. This choice is arbitrary but we should note that all other potentials in the network are to be defined relative to this reference node. The second step is to label the respective voltages of the remaining nodes $v_1(t)$, $v_2(t), \ldots$. Finally, it is necessary to assign arbitrarily the branch current directions between each node. The final step in the formulation procedure is to write Kirchhoff's current law for each node with each branch current being written in terms of the adjacent node voltages and the element relationships for each branch. This procedure will yield an integro-differential equation in the unknown node voltages for each node. Several examples should serve to clarify this method.

Example 3.11

Consider the circuit shown in Fig. 3.26, which is composed of a current source, a resistor, and a capacitor, each connected between the same pair of nodes or all in parallel. Formulate a node voltage equation governing this network, assuming no charge on the capacitor at time $t = 0$, that is, $v_c(0) = 0$. There are

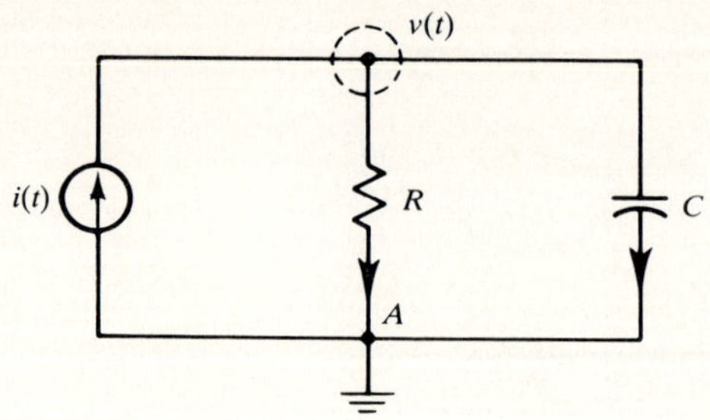

FIGURE 3.26 Single node circuit.

clearly two nodes and we shall assign node A as the reference node or ground. Label the remaining node $v(t)$ and assign the branch current directions. Here we have assumed that both of the passive element branch currents are downward, as indicated, but we could have assigned either, or both, to be upward. The branch current direction for the current source is in the direction of the current source but wouldn't have to be. Let us now apply Kirchhoff's current law to the small dashed boundary surrounding the node labeled $v(t)$. If we assume that current leaving the boundary is positive and that entering is negative and sum all currents to zero we get

$$-i(t) + \frac{v(t)}{R} + C\frac{dv}{dt} = 0$$

which is the equation describing the network. It could be solved by the transform method for $v(t)$ if $i(t)$ were known.

Example 3.12

Consider the earlier problem posed in Example 3.10, as shown in Fig. 3.27.

Assign the ground node arbitrarily; the choice for this problem is shown in Fig. 3.27. Note that the presence of the voltage source makes the voltage at the node labeled $v(t)$ a known quantity and, as a result, only two node voltage equations for the nodes labeled $v_1(t)$ and $v_2(t)$ are needed for a complete formulation of the problem. Now simply write Kirchhoff's current law for the node labeled $v_1(t)$ to give

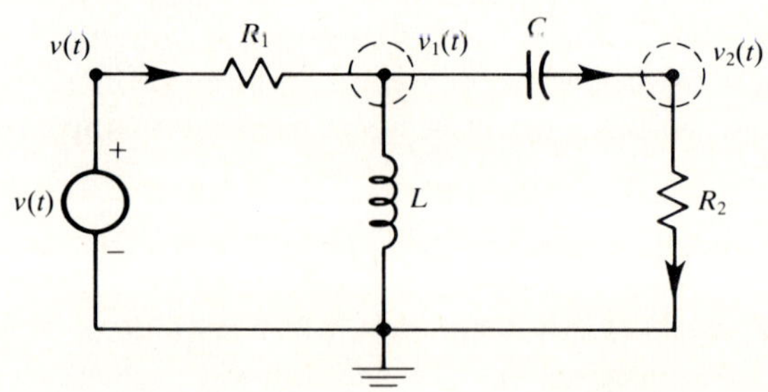

FIGURE 3.27 A two-loop network.

$$-\frac{v(t)-v_1(t)}{R_1}+\frac{1}{L}\int_0^t v_1\,dt+i_L(0)+C\frac{d}{dt}(v_1-v_2)=0$$

and similarly for the node labeled $v_2(t)$

$$-C\frac{d}{dt}(v_1-v_2)+\frac{v_2}{R_2}=0$$

This gives a pair of simultaneous integro-differential equations which can be solved by the Laplace transform method for $v_1(t)$ and $v_2(t)$. Let us do an analytical solution by assigning the following values for the source and circuit parameters and initial conditions of the network:

$$v(t)=10u(t)\text{ volts}$$

$$R_1=R_2=2\,\Omega$$

$$L=1\,\text{H}$$

$$C=2\,\text{F}$$

$$i_L(0)=v_c(0)=0$$

The circuit parameters given are not meant to be representative of usual physical values but rather are used for illustration. Transforming the differential equations with the parameter values given yields

$$\left(\frac{1}{2}+\frac{1}{s}+2s\right)V_1(s)-2sV_2(s)=\frac{5}{s}$$

$$-2sV_1(s)+\left(2s+\frac{1}{2}\right)V_2(s)=0$$

If we employ Cramer's rule to solve for $V_1(s)$, the result is

$$V_1(s)=\frac{\begin{vmatrix}\frac{5}{s} & -2s\\ 0 & 2s+\frac{1}{2}\end{vmatrix}}{\begin{vmatrix}\frac{1}{2}+\frac{1}{s}+2s & -2s\\ -2s & 2s+\frac{1}{2}\end{vmatrix}}$$

If we expand the determinants, the result is

$$V_1(s)=5\,\frac{s+0.25}{s^2+1.125s+0.25} \qquad \textbf{(a)}$$

If we apply the quadratic equation to the denominator, the roots are real and are $s_1=-0.3048$ and $s_2=-0.820$, so we can write Eq. **(a)** as

$$V_1(s) = 5\left[\frac{s + 0.25}{(s + 0.3048)(s + 0.820)}\right]$$

If we make a partial fraction expansion the result is

$$V_1(s) = 5\left[\frac{-0.1064}{s + 0.3048} + \frac{1.106}{s + 0.820}\right]$$

The inverse transform is then

$$v_1(t) = 5[-0.1064e^{-0.3048t} + 1.106e^{-0.820t}] \text{ volts}$$

This voltage response $v_1(t)$ is illustrated in Fig. 3.28; $v_2(t)$ could be found by a similar procedure.

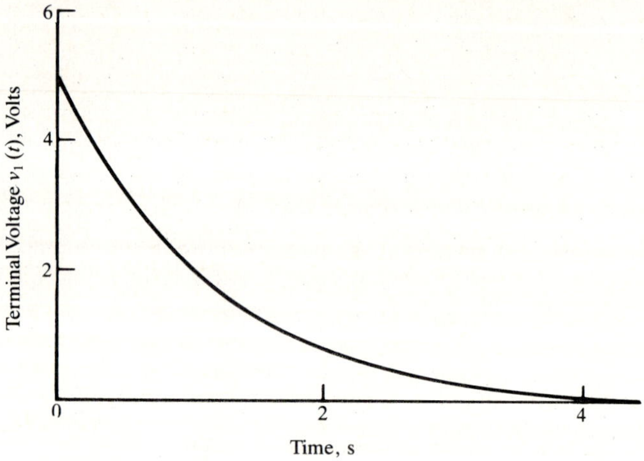

FIGURE 3.28 Node voltage at node one for the circuit of Fig. 3.27.

3.11 MUTUAL INDUCTANCE

In many electrical circuit problems there is a magnetic coupling, described by Faraday's law, between adjacent current-carrying conductors. This effect is especially significant when the conductors are in the form of adjacent, but not electrically connected, coils of wire wherein there are magnetic fields generated which can interact.

Consider the adjacent coils of Fig. 3.29, which are physically located such that the magnetic field generated by $i_1(t)$ in coil 1 is linked to coil 2 as shown. Faraday's law states that there is a potential induced in coil 2 due to the changing flux generated in coil 1, that is, $v_2(t) = \pm M(di_1/dt)$. The symbol M is used for this "inductance" which is known as the mutual

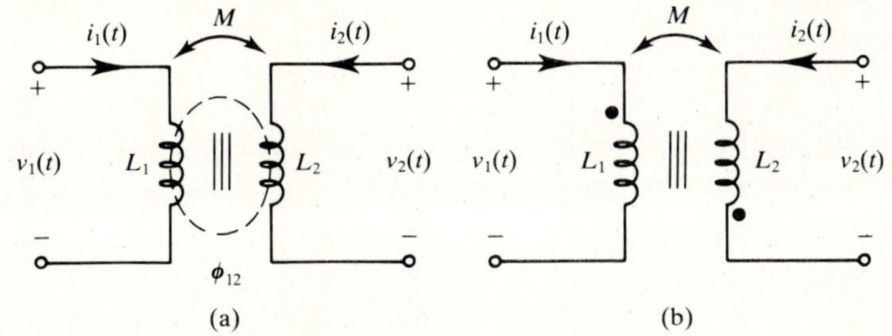

FIGURE 3.29 (a) Flux linkage between adjacent coils. (b) The dot notation for magnetic coupling.

inductance because it relates the voltage in one loop to the current change in another loop (as opposed to the symbol L which is used to relate the voltage in a loop to the current change in the same loop). If loop 2 also has a current $i_2(t)$, then the total voltage across the coil is the algebraic sum of the self-induced and the mutually induced voltages, that is, $v_2(t)$ between the terminals of coil $2 = L_2(di_2/dt) \pm M(di_1/dt)$.

By parallel reasoning, the equation for the terminal voltage $v_1(t)$ in loop 1 is

$$v_1(t) = L_1 \frac{di_1}{dt} \pm M \frac{di_2}{dt} \tag{3.11.1}$$

Inductances L_1 and L_2 are the respective self-inductances of the coils and M is the mutual inductance between these coils. The algebraic signs to be used on the $M(di/dt)$ terms depend on the geometry and the directions of winding of each coil. To account for these geometric properties, a dot notation is employed, as shown in Fig. 3.29(b). If *both* assumed loop currents enter or both leave a dotted terminal, indicating that the fluxes from the currents add in the coils, then the positive sign is used for the M term; otherwise the negative sign is used. For the network of Fig. 3.29(b) the appropriate equations are

$$v_1(t) = L_1 \frac{di_1}{dt} - M \frac{di_2}{dt} \tag{3.11.2}$$

and

$$v_2(t) = L_2 \frac{di_2}{dt} - M \frac{di_1}{dt} \tag{3.11.3}$$

Note that, in writing the loop equations for loops 1 and 2, it is possible to assign a voltage to "open" terminals. An example should clarify this concept.

Example 3.13

Formulate the loop equations for the network shown in Fig. 3.30 for the loop currents as shown. Note that $i_1(t)$ enters a dotted terminal but $i_2(t)$ leaves a dotted terminal. For loop 1 we apply Kirchhoff's voltage law to yield

$$-v(t) + R_1 i_1 + \left(L_1 \frac{di_1}{dt} - M \frac{di_2}{dt}\right) = 0$$

and for loop 2

$$R_2 i_2 + \frac{1}{C}\int i_2\, dt + \left(L_2 \frac{di_2}{dt} - M \frac{di_1}{dt}\right) = 0$$

These two equations could be solved simultaneously for $i_1(t)$ and $i_2(t)$.

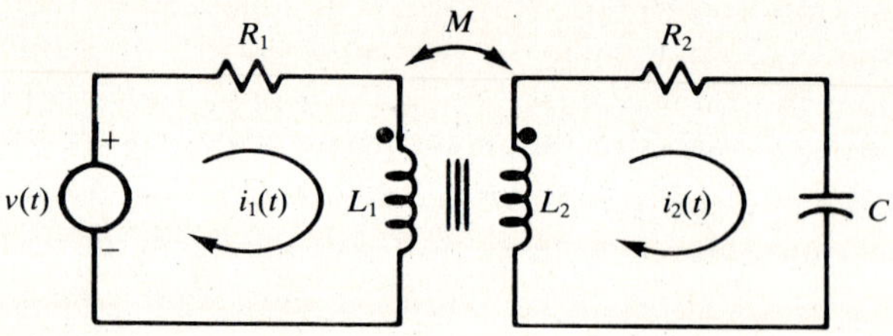

FIGURE 3.30 A two-loop circuit with mutual inductance.

3.12 THE IMPEDANCE CONCEPT

In our discussion of electric circuits, we have deliberately avoided the use of the impedance concept because of the desire to use first principles when formulating and solving electric circuit problems. We shall now attempt to tie the fundamental concepts exposed here to the concepts employed in the reader's previous exposure to electric circuit analysis.

Let us consider the single-loop circuit illustrated in Fig. 3.31. The loop equation of this circuit is given by applying Kirchhoff's voltage law around the loop to give

$$v(t) = R + \frac{1}{C}\int i\, dt + L\frac{di}{dt} \tag{3.12.1}$$

Let us now transform this equation to get, assuming zero initial conditions,

$$V(s) = \left(R + \frac{1}{sC} + Ls\right)I(s) \tag{3.12.2}$$

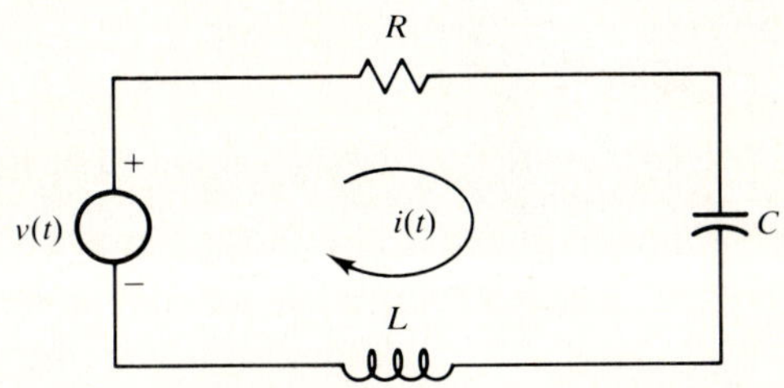

FIGURE 3.31 A single-loop network.

We shall define the term impedance, given the symbol $Z(s)$, to be the ratio of the s-domain voltage to s-domain current or

$$Z(s) = \frac{V(s)}{I(s)} \tag{3.12.3}$$

Note now that the terms in the parentheses of Eq. (3.12.2) are each a ratio of voltage to current and as such each term in the sum is an impedance. In fact, impedances connected in series, as in the loop in Fig. 3.31, are additive. The impedance of the three common passive circuit elements is given by transforming relations (3.9.3), (3.9.6), and (3.9.12) while ignoring any initial conditions present. For a resistor using Eq. (3.9.1) we get

$$Z_R(s) = R \tag{3.12.4}$$

For the inductor we get from Eq. (3.9.4)

$$Z_L(s) = sL \tag{3.12.5}$$

and for the capacitor we get, using relation (3.9.6),

$$Z_C(s) = \frac{1}{sC} \tag{3.12.6}$$

The reader should note the similarity between these impedances and those used for the solution of sinusoidal steady-state problems. The difference is that the familiar $j\omega$ has been replaced by the complex variable s. This replacement does not restrict us to the solution of sinusoidal steady-state problems as $j\omega$ does, but rather opens the door to a variety of transient problems as well. We shall further explore this connection between s and $j\omega$ in Chapter 4.

This impedance concept may be employed to formulate the governing equations for electrical networks directly without working in the time domain. An example of the formulation of circuit equations in the s domain should help the reader to bridge the gap between the formulations given earlier in this chapter and the methods outlined in a circuit analysis course.

Example 3.14

Formulate the s-domain loop equations for the network previously analyzed in Examples 3.10 and 3.12. We can draw the s-domain network and label the elements with the appropriate s-domain impedances, as illustrated in Fig. 3.32. Writing Kirchhoff's law for the left loop we obtain

$$R_1 I(s) + sL(I_1(s) - I_2(s)) = V(s)$$

and for the right loop we obtain

$$\frac{1}{sC} I_2(s) + R_2 I_2(s) + sL(I_2(s) - I_1(s)) = 0$$

These two equations are identical to what would be obtained by transforming Eqs. **(a)** and **(b)** in Example 3.10.

It should be noted in passing that, while the impedance concept was developed here for electrical circuits, in fact it applies equally well to mechanical, electromechanical, and any other systems that are described by integro-differential or differential equations. Such systems may relate responses to independent variables other than time, such as displacement.

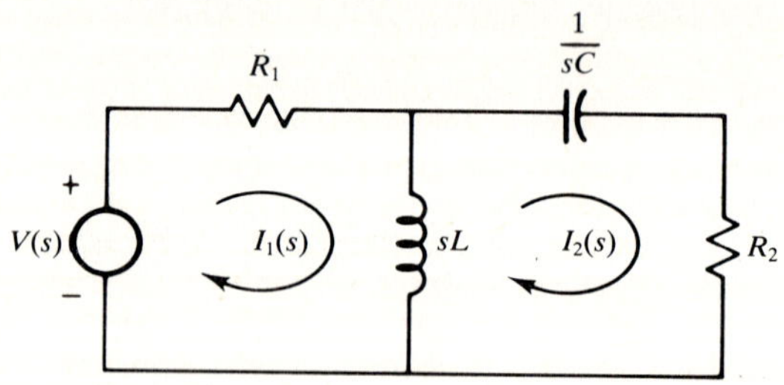

FIGURE 3.32 A two-loop electrical network.

3.13 ENERGY AND POWER IN SYSTEMS ELEMENTS

It is often important to be able to evaluate the energy stored or the power dissipated by an element in a mechanical system or an electrical network.

3.13.1 Energy Storage and Dissipation in Mechanical Elements

Let us consider the potential energy stored in the linear elastic spring shown in Fig. 3.33(a), which has the linear force-deflection relationship shown in Fig. 3.33(b).

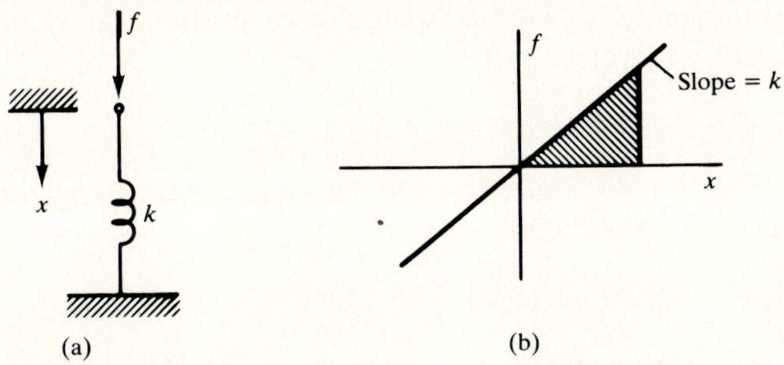

FIGURE 3.33 (a) Linear spring. (b) Force-deflection relationship.

The potential energy stored at any time is equal to the work done in compressing or stretching the spring to some net displacement, x, from equilibrium, or

$$W = \int_0^x f\,dx \tag{3.13.1}$$

which is the area under the force-deflection curve between the origin and x. For the spring $f = kx$,

$$W = \int_0^x kx\,dx = \frac{1}{2}kx^2 \tag{3.13.2}$$

The kinetic energy stored by a mass m in translation is the work done in accelerating the mass to the velocity u. The work done is

$$W = \int f\,dx \tag{3.13.3}$$

but the force and velocity are related by Newton's second law or

$$f = m\frac{du}{dt} \tag{3.13.4}$$

and then the work is

$$W = \int m\frac{du}{dt}\,dx = \int m\frac{dx}{dt}\,du \tag{3.13.5}$$

but dx/dt = the velocity, so

$$W = \int_0^u mu\,du = \frac{1}{2}mu^2 = \frac{1}{2}m\dot{x}^2 \tag{3.13.6}$$

which is the usual well-known result from mechanics.

The instantaneous power dissipated in a mechanical element is the product of force and velocity or

$$p(t) = fu \tag{3.13.7}$$

For the linear dashpot the force is $f = bu$ or

$$p(t) = bu^2 = b\dot{x}^2 \tag{3.13.8}$$

There are similar relationships for energies and power in torsional system elements.

3.13.2 Energy Storage and Dissipation in Electric Circuit Elements

The energy stored at any time t in an electrical element is

$$W = \int v \, dq \tag{3.13.9}$$

where dq is an element of charge transported and v is the instantaneous terminal voltage. For a capacitor, the terminal voltage is related to the charge by the capacitance or

$$v(t) = \frac{q(t)}{C} \tag{3.13.10}$$

The energy stored in the electric field of the capacitor is then

$$W = \frac{1}{C} \int_0^q q \, dq = \frac{1}{2C} q^2 = \frac{1}{2} Cv^2 \tag{3.13.11}$$

Similarly, the energy stored in the magnetic field of an inductor can be given by use of Eq. (3.13.9) and the relation for the terminal voltage of an inductor which is

$$v = L \frac{di}{dt} \tag{3.13.12}$$

Then, from Eq. (3.13.9), the energy is

$$W = \int L \frac{di}{dt} \, dq \tag{3.13.13}$$

but $dq/dt = i$, so

$$W = \int_0^i Li\, di = \frac{1}{2} Li^2 \tag{3.13.14}$$

The instantaneous power supplied to any circuit element is given by

$$p(t) = v(t)i(t) \tag{3.13.15}$$

For a resistor the voltage-current relationship is given by Ohm's law as $v(t) = Ri(t)$, so the instantaneous power is

$$p(t) = Ri^2(t) = \frac{v^2(t)}{R} \tag{3.13.16}$$

which is known to be total energy dissipation, i.e., no energy is stored in a resistor. While similar expressions for the instantaneous power can be developed for inductance and capacitance, it is clear that, just as there is no energy stored in a resistor, there is no energy dissipated in an ideal capacitor or an ideal inductor. Thus, $p(t)$ for an inductor is $Li(di/dt)$ and $p(t)$ for a capacitor is $Cv(dv/dt)$, both representing the *rate* at which energy is *stored* in the device.

3.14 SUMMARY

In this chapter we have explored the ideas of modeling dynamic mechanical systems and electrical networks. The formulation of such models is based on well-known physical principles such as Newton's second law, Kirchhoff's circuit laws, and Faraday's law. A high degree of physical understanding was encouraged for the application of these fundamental principles. In addition to these modeling concepts, the tool of the Laplace transformation was employed to calculate the temporal response of such systems.

REFERENCES

1. Rosenberg, R. C., and Karnopp, D. C. *Introduction to Physical System Dynamics*. McGraw-Hill, New York, 1983.
2. Cheng, D. K. *Analysis of Linear Systems*. Addison-Wesley, Reading, MA, 1959.
3. Ogata, K. *System Dynamics*. Prentice-Hall, Englewood Cliffs, NJ, 1978.

PROBLEMS

3-1. Employing Newton's second law, write the equations of motion for the three systems, assuming that force $f(t)$ and displacement $z(t)$ are known functions of time.

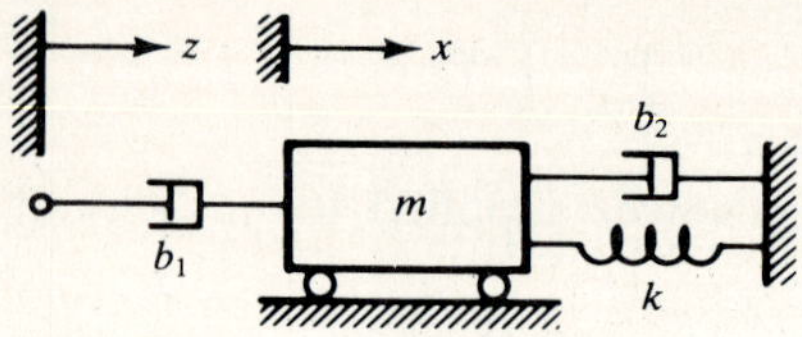

(a)

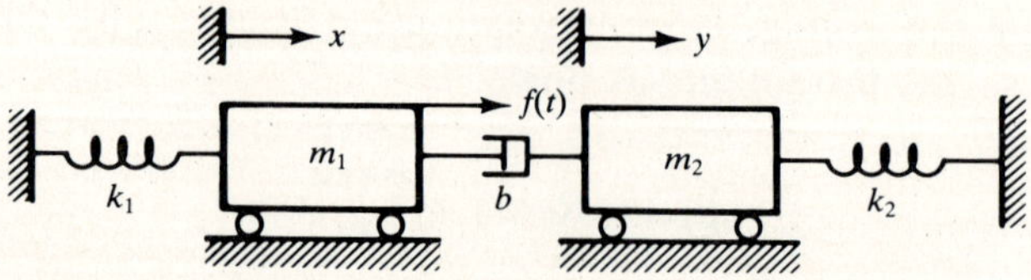

(b)

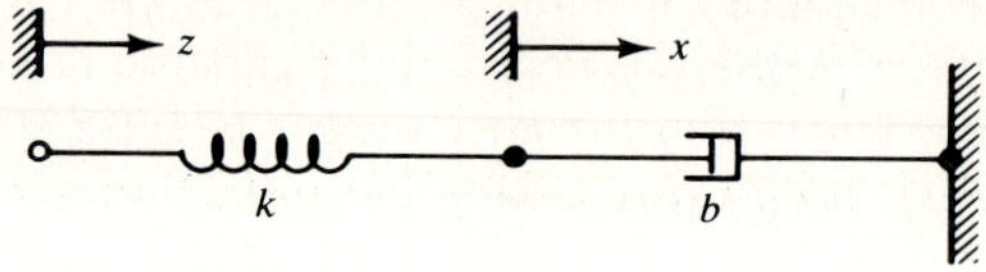

(c)

3-2. The torsional system below is equivalent to the crankshaft for a two-cylinder engine.

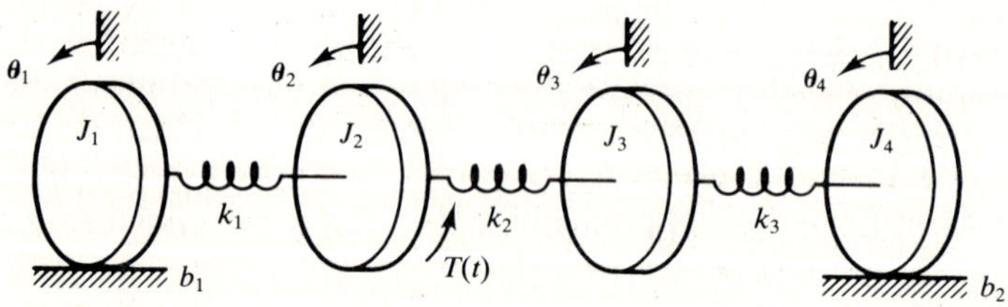

Write four differential equations assuming that torque $T(t)$ is known. Assume that b_1 and b_2 represent viscous friction between the respective inertia and ground.

3-3. A vibration isolation system for an electronics package is shown below.

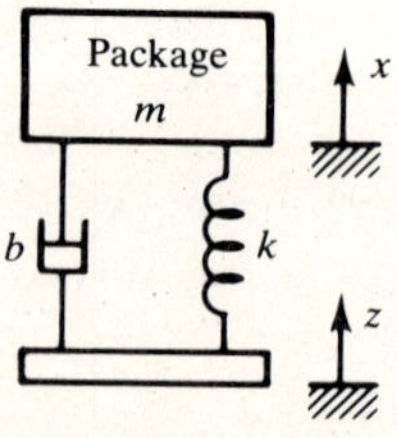

(a) Write the differential equation of motion for the package assuming that the motion of the base $z(t)$ is known.

(b) Let the relative displacement be $y(t) = x(t) - z(t)$. Using this relation solved for $x(t)$ and its derivatives in the equation of motion, obtain the equation of motion in terms of y and z. Note how the forcing function is now the acceleration of the base $\ddot{z}$.

3-4. In Problem 3-3(b), the following system parameters apply:

$$m = 5\ \text{kg},\ b = 8\ \text{N}\cdot\text{s/m},\ k = 100\ \text{N/m}$$

and the base acceleration is

$$\ddot{z} = 4u(t)\ \text{m/s}^2$$

Assuming that the system starts from rest, find $y(t)$.

3-5. The system below is a simple approximation to an energy-absorbing automobile bumper system. Write the equation of motion assuming $z(t)$ to be a known function. If the system impacts a rigid wall, what is $z(t)$?

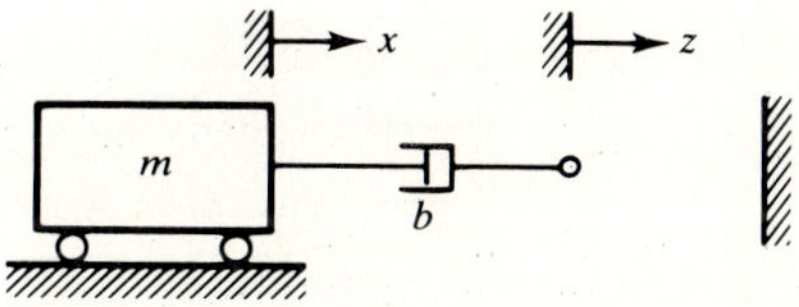

3-6. In Problem 3-5, the mass of the car is $m = 1000$ kg and the absorption coefficient $b = 400\ \text{N}\cdot\text{s/m}$. Find $x(t)$ if $x(0) = z(0) = 0$ and $\dot{x}(0) = 30\ \text{m/s}$.

3-7. The system shown below represents a deep mine elevator system with m representing the transport car and J the capstan which lifts the car. Here we shall consider the cable to be inextensible and under tension $F(t)$. The torque applied to the capstan will be represented by $T(t)$.

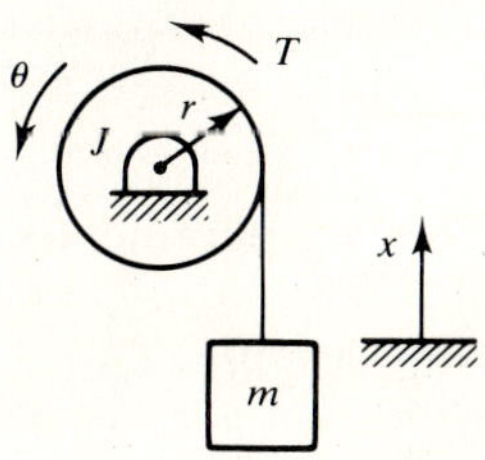

(a) Draw the free-body diagram for both m and J, including the effect of gravity.

(b) Write the equations of motion for the capstan and the car using the coordinates indicated.

3-8. For the system of Problem 3-7 the following parameters are given:

$$m = 400 \text{ kg}, J = 300 \text{ kg} \cdot \text{m}^2, r = 4 \text{ m}$$

$$T = 20000 \cdot u(t) \text{ N} \cdot \text{m}$$

If the system starts from rest, find the motion $x(t)$.

3-9. The system given in Problem 3-7 may be further complicated if now the cable may be represented as an elastic spring of modulus k, as shown below. Write the two equations of motion including the effect of gravity.

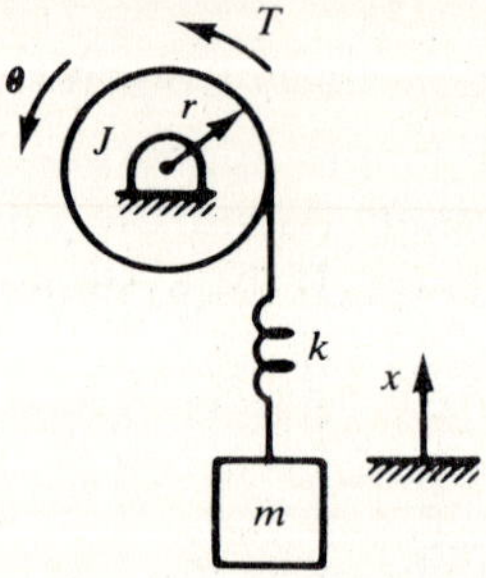

3-10. In the composite mechanical system shown below, k_1 and k_2 are both rectilinear springs and J is the mass moment of inertia about the pivot.
(a) Draw the free-body diagrams for m and J.
(b) Write the two differential equations of motion for m and J.

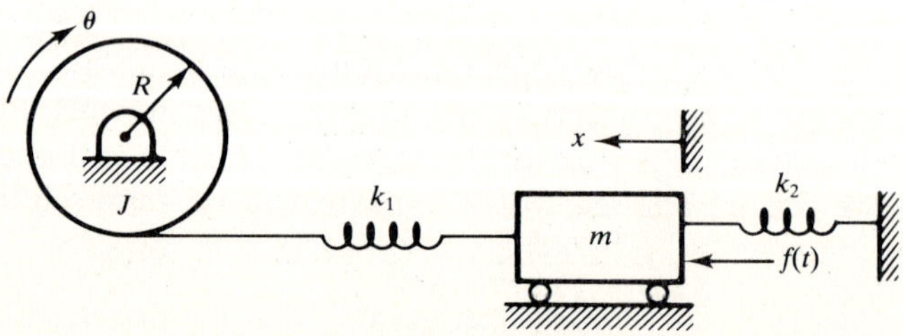

3-11. In Problem 3-10 the following parameters are given:

$$J = 10 \text{ kg} \cdot \text{m}^2, m = 5 \text{ kg}$$

$$k_1 = 10 \text{ N/m}, k_2 = 40 \text{ N/m}, r = 1 \text{ m}$$

The forcing function is $f(t) = 10t \cdot u(t)$ N. Solve for $X(s)$, assuming that the system starts from rest.

3-12. Draw the free-body diagram and write the equation of motion for *each* mass. Assume that the lever is massless and undergoes small rotations. Neglect friction.

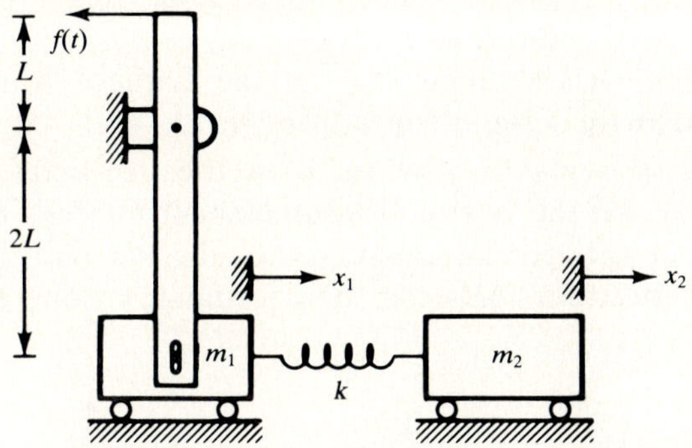

3-13. **(a)** Draw and correctly label all pertinent forces on the free-body diagrams for masses m_1 and m_2.
(b) Write the two differential equations of equilibrium.

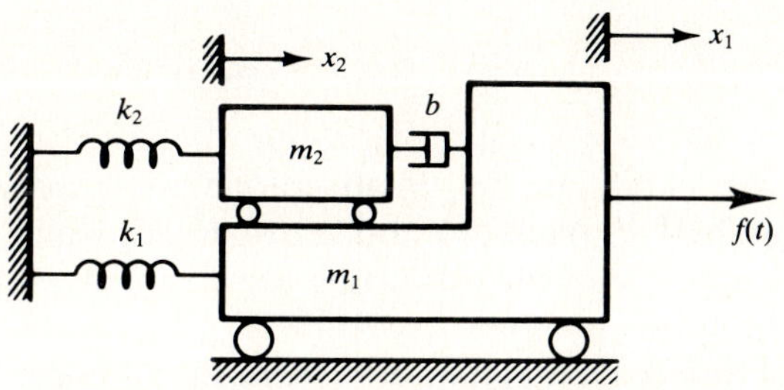

3-14. For small angles of deflection θ for the system shown, use $\Sigma\, T = J\ddot{\theta}$ to generate the equation of motion. Evaluate J in terms of m_1 and m_2.

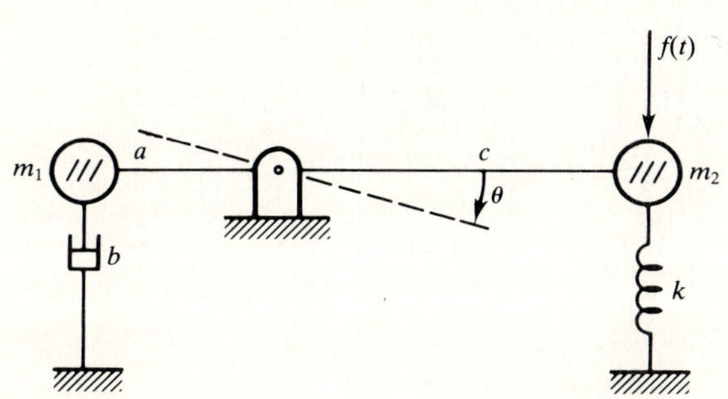

3-15. In the system of Problem 3-14 use the following system parameters:

$$a = c = 1\text{ m},\ m_1 = 1\text{ kg},\ m_2 = 2\text{ kg}$$

$$k = 40\text{ N/m},\ b = 5\text{ N}\cdot\text{s/m}$$

and $f(t) = 10 \cdot u(t)$ N. Find $\theta(t)$ by the Laplace transform method, assuming that the system starts from rest.

3-16. **(a)** For the system shown below, draw the free-body diagrams of the mass m_2 and the bar m_1. Assume small motions θ.

(b) Write the differential equations for the bar and the mass.

(c) Find a constraint between the coordinates x and θ.

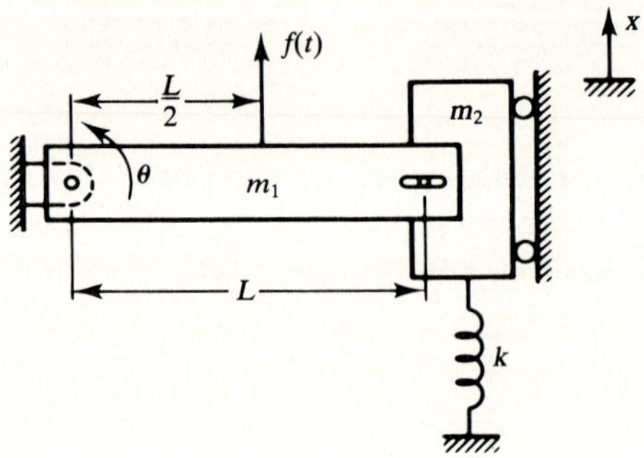

3-17. Consider the planar model of an automobile chassis shown below, which is allowed to pitch (θ) and translate vertically (z). The known motions of the front and rear axles are denoted as $x(t)$ and $y(t)$, as illustrated.

(a) Draw a free-body diagram of the chassis with only changes in force from static equilibrium considered; thus gravity forces may be omitted.

(b) Write the two governing differential equations of motion.

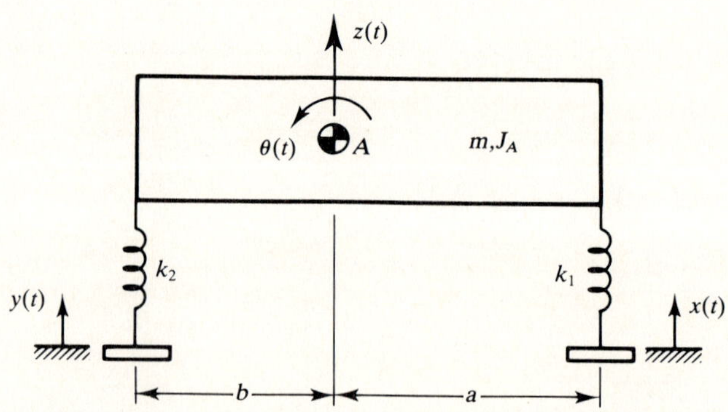

3-18. Determine the magnitude and proper polarity of the voltage across the 10 Ω resistor by the node voltage method (KCL). What is the current through the 10 Ω resistor?

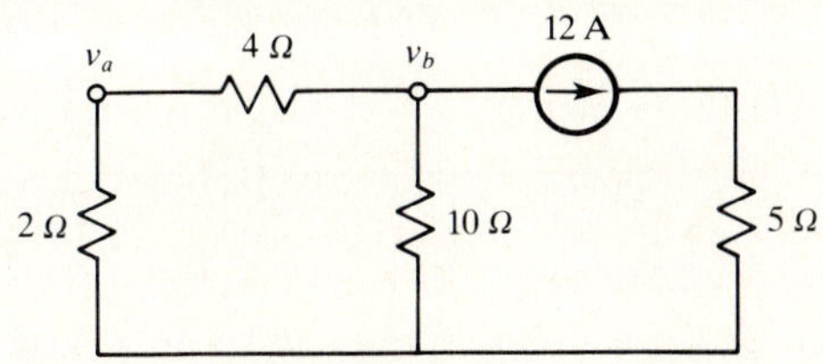

3-19. Write the node voltage equations employing Kirchhoff's current law for the nodes labeled v_1 and v_2 in the network shown.

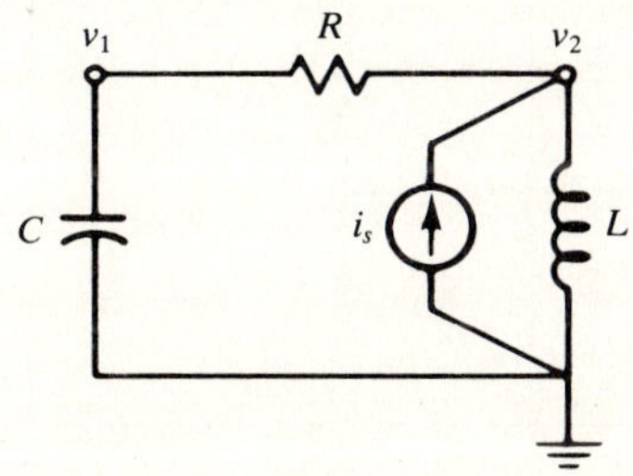

3-20. Write the loop equations using Kirchhoff's voltage law for loops 1 and 2 of the network shown.

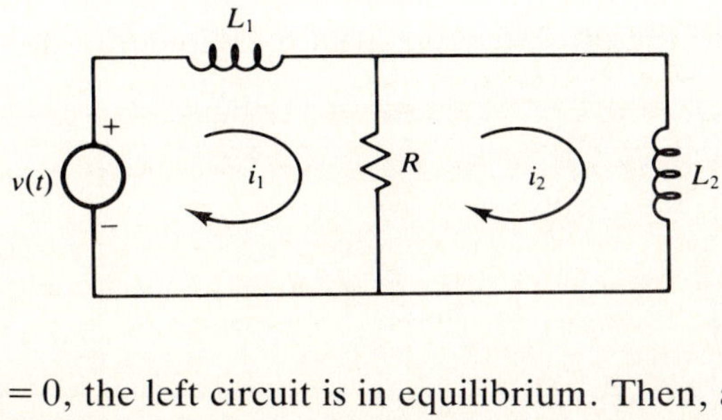

3-21. Prior to $t = 0$, the left circuit is in equilibrium. Then, at $t = 0$, switch S_1 opens and switch S_2 closes. Since the current in the inductor cannot change instantaneously, it will be the same after the switching operation as prior to it.

(a) Write the governing differential equation for the right loop.

(b) Transform this differential equation term by term and solve for $I(s)$.

(c) Invert this solution to give $i(t)$.

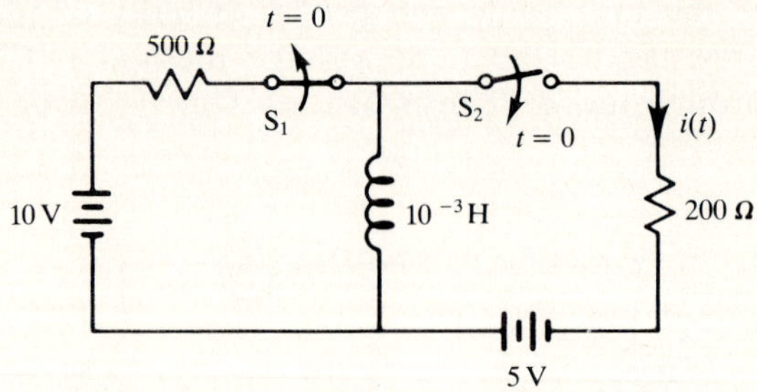

3-22. At $t = 0$ switch S_1 opens, removing the battery from the circuit which, up to that time, is in steady state.

(a) Write the loop equation in the time domain for $t \geq 0$.

(b) Solve this loop equation using Laplace transforms.

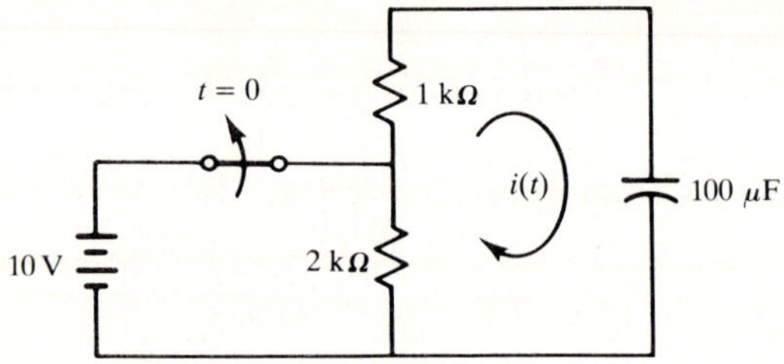

3-23. At $t = 0$ switch S_1 closes, completing the loop in the circuit shown below.

(a) Write the loop equation in the time domain for $t \geq 0$.

(b) Assuming no initial charge on the capacitor *and* no initial inductor current, find $i(t)$ by means of Laplace transforms.

(c) By any method, find the capacitor voltage $v_c(t)$.

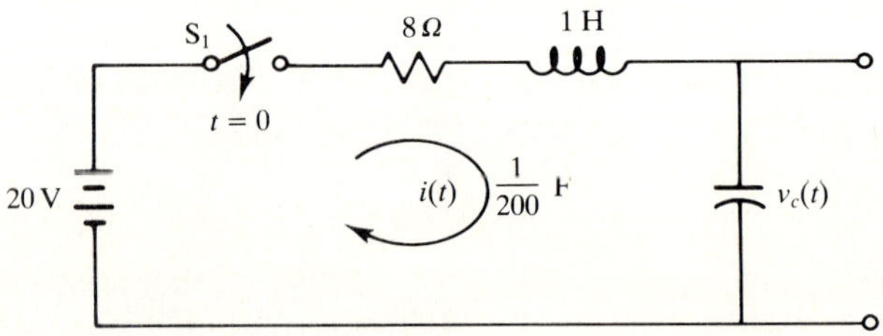

3-24. For the circuit given in Problem 3-21 the following data are given:

$$v(t) = e^{-t}$$

$$L_1 = 1\ \text{H}$$

$$L_2 = 1\ \text{H}$$

$$R = 1\ \Omega$$

The initial current in L_2 is 1 A directed upward. Find $i_2(t)$ if the initial current in L_1 is zero.

3-25. Write the node voltage equations for the nodes v_2 and v_3 for the network shown.

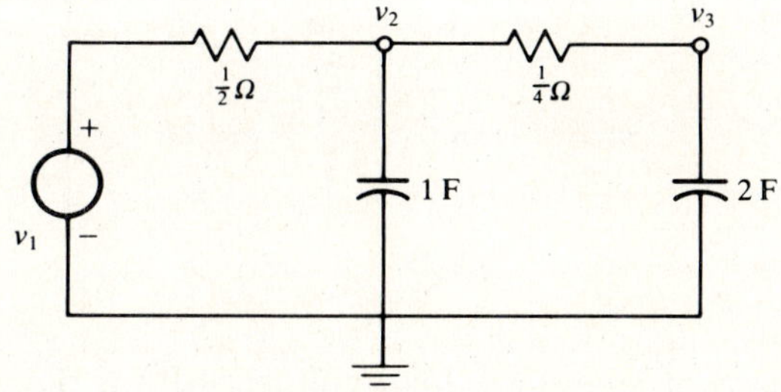

3-26. For the circuit of Problem 3-23 the source is $v_1(t) = tu(t)$. Find node voltage $v_2(t)$.

3-27. For the impedance bridge below, write the node voltage equations (Kirchhoff's current law) for nodes v_1, v_2, and v_3.

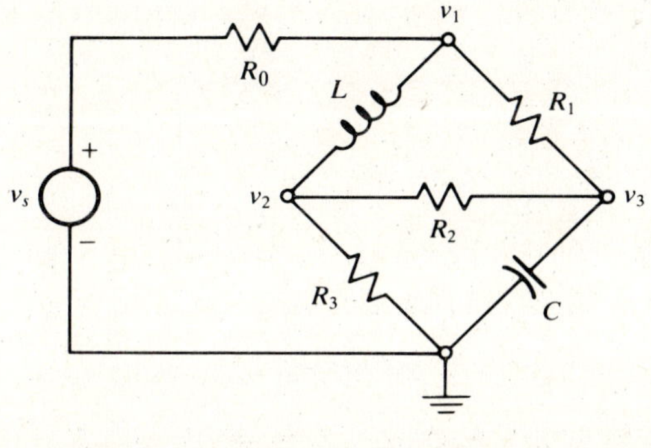

3-28. Write and solve the loop equations for the network with mutual inductance shown. The mutual inductance is $M = 10^{-2}$ H.

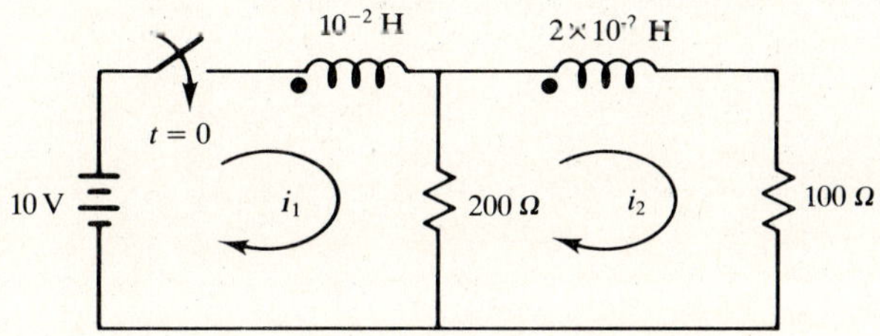

Assume that initial currents are zero.

3-29. The circuit shown is a model of a transformer driving the 5 Ω load resistance.

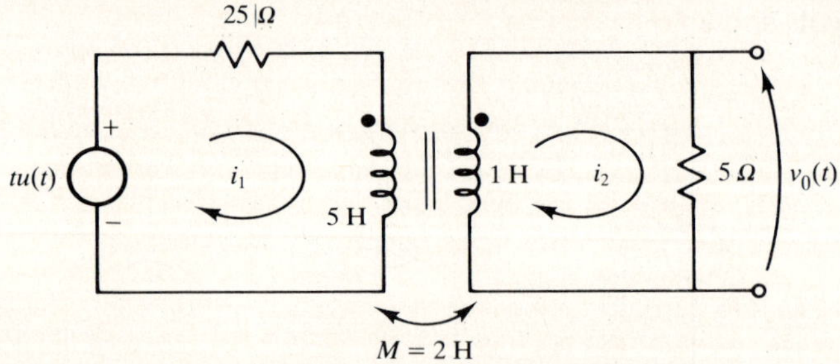

Find the current i_2 and the load voltage $v_0(t)$ if the initial currents in the inductors are zero.

3-30. For the circuit shown, the loop current $i(t)$ at the instant of consideration is 2 amperes and the capacitor voltage is 100 volts.

(a) What is the total energy stored at this instant?

(b) What is a quick estimate of the maximum energy stored by the capacitor at any time in the future?

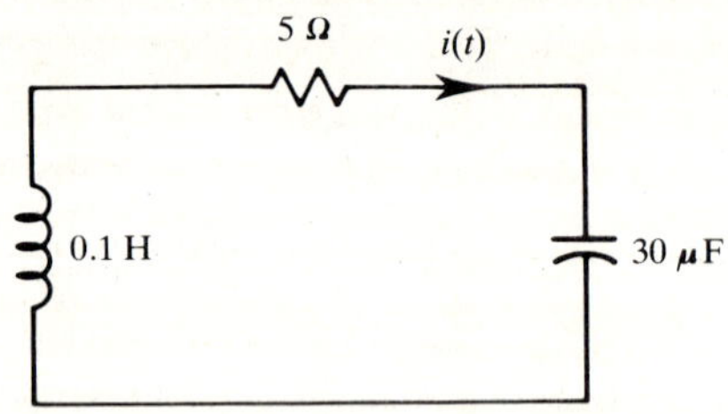

4

APPLICATION OF TRANSFORM CONCEPTS TO ENGINEERING SYSTEMS

4.1 INTRODUCTION

In many fields where analytical techniques are employed, the concept of the input-output model has proved useful. This idea is particularly valuable in such fields as electronics, chemical process design, industrial engineering, and economics. In this chapter we shall explore the concept of a "black box" having inputs and outputs as a representation of an engineering system. When we discuss the ideas of input and output we do not necessarily mean that there is a flow of some physical quantity into and out of the system boundary but rather that flows of information or signals exist. This does not preclude the flow of mass or energy in a systems problem but the physical flows may or may not be the same as the information or signal flows.

4.2 TRANSFER FUNCTIONS FOR SINGLE-INPUT–SINGLE-OUTPUT SYSTEMS

Consider the linear time invariant system of Fig. 4.1, which can be thought of as a "black box," the dynamics of which act upon the input or stimulus to produce the output or response. We are interested in the relationship between the stimulus and the response. This may be complex because the system is a dynamic system and these dynamics modify the input to produce the output. This system could be an electronic amplifier, an electrical filter, an electric motor, a vibratory system, a chemical process, or some other dynamic system.

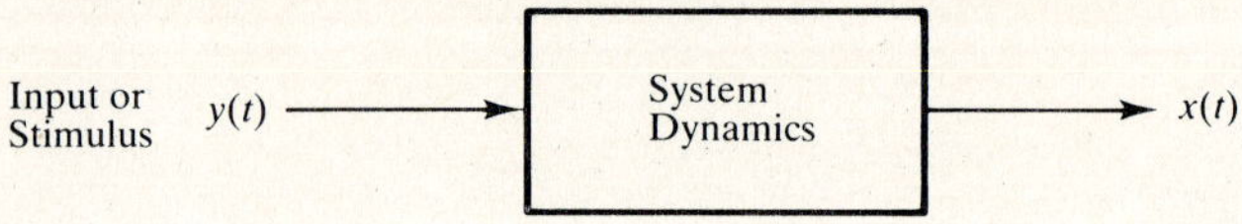

FIGURE 4.1 "Black box" representation of a dynamic single-input–single-output system.

We shall now define a *causal system* to be one in which the response $x(t)$ is caused by stimulus $y(t)$ and response cannot occur until or after $y(t)$ occurs. In other words, the system output cannot anticipate the input.

The question arises as to how we might analytically represent the input-output relations of the system. We have already seen that many interesting systems can be described by linear algebraic and integro-differential equations. These equations will, in general, be coupled together in such a manner that they will not be solvable one at a time but rather must be solved simultaneously.

We find that, if we perform the Laplace transformation on these describing equations, the result will be a set of coupled linear algebraic equations in the transforms of the problem variables. These equations may be solved for the output (response) variables in terms of the input (stimulus) variables by Cramer's rule, substitution, or some other standard method.

For the single-input–single-output system of Fig. 4.1, let us define the transfer function for the system, $G(s)$, to be the ratio of the output variable-to-the input variable in the s domain if all initial conditions are assumed to be zero. Please note that the transfer function $G(s)$ will depend greatly on which system variables are chosen as input and output. The mathematical definition of the transfer function for the system of Fig. 4.1 is then

$$G(s) = \frac{X(s)}{Y(s)} = \frac{\mathscr{L}[x(t)]}{\mathscr{L}[y(t)]} \quad \text{for zero initial conditions} \tag{4.2.1}$$

We immediately see that, given any input function $Y(s)$ and zero initial conditions, the response function $X(s)$ may be obtained from

$$X(s) = G(s) \cdot Y(s) \tag{4.2.2}$$

The real utility of a transfer function lies in the fact that it is possible to invert the transforms given in expression (4.2.2) to get the time-domain response $x(t)$.

Example 4.1

Consider the system shown in Fig. 4.2, wherein the driving function or stimulus is $y(t)$ and the response is displacement $x(t)$. Find the transfer

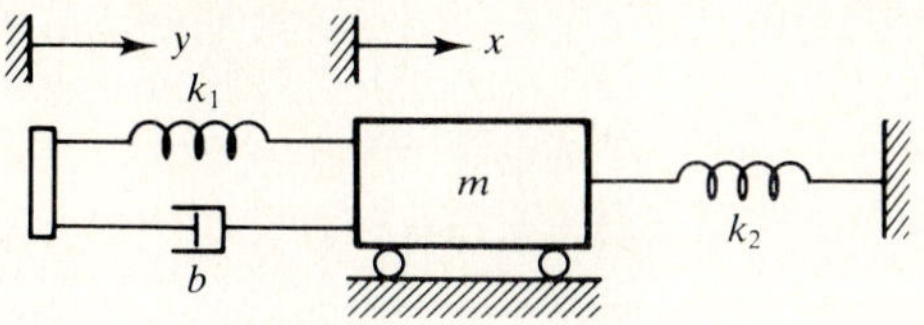

FIGURE 4.2 A linear mechanical system.

function $G(s) = X(s)/Y(s)$. Using a free-body diagram for the mass, the differential equation of motion is

$$m\frac{d^2x}{dt^2} + b\frac{dx}{dt} + (k_1 + k_2)x = b\frac{dy}{dt} + k_1 y$$

If we take the Laplace transform, the result for zero initial conditions is

$$(ms^2 + bs + k_1 + k_2)X(s) = (bs + k_1)Y(s)$$

Then the transfer function is

$$G(s) = \frac{X(s)}{Y(s)} = \frac{bs + k_1}{ms^2 + bs + k_1 + k_2}$$

which can be used to represent the system in block diagram form as illustrated in Fig. 4.3. Now let us choose the system parameters to be

$$m = 5\ \text{kg}$$

$$k_1 = k_2 = 500\ \text{N/m}$$

$$b = 50\ \text{N}\cdot\text{s/m}$$

which gives a transfer function of

$$G(s) = \frac{X(s)}{Y(s)} = \frac{50s + 500}{5s^2 + 50s + 1000}$$

which can be reduced to

$$G(s) = \frac{10s + 100}{s^2 + 10s + 200}$$

Now let us investigate the response of this system to a sudden displacement of 0.1 m, a step input in $y(t)$, so

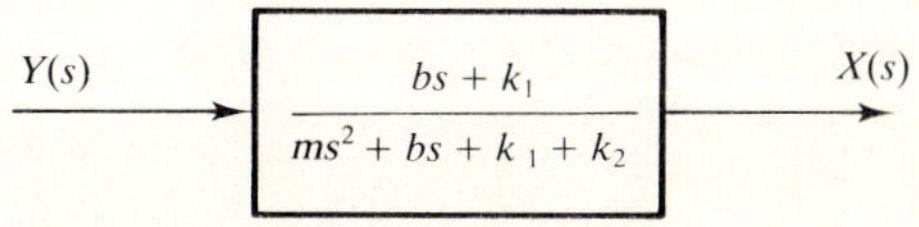

FIGURE 4.3 Block diagram for the mechanical system of Fig. 4.2.

$$y(t) = 0.1\, u(t) \text{ and } Y(s) = 0.1/s$$

The response is then

$$X(s) = \frac{s + 10}{s(s^2 + 10s + 200)}$$

If we check the roots of the denominator, we find that they are complex and hence we should complete the square in the quadratic term to give

$$X(s) = \frac{s + 10}{s[(s + 5)^2 + 13.23^2]} \qquad \textbf{(a)}$$

A partial fraction expansion will be of the form

$$X(s) = \frac{A}{s} + \frac{B(s + 5)}{(s + 5)^2 + 13.23^2} + \frac{C(13.23)}{(s + 5)^2 + 13.23^2} \qquad \textbf{(b)}$$

If we evaluate A by the Heaviside method, we may then evaluate B and C by finding a common denominator and equating the numerators of Eqs. **(a)** and **(b)** to give

$$A = -B = 0.05$$

$$C = 0.0567$$

The time response is then found by inversion to be

$$x(t) = 0.05\, u(t) - 0.05e^{-5t} \cos(13.23t) + 0.0567e^{-5t} \sin(13.23t) \text{ m}$$

Thus, a step input in $y(t)$ yields an oscillatory response which settles to a deplacement of 0.05 m for $x(t)$ which is the same steady-state displacement as at the support.

Example 4.2

Consider the two-loop electrical circuit shown in Fig. 4.4. Find the specific transfer function $G(s) = I_2(s)/V(s)$. (It is recognized that there are many other transfer functions which could be evaluated for this network.) By the methods given in Chapter 3, the two loop equations for zero initial conditions are

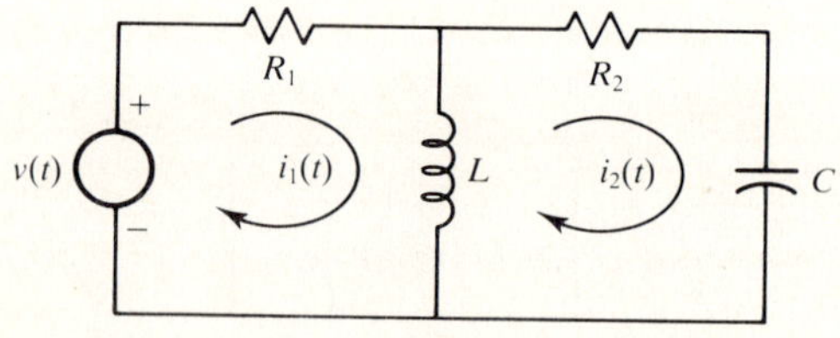

FIGURE 4.4 A two-loop electrical network.

$$R_1 i_1 + L \frac{d}{dt}(i_1 - i_2) = v(t)$$

$$L \frac{d}{dt}(i_2 - i_1) + R_2 i_2 + \frac{1}{C}\int i_2 \, dt = 0$$

If we take the Laplace transform of these equations, we get

$$(Ls + R_1)I_1(s) - Ls\, I_2(s) = V(s)$$

$$-Ls\, I_1(s) + \left(Ls + R_2 + \frac{1}{sC}\right)I_2(s) = 0$$

Although we could solve these two equations by the substitution method, the best way to avoid sign errors is to use Cramer's rule:

$$I_2(s) = \frac{\begin{vmatrix} Ls + R_1 & V(s) \\ -Ls & 0 \end{vmatrix}}{\begin{vmatrix} Ls + R_1 & -Ls \\ -Ls & Ls + R_2 + \dfrac{1}{sC} \end{vmatrix}}$$

Upon evaluation of the determinants, the result is

$$I_2(s) = \frac{V(s)Ls}{(Ls + R_1)\left(Ls + R_2 + \dfrac{1}{sC}\right) - L^2 s^2}$$

After some algebra, the resulting transfer function is

$$\frac{I_2(s)}{V(s)} = \frac{LCs^2}{(R_1 + R_2)LCs^2 + (R_1 R_2 C + L)s + R_1}$$

This transfer function could also be represented by a block diagram as in the previous example.

4.3 THE TRANSFER FUNCTION AND SYSTEM STABILITY

For our purposes, a single-input–single-output linear system can always be represented by a transfer function, which is the ratio of two polynomials in the complex variable s:

$$G(s) = \frac{N(s)}{D(s)} = \frac{b_m s^m + b_{m-1}s^{m-1} + \cdots + b_1 s + b_0}{s^n + a_{n-1}s^{n-1} + \cdots + a_1 s + a_0} \qquad \textbf{(4.3.1)}$$

The numerator and denominator polynomials can be written in factored form to give

$$G(s) = \frac{b_m(s - z_1)(s - z_2) \ldots (s - z_m)}{(s - p_1)(s - p_2) \ldots (s - p_{n-1})(s - p_n)} \qquad n > m \qquad \textbf{(4.3.2)}$$

In Chapter 2 we discussed the concepts of poles and zeros of s-domain functions so these ideas should not be new to the reader of this chapter. The roots of the numerator $z_1, z_2, \ldots, z_m$ are commonly referred to as the *zeros* of $G(s)$, while the roots of the denominator $p_1, p_2, \ldots, p_n$ are commonly called the *poles* of $G(s)$. Recall that the z_i and p_i may be real or occur in complex conjugate pairs.

When the system represented by $G(s)$ is subjected to an input, the response solution will have, in addition to other terms, terms which arise from the poles of $G(s)$. Examination of the locations of the poles $p_1, \ldots, p_n$ in the s plane shows certain things about the system response in the time domain. Poles on the negative real axis contribute responses which die out for large values of time. Poles on the positive real axis contribute terms which increase without bound for large time. A complex pair of poles with positive real parts yields oscillatory responses which increase without bound, while a complex pair of poles with negative real parts yields decaying oscillatory responses. Pairs of purely imaginary poles give oscillations which neither increase nor decrease as time evolves. Poles at the origin of the s plane contribute polynomial type functions of time. These are summarized in Fig. 4.5.

Let us now state some definitions regarding the stability of a system with a transfer function $G(s)$.

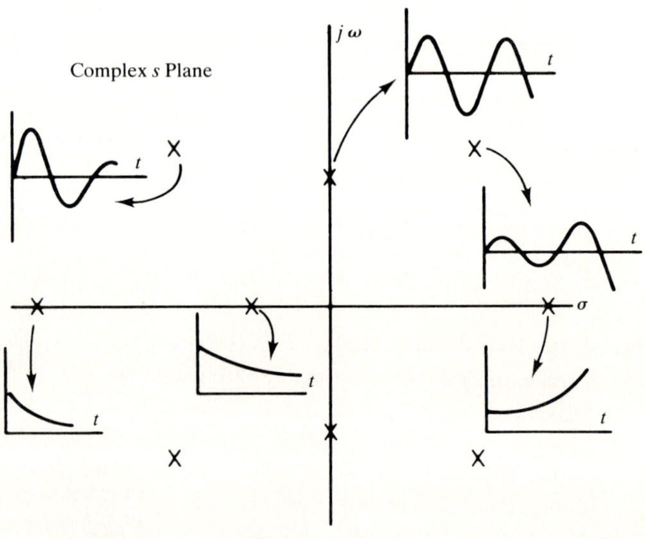

FIGURE 4.5 Pole locations and associated responses.

Definition: A linear system represented by transfer function $G(s)$ is said to be *stable* if all the poles of $G(s)$ lie in the left half of the complex s plane.

Definition: A linear system is said to be *marginally stable* if the poles of $G(s)$ lie in the left half of the complex s plane and at least one complex pair on the imaginary axis.

Definition: A linear system is said to be *unstable* if any of the poles of $G(s)$ lie in the right half of the complex s plane.

Example 4.3

Consider the system of Example 4.1, in which we found the transfer function to be

$$G(s) = \frac{10(s+10)}{s^2 + 10s + 200}$$

If we examine the roots of the quadratic denominator, we find that they are at $s = -5 \pm j13.23$ and in factored form the transfer function is

$$G(s) = \frac{10(s+10)}{(s+5+j13.23)(s+5-j13.23)}$$

From this we may plot the pole-zero diagram of $G(s)$ to yield that of Fig. 4.6. Since the poles lie in the left half of the s plane the system is stable.

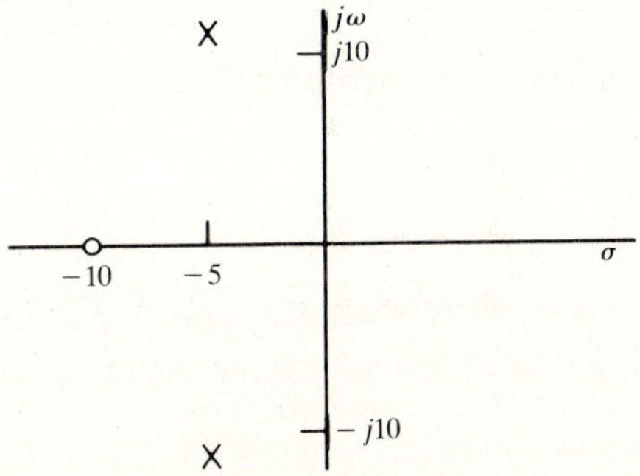

FIGURE 4.6 Pole-zero diagram for a transfer function in the complex s plane.

4.4 STEADY-STATE RESPONSE OF STABLE SYSTEMS TO SINUSOIDAL INPUTS

Consider the stable single-input–single-output system represented by transfer function $G(s)$ to be forced by a sinusoidal forcing function $y(t) = A \cos \omega t$, as shown in Fig. 4.7. The transform of the input is

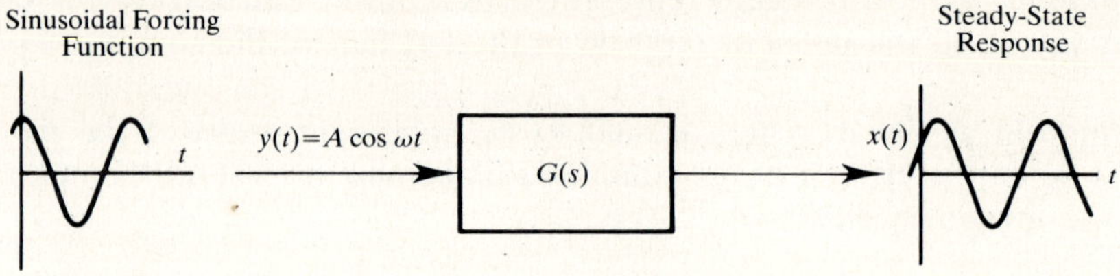

FIGURE 4.7 A sinusoidally forced system.

$$Y(s) = \frac{As}{s^2 + \omega^2} \tag{4.4.1}$$

The response in the s domain can be written as the product of the transfer function and the transform of the input function.

$$X(s) = G(s)\,\frac{As}{s^2 + \omega^2} \tag{4.4.2}$$

But $G(s)$ can be written with a factored denominator as

$$G(s) = \frac{N(s)}{(s - p_1)(s - p_2) \ldots (s - p_n)} \tag{4.4.3}$$

where, since the system is stable, the real parts of the p_i are all negative. After factoring the quadratic denominator, the transform of the response $X(s)$ can be written as

$$X(s) = A\left[\frac{sN(s)}{(s + j\omega)(s - j\omega)(s - p_1) \ldots (s - p_n)}\right] \tag{4.4.4}$$

Let us now write this as a partial fraction expansion:

$$X(s) = A\left[\frac{B}{s + j\omega} + \frac{C}{s - j\omega} + \sum_{i=1}^{n} \frac{K_i}{s - p_i}\right] \tag{4.4.5}$$

If the coefficients B, C, $K_1, \ldots, K_n$ were known, the time-domain solution would be

$$x(t) = A\left(Be^{-j\omega t} + Ce^{j\omega t} + \sum_{i=1}^{n} K_i e^{p_i t}\right) \tag{4.4.6}$$

If the system is stable, all the terms $e^{p_i t}$ approach zero as t becomes large, so the steady-state response is given by the first two terms or

$$x_{ss}(t) = A(Be^{-j\omega t} + Ce^{j\omega t}) \tag{4.4.7}$$

where B and C are still unknown but are given as usual by the method of partial fractions to be

$$B = \left.\frac{G(s)s}{s - j\omega}\right|_{s=-j\omega} = \frac{G(-j\omega)}{2} \tag{4.4.8}$$

and

$$C = \left.\frac{G(s)s}{s + j\omega}\right|_{s=j\omega} = \frac{G(j\omega)}{2} \tag{4.4.9}$$

Since $G(j\omega)$ is a complex quantity, it could be written in polar form as

$$G(j\omega) = |G(j\omega)|e^{j\phi} \tag{4.4.10}$$

where

$$\phi = \angle G(j\omega) \tag{4.4.11}$$

and $G(-j\omega)$ can be written as a quantity with the same magnitude and the conjugate angle

$$G(-j\omega) = |G(j\omega)|e^{-j\phi} \tag{4.4.12}$$

The steady-state response of Eq. (4.4.7) can now be written as

$$x_{ss}(t) = A|G(j\omega)|\left[\frac{e^{-j\omega t}e^{-j\phi}}{2} + \frac{e^{j\omega t}e^{j\phi}}{2}\right] \tag{4.4.13}$$

$$= A|G(j\omega)|\left[\frac{e^{-j(\omega t+\phi)} + e^{j(\omega t+\phi)}}{2}\right] \tag{4.4.14}$$

Using the Euler identity, we recognize that the quantity in the brackets is a cosine function. Then the steady-state response can be written

$$x_{ss}(t) = A|G(j\omega)|\cos(\omega t + \phi) \tag{4.4.15}$$

The function $G(j\omega)$ is often referred to as the *frequency response function* or the *harmonic response function*.

We can write the frequency response function as the ratio of two complex numbers:

$$G(j\omega) = \frac{N(j\omega)}{D(j\omega)} \tag{4.4.16}$$

Then

$$\phi = \angle G(j\omega) = \angle N(j\omega) - \angle D(j\omega) \tag{4.4.17}$$

Frequency response data for a system are commonly presented in the form of plots of $|G(j\omega)|$ as a function of frequency ω and $\angle G(j\omega)$ as a function of ω.

Note that, when the driving function is a sine rather than a cosine, the output expression [Eq. (4.4.15)] should be written with a sine rather than a cosine. Also, if the system is driven by a sum of two sinusoids of different frequencies, the principle of superposition is the appropriate tool. In other words, treat each piece of the driving function separately to calculate its respective response, then sum the responses.

Example 4.4

Let us consider the system of Examples 4.1 and 4.3, wherein the transfer function between the input displacement $y(t)$ and the displacement of the mass $x(t)$ is

$$G(s) = \frac{X(s)}{Y(s)} = \frac{10s + 100}{s^2 + 10s + 200}$$

For the driving displacement of $y(t) = 0.02\cos(20t)$ m, find the steady-state response of the mass.

At radian frequency ω, the frequency response function is

$$G(j\omega) = \frac{10j\omega + 100}{-\omega^2 + j10\omega + 200}$$

The magnitude and phase character of this function are sketched in Fig. 4.8. If we evaluate this function at a frequency of 20 r/s we get

$$G(j20) = \frac{100 + j200}{-400 + j200 + 200}$$

$$= \frac{100 + j200}{-200 + j200} = \frac{2.24 \angle 63.4^\circ}{2.828 \angle 135^\circ}$$

$$= 0.792 \angle -71.6^\circ$$

The steady-state response is then

$$x(t) = 0.0158\cos\left(20t - \frac{71.6^\circ}{57.3^\circ/r}\right)\text{m}$$

Another way to present the data representing the frequency response of a system is in terms of a Bode plot, wherein the frequency response magnitude function in decibels† (dB) is plotted as a function of the logarithm of

---

†The term decibel has acoustic origins to be used later in electronic amplifier design relative to transcontinental telephony. The definition given in expression (4.4.18) is now commonly used in system analysis and will, in this context, serve as the origin.

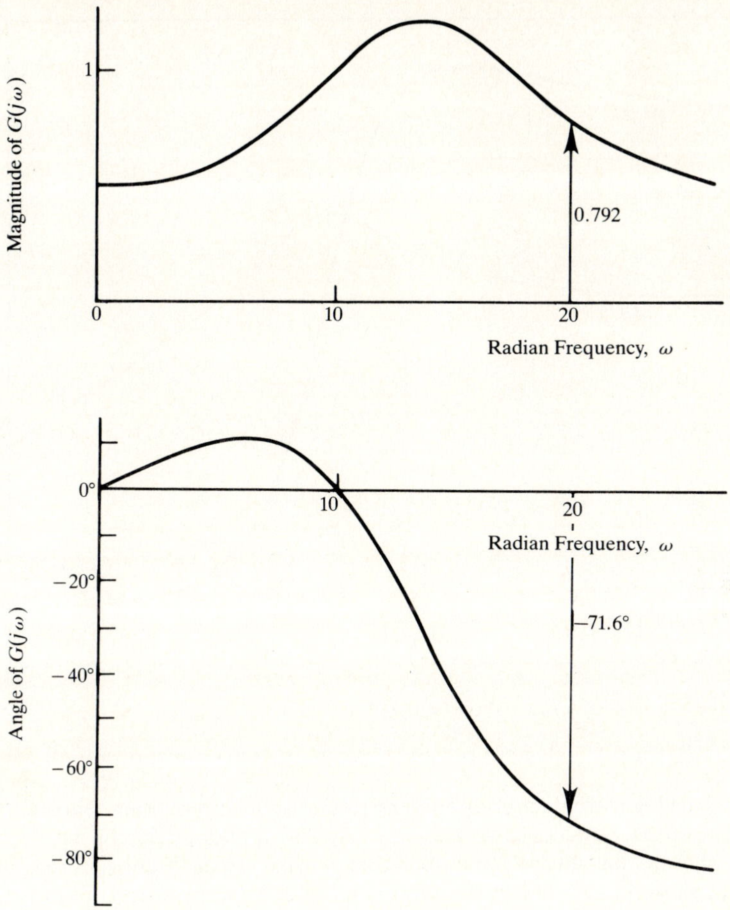

FIGURE 4.8 Magnitude and phase character of the frequency response function for Example 4.4.

frequency. The gain of the system in decibels is defined as

$$\mathrm{dB}(\omega) = 20 \log_{10}(|G(j\omega)|) \tag{4.4.18}$$

It is also common to plot phase as a function of the logarithm of frequency.

Example 4.5

Present the data generated in Example 4.4 in the form of decibels and phase versus frequency on a logarithmic scale.

Using relation (4.4.18), the data of Fig. 4.8 can be converted to the plots illustrated in Fig. 4.9.

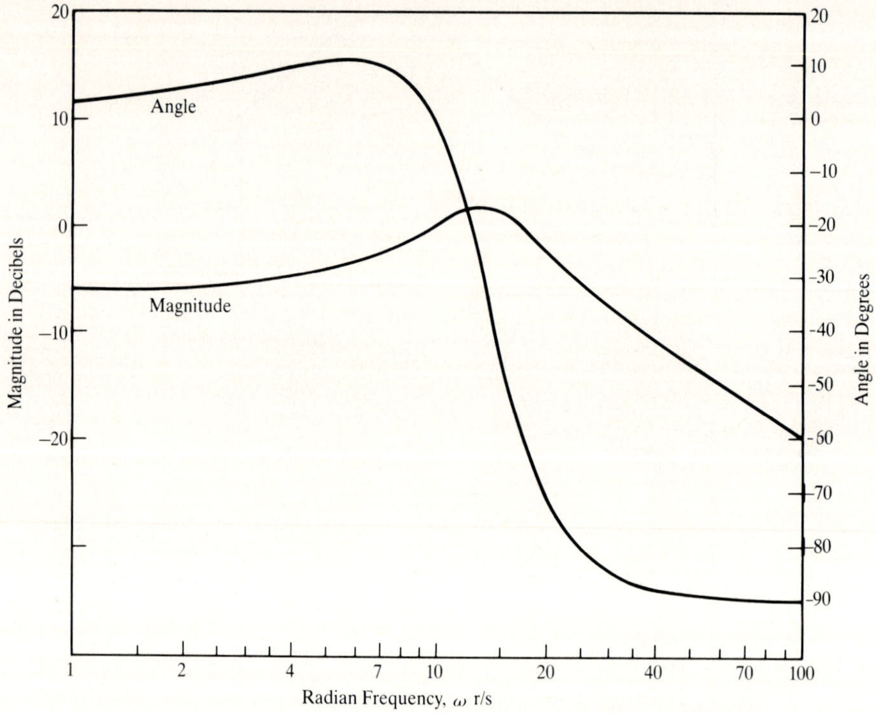

FIGURE 4.9 Decibel presentation of frequency response data.

This logarithmic presentation of frequency response data tends to emphasize the frequency response character (lowpass, bandpass, highpass, notch, etc.) but one should always remember that regions of the frequency response are never as well defined as one might be led to believe by a Bode plot. Frequency response character for typical systems is shown in both rectangular and logarithmic presentation in Fig. 4.10 for lowpass, bandpass, and highpass cases. Note that, if the magnitude of $G(j\omega)$ is less than unity, the gain in dB will be negative.

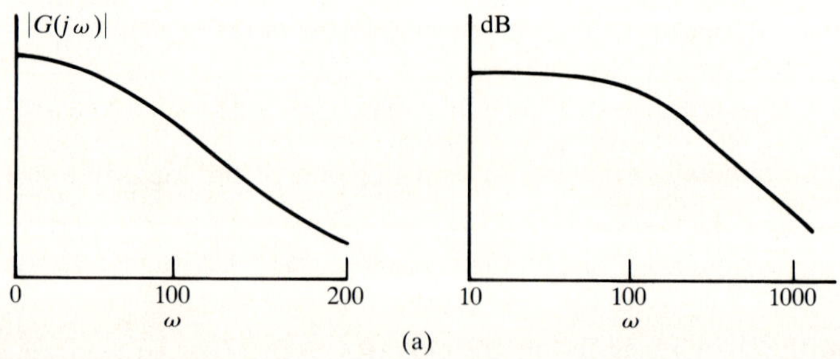

FIGURE 4.10 (a) Lowpass frequency response character. (b) Bandpass frequency response character. (c) Highpass frequency response character.

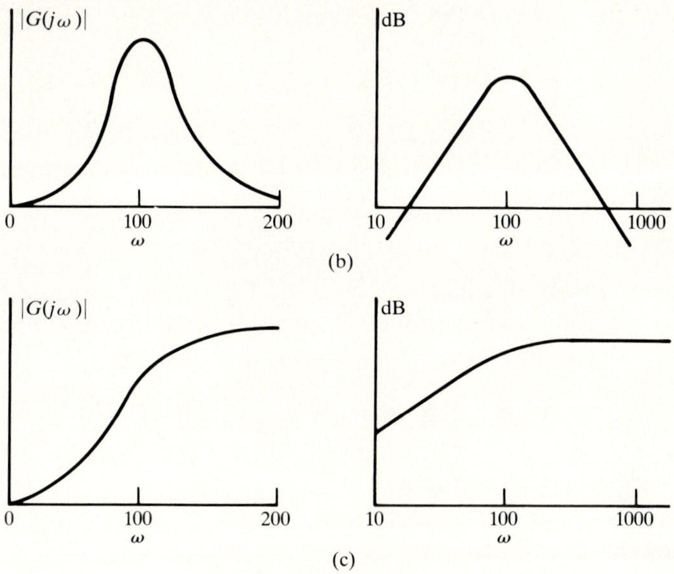

FIGURE 4.10 (*Continued*).

4.5 GENERALIZATION FOR SIMPLE SYSTEMS

In this section we shall examine, in a general fashion, the time and frequency response character of systems described by first- and second-order differential equations.

4.5.1 First-Order Systems

The student is certainly familiar with physical problems which are described by first-order linear ordinary differential equations such as electric circuits containing resistors and capacitors or resistors and inductors in combination with sources and/or initial stored energy. The solutions of such systems are characterized in part by a single exponential term of the form $e^{-t/\tau}$, where the parameter τ, called the *time constant*, has units of seconds and is a measure of the speed with which the system responds. In other words, the smaller the time constant τ, the faster the system responds. Simple problems involving hydraulic systems, thermal systems, and chemical kinetics often result in first-order systems, and it is this wide scope of problems which makes the generalization for first-order systems desirable.

Consider the system governed by the first-order differential equation

$$\frac{dx}{dt} + \beta x = \beta y \tag{4.5.1}$$

Transforming, and ignoring the initial conditions, we get

$$sX(s) + \beta X(s) = \beta Y(s) \tag{4.5.2}$$

and solving for the transfer function gives

$$G(s) = \frac{X(s)}{Y(s)} = \frac{\beta}{s + \beta} \tag{4.5.3}$$

This may be thought of conceptually as in Fig. 4.11. There is only one system parameter, β. This parameter appears in time-domain responses of the form $e^{-\beta t}$, so we may draw the following conclusions:

1. There is only one system pole located at $s = -\beta$.
2. The parameter β is the reciprocal of the system time constant, which is a measure of the speed of response.

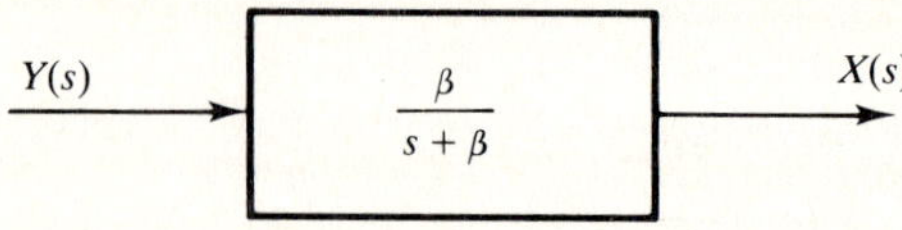

FIGURE 4.11 First-order system block diagram.

Let us now examine the system step response where

$$y(t) = Au(t) \tag{4.5.4}$$

where $u(t)$ is a unit step function; this expression has Laplace transform

$$Y(s) = \frac{A}{s} \tag{4.5.5}$$

Then the system response is

$$X(s) = G(s)Y(s) = A\frac{\beta}{s(s + \beta)} \tag{4.5.6}$$

and, upon inverting the transform, we get

$$x(t) = A(1 - e^{-\beta t}) \tag{4.5.7}$$

which is illustrated in Fig. 4.12.

Let us now consider the frequency response of the first-order system which we get by letting s be $j\omega$ in the transfer function (Eq. (4.5.3)) or

$$G(j\omega) = \frac{\beta}{j\omega + \beta} \tag{4.5.8}$$

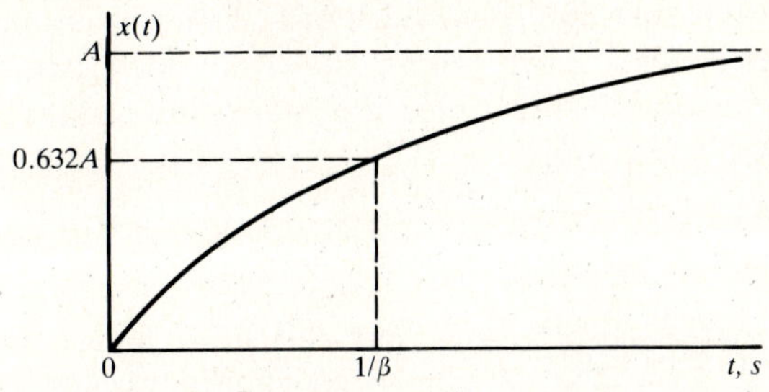

FIGURE 4.12 First-order system step response.

We are usually interested in the magnitude and phase information associated with the frequency response functions which in this case are, respectively,

$$|G(j\omega)| = \frac{\beta}{\sqrt{\omega^2 + \beta^2}} = \frac{1}{\sqrt{1 + (\omega/\beta)^2}} \tag{4.5.9}$$

and

$$\angle G(j\omega) = -\tan^{-1}\left(\frac{\omega}{\beta}\right) \tag{4.5.10}$$

If we plot these data, we get the frequency response plots of Fig. 4.13. From Fig. 4.13 we see that the parameter β is representative of the width of the

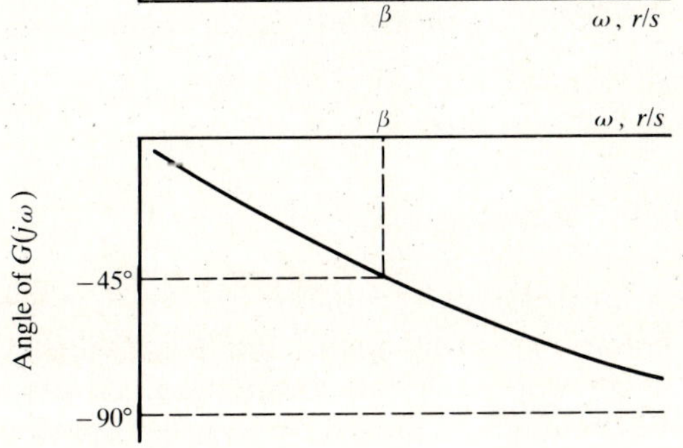

FIGURE 4.13 Frequency response data for a first-order system.

pass band of the system. The bandwidth of a system is commonly defined as the span of frequencies over which the frequency response magnitude function is greater than $1/\sqrt{2}$ times the maximum value. In this case the bandwidth is β r/s.

Example 4.6

Consider the mechanical system consisting of a damper and spring as illustrated in Fig. 4.14. The differential equation governing the motion of the junction of the two mechanical elements is

$$b\frac{dx}{dt} + kx = f(t)$$

or

$$\frac{dx}{dt} + \frac{k}{b}x = \frac{1}{b}f(t)$$

Comparing this equation with Eq. (4.5.1) indicates that the parameter $\beta = k/b$ or the time constant is $\tau = b/k$. This makes sense, since large b (a sluggish damper) will cause slow response. Likewise, the higher the spring stiffness k, the faster the system response. If we use the following numerical values for k and b

$$k = 3 \times 10^3 \text{ N/m}$$

$$b = 20 \text{ N}\cdot\text{s/m}$$

the time constant τ is 6.67×10^{-3}s and the bandwidth β is 150 r/s.

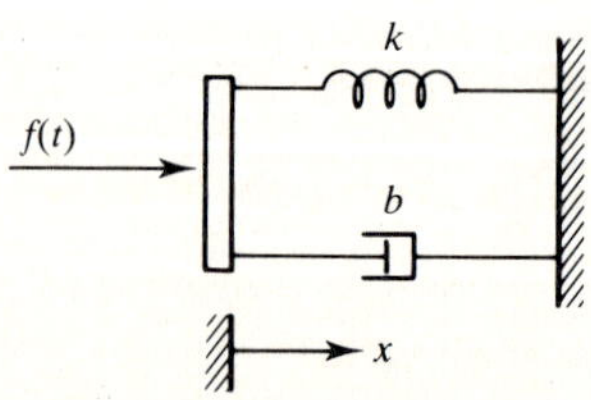

FIGURE 4.14 Spring-damper shock absorbing system.

4.5.2 Second-Order Systems

A number of important physical problems can be represented by second-order differential equations. Problems of this type which we have encountered thus far are those of mechanical vibration and electric circuit problems in which both inductors and capacitors are incorporated. The natural responses of such systems are generally damped or undamped oscillations and, when driven by a sinusoid at a critical frequency, the system will tend to oscillate with large amplitude (or resonate). This particular frequency is determined by the system parameters. The responses of this type of system differ radically from those of the simple first-order system which we have just examined. Consider the system governed by the differential equation

$$\frac{d^2x}{dt^2} + 2\zeta\omega_n \frac{dx}{dt} + \omega_n^2 x = \omega_n^2 y \qquad \textbf{(4.5.11)}$$

where the parameters ω_n and ζ are both positive and $y(t)$ is an arbitrary driving function. Taking the Laplace transform and ignoring initial conditions yields

$$(s^2 + 2\zeta\omega_n s + \omega_n^2)X(s) = \omega_n^2 Y(s) \qquad \textbf{(4.5.12)}$$

and solving for the transfer function gives

$$G(s) = \frac{X(s)}{Y(s)} = \frac{\omega_n^2}{s^2 + 2\zeta\omega_n s + \omega_n^2} \qquad \textbf{(4.5.13)}$$

Clearly, this system has two parameters, ω_n and ζ, which influence the location of the poles of $G(s)$. The parameter ω_n is commonly called the *undamped natural frequency*, while the parameter ζ is called the *damping ratio*. The denominator polynomial or system characteristic equation is

$$s^2 + 2\zeta\omega_n s + \omega_n^2 = 0 \qquad \textbf{(4.5.14)}$$

There are a number of possibilities for the location of the roots of this equation, depending on the relative magnitudes of ω_n and ζ. Let us complete the square in Eq. (4.5.14) to yield

$$(s + \zeta\omega_n)^2 + \omega_n^2(1 - \zeta^2) = 0 \qquad \textbf{(4.5.15)}$$

We shall now consider the three possible cases separately.

Case I $(0 \le \zeta \le 1)$ Clearly, if $\zeta < 1$, the roots are a complex conjugate pair and are

$$s_{1,2} = -\zeta\omega_n \pm j\omega_n\sqrt{1 - \zeta^2} \qquad \textbf{(4.5.16)}$$

These roots are located as indicated in Fig. 4.15. This case is often referred to as the *underdamped* case. The solutions associated with the roots of Eq. (4.5.14) are

$$e^{-\zeta\omega_n t} \cos \omega_n\sqrt{1 - \zeta^2}\, t,\; e^{-\zeta\omega_n t} \sin \omega_n\sqrt{1 - \zeta^2}\, t \qquad \textbf{(4.5.17)}$$

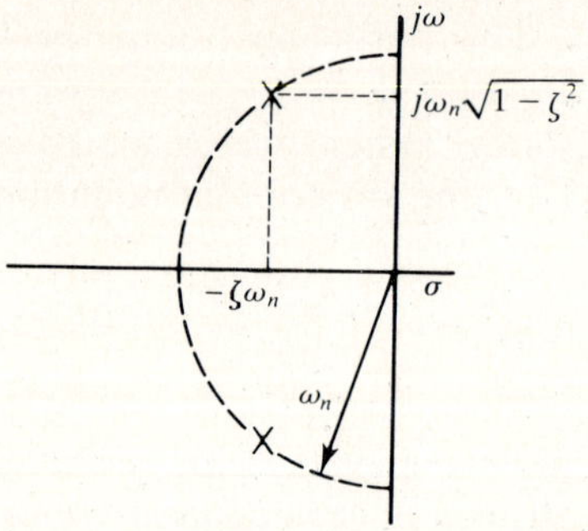

FIGURE 4.15 Underdamped pole locations ($\zeta < 1$).

The frequency $\omega_n\sqrt{1-\zeta^2}$ which appears in the sinusoids is referred to as the *damped natural frequency* and is commonly given the symbol ω_d. Also, it is interesting to note that if $\zeta = 0$ the roots are purely imaginary and the solutions are sustained sines and cosines.

Case II ($\zeta = 1$) In this case the roots of Eq. (4.5.14) are real and equal or

$$s_{1,2} = -\omega_n \qquad \textbf{(4.5.18)}$$

which means that they are coincident in the s plane as shown in Fig. 4.16. This is referred to as the *critically damped case* and the solutions associated with the poles of $G(s)$ are given by the repeated real root case for Laplace transforms to be

$$te^{-\omega_n t},\ e^{-\omega_n t} \qquad \textbf{(4.5.19)}$$

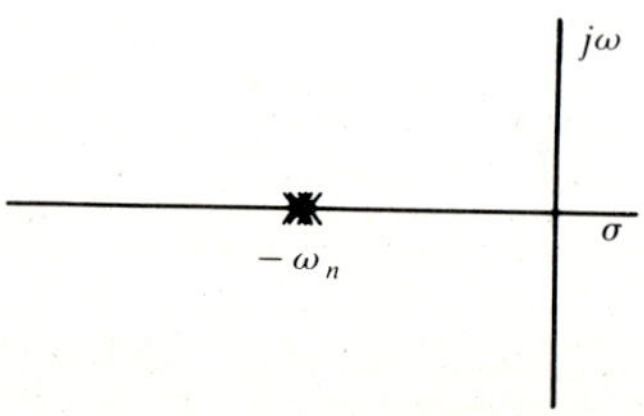

FIGURE 4.16 Critically damped pole locations ($\zeta = 1$).

Case III ($\zeta > 1$) In this case the roots of Eq. (4.5.14) are real, negative, and distinct or

$$s_{1,2} = -\zeta\omega_n \pm \omega_n\sqrt{\zeta^2 - 1} \qquad \textbf{(4.5.20)}$$

and these roots are located symmetrically about $s = -\zeta\omega_n$ and are shown in Fig. 4.17. This case is commonly referred to as the *overdamped case* and the solutions associated with these poles are

$$e^{-(\zeta\omega_n + \omega_n\sqrt{\zeta^2-1})t},\ e^{-(\zeta\omega_n - \omega_n\sqrt{\zeta^2-1})t} \qquad \textbf{(4.5.21)}$$

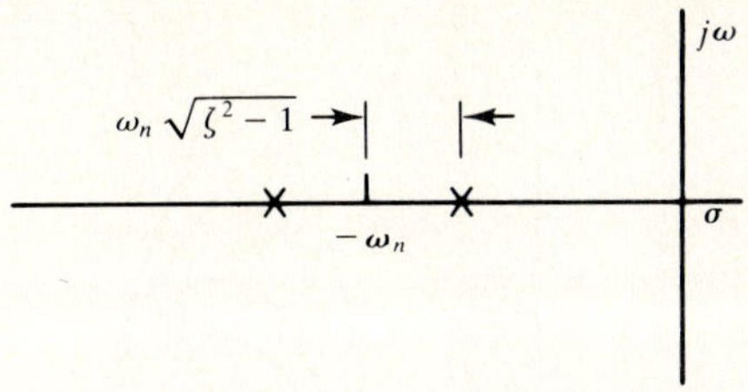

FIGURE 4.17 Overdamped pole locations.

Second-Order System Step Response Let us consider a system with a transfer function given by Eq. (4.5.13) which is forced from rest (zero initial conditions) by a step forcing function $y(t)$ of magnitude A which has transform $Y(s) = A/s$. The transform of the response is given by

$$X(s) = \frac{A\omega_n^2}{s(s^2 + 2\zeta\omega_n s + \omega_n^2)} \tag{4.5.22}$$

If we perform a partial fraction expansion, the resulting response is composed of a constant term and either exponential terms or damped sines and cosines. We must then perform the partial fraction expansion for the three cases of damping:

$$x(t) = \begin{cases} A\left[1 - e^{-\zeta\omega_n t}\left(\cos\omega_d t + \dfrac{\zeta}{\sqrt{1-\zeta^2}}\sin\omega_d t\right)\right]u(t) \quad 0 < \zeta \le 1 & \text{(4.5.23a)} \\ A[1 - (1 + \omega_n t)e^{-\omega_n t}]u(t) \quad \zeta = 1 & \text{(4.5.23b)} \\ A\left[1 + \dfrac{1}{2(\zeta^2-1) - 2\zeta\sqrt{\zeta^2-1}}\, e^{-\omega_n(\zeta - \sqrt{\zeta^2-1})t} \right. & \\ \quad + \dfrac{1}{2(\zeta^2-1) + 2\zeta\sqrt{\zeta^2-1}} & \\ \quad \left. \times\, e^{-\omega_n(\zeta + \sqrt{\zeta^2-1})t}\right]u(t) \quad \zeta > 1 & \text{(4.5.23c)} \end{cases}$$

These step responses are shown plotted in dimensionless form in Fig. 4.18 for a variety of values of the damping ratio ζ. Note that for the underdamped case ($0 < \zeta < 1$), the response overshoots the value which it reaches for large time. If we evaluate expression (4.5.23a) for $\omega_d t = \pi$ the peak value will be given and that is

$$x_p = A\left[1 + \exp\left(\frac{-\zeta\pi}{\sqrt{1-\zeta^2}}\right)\right] \tag{4.5.24}$$

and the percent overshoot is then

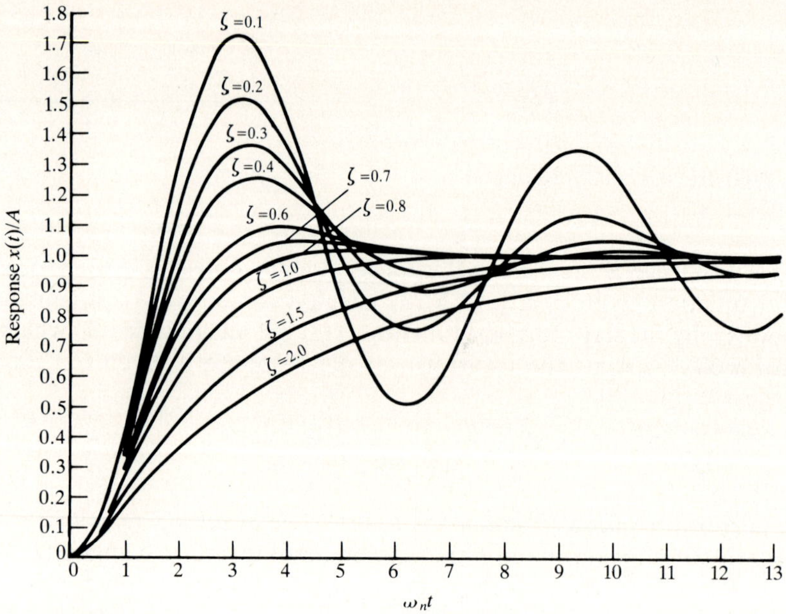

FIGURE 4.18 Second-order system step response for various values of the damping ratio.

$$PO = \exp\left(\frac{-\zeta\pi}{\sqrt{1-\zeta^2}}\right) \times 100 \tag{4.5.25}$$

The percent overshoot is shown in Fig. 4.19 as a function of the damping ratio ζ.

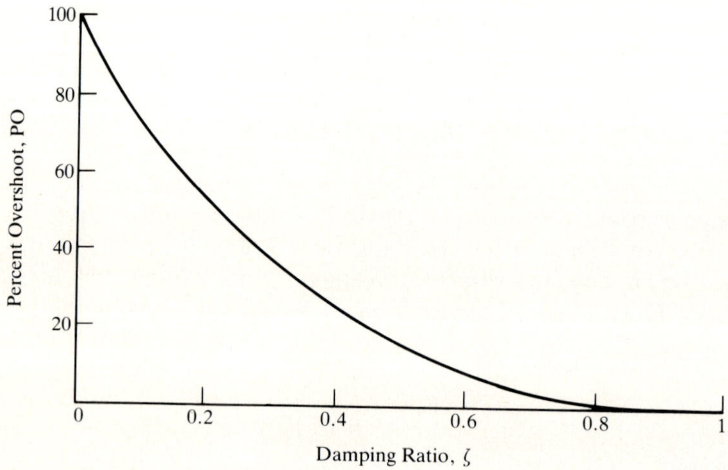

FIGURE 4.19 Variation of percent overshoot with damping ratio.

Frequency Response of Second-Order Systems If the input to the second-order system with the transfer function of expression (4.5.13) were sinusoidal, we would be interested in the character (amplitude and phase) of the steady-state response to the oscillatory forcing function. This is reflected in the harmonic response function for the system $G(j\omega)$ which is given by letting s be $j\omega$ in Eq. (4.5.13) or

$$G(j\omega) = \frac{\omega_n^2}{-\omega^2 + j2\zeta\omega_n\omega + \omega_n^2} \tag{4.5.26}$$

If we divide both numerator and denominator by ω_n^2 we get

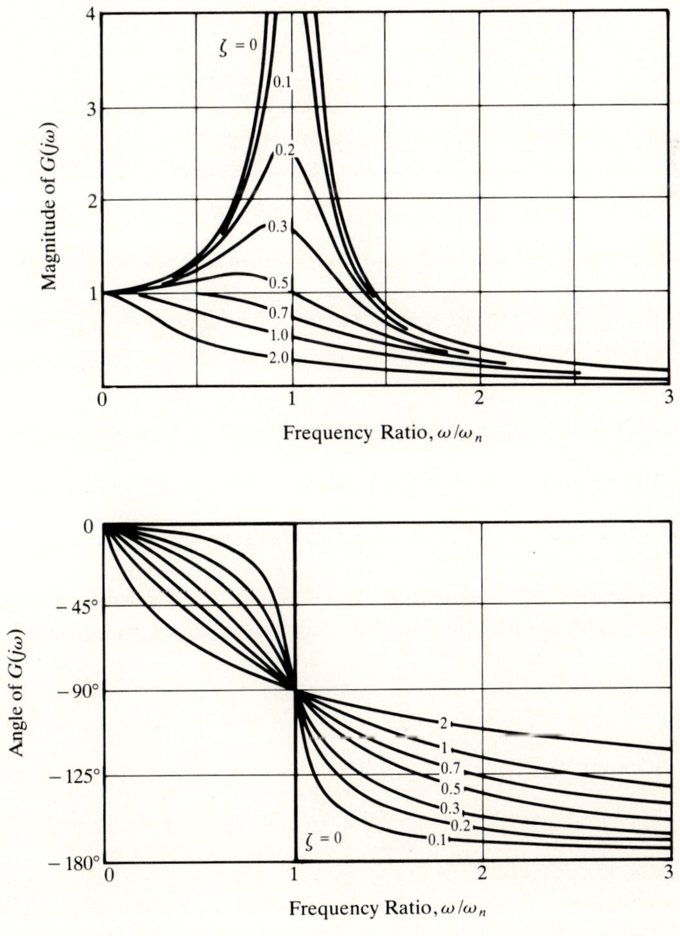

FIGURE 4.20 Magnitude and phase character of a second-order system for various damping ratios.

$$G(j\omega) = \frac{1}{1 - \left(\dfrac{\omega}{\omega_n}\right)^2 + j2\zeta\left(\dfrac{\omega}{\omega_n}\right)} \qquad \textbf{(4.5.27)}$$

The magnitude of $G(j\omega)$ is given easily by

$$|G(j\omega)| = \frac{1}{\sqrt{\left[1 - \left(\dfrac{\omega}{\omega_n}\right)\right]^2 + \left(2\zeta\,\dfrac{\omega}{\omega_n}\right)^2}} \qquad \textbf{(4.5.28)}$$

and the angle of $G(j\omega)$ is

$$\angle G(j\omega) = \tan^{-1}\left(\frac{2\zeta\,\dfrac{\omega}{\omega_n}}{1 - \left(\dfrac{\omega}{\omega_n}\right)^2}\right) \qquad \textbf{(4.5.29)}$$

The magnitude and phase for various values of the damping ratio ζ are given in Fig. 4.20.

Example 4.7

Consider the mechanical system examined in Examples 4.1 and 4.3 in which the transfer function was

$$G(s) = \frac{X(s)}{Y(s)} = \frac{10s + 100}{s^2 + 10s + 200}$$

The characteristic equation for this second-order system is

$$s^2 + 10s + 200 = 0$$

Comparing this expression with Eq. (4.5.14), the following conclusions can be drawn. The undamped natural frequency ω_n is given by

$$\omega_n^2 = 200$$

Thus

$$\omega_n = 14.14 \text{ r/s}$$

and from the middle term

$$2\zeta\omega_n = 10$$

Then the damping ratio ζ is

$$\zeta = 0.354$$

and the system is underdamped. Note that this system has an s in the numerator of the transfer function and thus this second-order system is *not* the same as that discussed earlier in this section. The results concerning step response in Figs. 4.18 and 4.19 are also not applicable to this problem. Also, the frequency response data of Fig. 4.20 are not applicable to this system, as one can see by comparison of Fig. 4.8 and Fig. 4.19 for a damping ratio of 0.5.

We have given the frequency response curves for first- and second-order systems in Figs. 4.13 and 4.20. We must recall that the second-order curves of Fig. 4.20 are restricted to a system with a constant transfer function numerator.

4.6 THE CONVOLUTION THEOREM AND THE IMPULSE RESPONSE

An important theorem in the theory of the Laplace transform is the convolution theorem. This theorem has important applications in system theory, communication theory, solid mechanics, and probability theory.

Given two Laplace transformable functions $f_1(t)$ and $f_2(t)$ which are each zero for $t<0$ and which have respective transforms $F_1(s)$ and $F_2(s)$, the inverse transform of the product of $F_1(s)$ and $F_2(s)$ is given by the following integral:

$$\mathscr{L}^{-1}[F_1(s)F_2(s)] = \int_0^t f_1(\tau)f_2(t-\tau)\,d\tau \tag{4.6.1}$$

We shall accomplish the proof by considering the forward transform of Eq. (4.6.1) or proving that

$$F_1(s)F_2(s) = \mathscr{L}\left[\int_0^t f_1(\tau)f_2(t-\tau)\,d\tau\right] \tag{4.6.2}$$

Let us accomplish the forward transform indicated in Eq. (4.6.2). We shall refer to the integral as I, so, from the transform definition,

$$\mathscr{L}[I] = \int_0^\infty \int_0^t f_1(\tau)f_2(t-\tau)\,d\tau\, e^{-st}\,dt \tag{4.6.3}$$

Since $f_2(t-\tau)$ vanishes for negative arguments, $f_2(t-\tau)=0$ for $\tau>t$, so the upper limit may be written as infinity. Then

$$\mathscr{L}[I] = \int_0^\infty \int_0^\infty f_1(\tau)f_2(t-\tau)\,d\tau\, e^{-st}\,dt \tag{4.6.4}$$

Let us now make a change of variable such that $t-\tau=\alpha$ so that, when $t=\infty$, $\alpha=\infty$, and when $t=0$, $\alpha=-\tau$ and $dt=d\alpha$. Then Eq. (4.6.4) yields

$$\mathscr{L}[I] = \int_{-\tau}^\infty \int_0^\infty f_1(\tau)f_2(\alpha)e^{-s(\alpha+\tau)}\,d\tau\,d\alpha \tag{4.6.5}$$

If $f_1(\alpha)$ vanishes for negative α, then the lower limit on the α integral may be made zero; so we may now change the order of integration to give

$$\mathscr{L}[I] = \int_0^\infty f_1(\tau)e^{-s\tau}\, d\tau \int_0^\infty f_2(\alpha)e^{-s\alpha}\, d\alpha \qquad \textbf{(4.6.6)}$$

It is now easy to recognize each separate integral as the respective Laplace transforms of $f_2(\alpha)$ and $f_1(\tau)$; then

$$\mathscr{L}\left[\int_0^t f_1(\tau)f_2(t-\tau)\, d\tau\right] = F_1(s)F_2(s) \qquad \textbf{(4.6.7)}$$

and the proof of Eq. (4.6.1) is complete by taking the inverse transform of both sides of Eq. (4.6.7) or

$$\mathscr{L}^{-1}[F_1(s)F_2(s)] = \int_0^t f_1(\tau)f_2(t-\tau)\, d\tau \qquad \textbf{(4.6.8)}$$

We should note that the arguments of $f_1(.)$ and $f_2(.)$ may be interchanged without changing the result.

Example 4.8

Use the convolution theorem to evaluate the inverse transform of the product of the two s-domain functions given below:

$$F(s) = F_1(s)F_2(s) = \frac{1}{s+a}\,\frac{1}{s} \qquad (*)$$

By the convolution theorem, the inverse transform is

$$f(t) = \int_0^t e^{-a\tau}u(t-\tau)\, d\tau$$

By definition of the unit step function $u(t-\tau) = 1$ for $\tau < t$, so

$$f(t) = \int_0^t e^{-a\tau}\, d\tau$$

$$= \frac{-1}{a}\, e^{-a\tau}\Big|_0^t = -\frac{1}{a}\,(e^{-at} - 1)$$

which is the same answer which would be given by a partial fraction expression of Eq. (*). One error commonly made by students studying the convolution theorem is to say that the inverse of the product of two s-domain functions is the product of the individual inverses. This is not true and may be easily verified for the particular functions involved in Example 4.7. We may thus write in general

$$\mathscr{L}^{-1}[F_1(s)F_2(s)] \neq f_1(t)f_2(t) \qquad \textbf{(4.6.9)}$$

4.6.1 Application of the Convolution Theorem to System Response

One of the important applications of the convolution theorem is in the calculation of the forced response of a linear system. Let us consider the system of Fig. 4.21 with a transfer function $G(s)$ and all zero initial conditions. Let us examine the case where the forcing function $y(t)$ is a Dirac delta or impulse function or $y(t) = \delta(t)$. The transform of the input is unity or

$$\mathscr{L}[\delta(t)] = 1 \tag{4.6.10}$$

and thus the s-domain response is

$$X(s) = 1 \cdot G(s) = G(s) \tag{4.6.11}$$

and in the time domain

$$x(t) = g(t) \tag{4.6.12}$$

where $g(t)$ is the inverse transform of $G(s)$ and is commonly referred to as the *impulse response function* of the system.

Now let us consider the response of the system to some arbitrary forcing function $y(t)$. If we find the response by employing the transfer function we get

$$X(s) = G(s)\,Y(s) \tag{4.6.13}$$

This is exactly the form to which the convolution theorem is applicable. If we apply expression (4.6.8) we get

$$x(t) = \int_0^t g(\tau)y(t-\tau)\,d\tau \tag{4.6.14}$$

or

$$x(t) = \int_0^t g(t-\tau)y(\tau)\,d\tau \tag{4.6.15}$$

Often it is easier conceptually to write the convolution integral than to perform the actual integration; however, if one is careful with the arguments

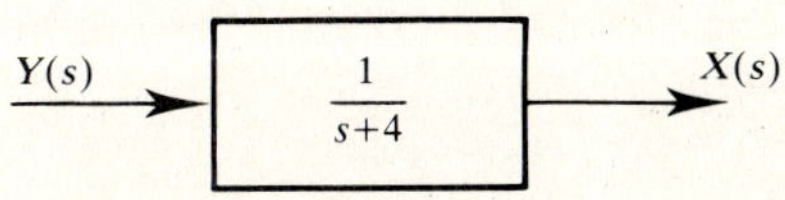

FIGURE 4.21 Linear system.

and limits of integration, then the correct response is given. To interpret graphically the elements of the integrand in expression (4.6.14), let us consider the step response of a lightly damped second-order system.

The input $y(t)$ and the impulse response $g(t)$ functions are shown in Figs. 4.22(a) and (b), respectively. Note that the variable of integration in Eq. (4.6.14) is τ and not t. For some particular time $t = t_1$, both elements of the integrand are shown in Fig. 4.22(c). The product of these functions is illustrated in Fig. 4.22(d); the area beneath this curve represents the response $x(t_1)$. The situation for some $t_2 > t_1$ is illustrated in Fig. 4.22(e) and the total response curve $x(t)$ is plotted in Fig. 4.22(f).

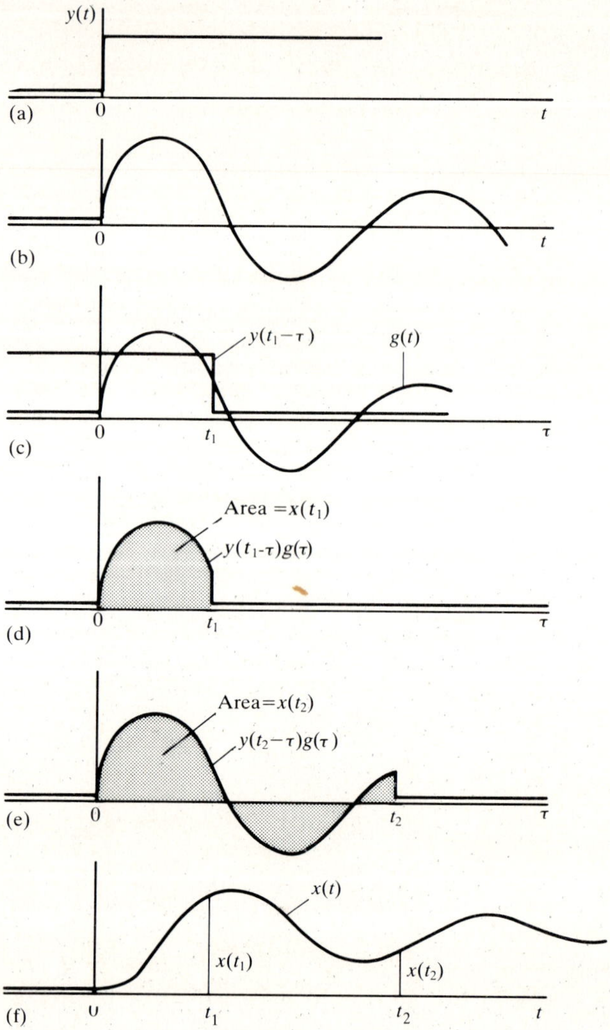

FIGURE 4.22 Graphical interpretation of second-order system step response.

Example 4.9

A system with a transfer function of

$$G(s) = \frac{1}{s+1}$$

is driven from rest by an exponential function

$$y(t) = e^{-3t}$$

Find the system response $x(t)$ by application of the convolution integral. The system impulse response is

$$g(t) = \begin{cases} 0 & t<0 \\ e^{-t} & t\geq 0 \end{cases}$$

Then the response is given by Eq. (4.6.14) to be

$$x(t) = \int_0^t e^{-\tau} e^{-3(t-\tau)}\, d\tau \qquad (*)$$

Let us now complete the integration indicated in Eq. (*) to give

$$\begin{aligned} x(t) &= e^{-3t} \int_0^t e^{2\tau}\, d\tau \\ &= e^{-3t} \frac{1}{2} e^{2\tau}\Big|_0^t \\ &= \frac{e^{-3t}}{2} (e^{2t} - 1) \\ &= \frac{1}{2} (e^{-t} - e^{-3t}) \end{aligned}$$

which is the same result which would be given by the inversion of the resulting product of transforms.

4.7 SYSTEM BLOCK DIAGRAMS

A system composed of more than one element can also be characterized by a composite block diagram of transfer functions and signal summers. These two elements are shown in Fig. 4.23 along with the governing equation.

Any linear system can be thought of as composed of combinations of transfer functions as in Fig. 4.23(a) and summers as in 4.23(b). A summer may have any number of inputs and only one output. The signs present at each input determine whether that input is added or subtracted to produce the output. When two blocks are in cascade, connected as shown in Fig.

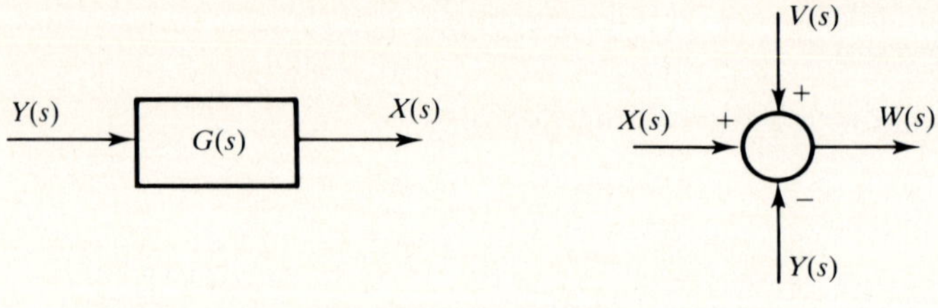

$X(s) = G(s)Y(s)$

(a)

$W(s) = V(s) + X(s) - Y(s)$

(b)

FIGURE 4.23 System block diagram elements.

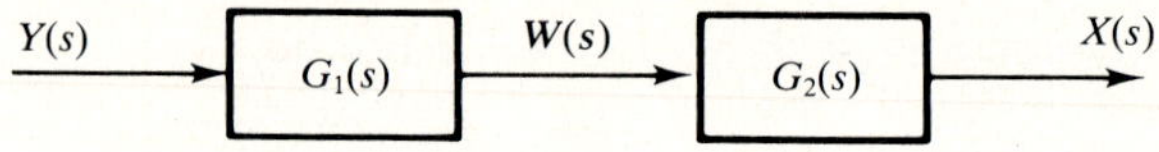

FIGURE 4.24 Cascaded transfer functions.

4.24, it will be assumed that the output of block one is not affected by the attachment of block two. In other words, block two does not load and therefore change the output of block one. For this configuration the equivalent input-output relation is

$$G(s) = \frac{X(s)}{Y(s)} = G_1(s)G_2(s) \tag{4.7.1}$$

The input-output relation for the system may be obtained by writing the governing equations for each system element and combining them. This procedure is best illustrated by the example which follows.

Example 4.10

Find the transfer function $X(s)/Y(s)$ for the complex block diagram of Fig. 4.25, where it is implied that both transfer functions and signals are functions of the variable s. The first step in the solution is to label all signals V, W, Z, and Q, as shown in the figure. Let us now write the equations for each element in the block diagram. For the left summer

$$W = Y - V \tag{a}$$

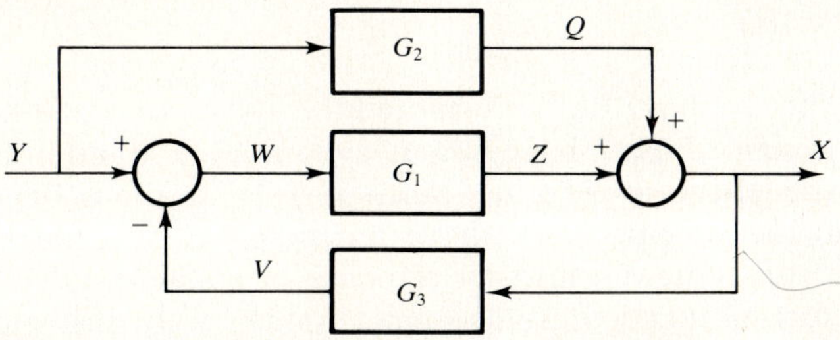

FIGURE 4.25 A block diagram with feed forward and feedback paths.

For G_1

$$Z = G_1 W \tag{b}$$

For G_2

$$Q = G_2 Y \tag{c}$$

For G_3

$$V = G_3 X \tag{d}$$

And for the right summer

$$X = Z + Q \tag{e}$$

Substitution of Eqs. **(b)** and **(c)** for Z and Q, respectively, gives

$$X = G_1 W + G_2 Y$$

Now substitute Eq. **(a)** for W to give

$$X = G_1(Y - V) + G_2 Y$$

and finally substitute Eq. **(d)** for V to give

$$X = G_1(Y - G_3 X) + G_2 Y$$

Now isolate X and Y to give

$$X(1 + G_1 G_3) = Y(G_1 + G_2)$$

The input-output transfer function is

$$\frac{X}{Y} = \frac{G_1 + G_2}{1 + G_1 G_3}$$

4.8 SUMMARY

In this chapter we have further explored the system idea and how we could represent a single-input–single-output system by a transfer function. We discussed the effects of transfer function pole locations on the dynamic response to the system. The transfer function was then modified so as to predict the steady-state sinusoidal response of a stable system driven by a sinusoidal forcing function. Some simple systems were characterized generally, and time and frequency response characteristics of first- and second-order systems were explored. The convolution theorem was developed and its application to system forced response was explored. The representation of systems by block diagrams was then discussed and the reduction of complex block diagrams to less complex diagrams was undertaken.

REFERENCES

1. Cheng, D. K. *Analysis of Linear Systems*. Addison-Wesley, Reading, MA, 1959.
2. Dorf, R. C. *Modern Control Systems*, 3d. ed. Addison-Wesley, Reading, MA, 1980.
3. D'Azzo, J. J., and C. H. Houpis. *Feedback Control System Analysis and Synthesis*. 2d ed. McGraw-Hill, New York, 1966.

PROBLEMS

4-1. For the mechanical systems shown below, find the transfer function $X(s)/Y(s)$. Assume that $y(t)$ is a known function in each case.

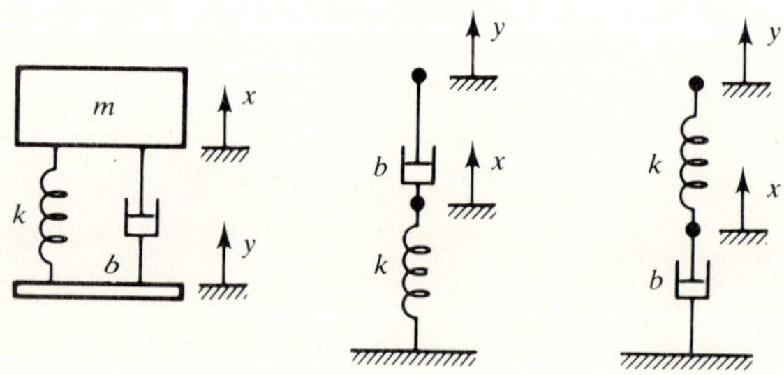

(a) (b) (c)

4-2. Find the transfer function $V_0(s)/V_i(s)$ for the three networks given below. Assume that no current exists in the output circuit branches.

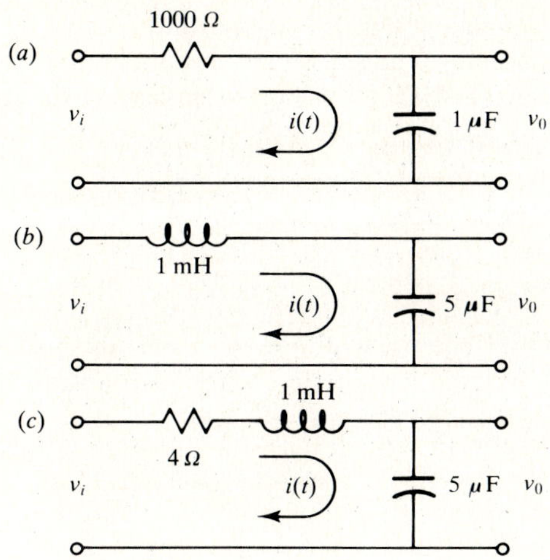

4-3. For the electrical network shown, find the transfer function $G(s) = V_2(s)/V_1(s)$.

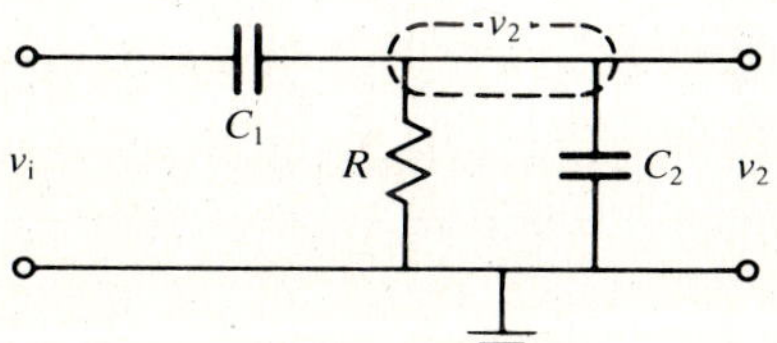

4-4. For the network shown, find the transfer function $G_1(s) = I_1(s)/V(s)$.

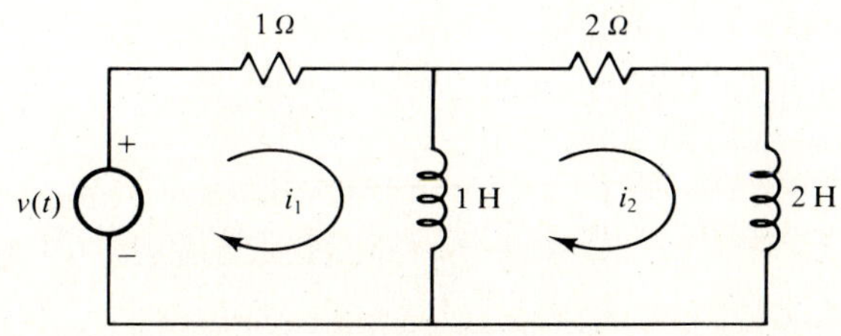

4-5. For the network shown, find the transfer function $V_0(s)/V_i(s)$.

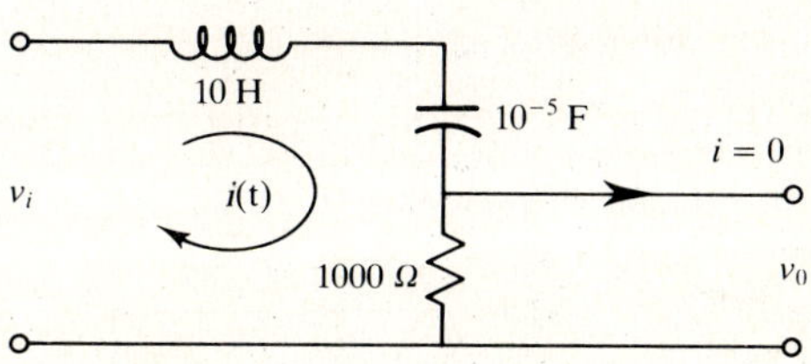

4-6. A linear system is shown below and the time-domain input and output are as shown. (*a*) Assuming that the initial conditions are zero, find the transfer function $G(s) = X(s)/Y(s)$. (*b*) Find the governing differential equation.

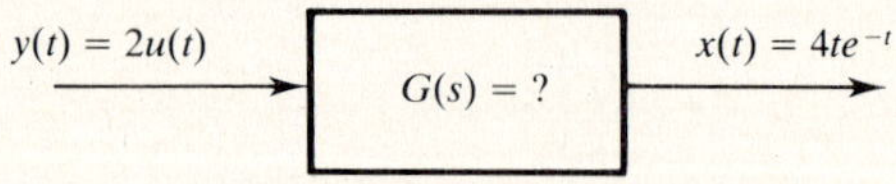

4-7. **(a)** For the network shown, write the loop equation for the closed loop shown. **(b)** Write the relation between the current $i(t)$ and the output voltage $v_0(t)$. **(c)** Transform these relations and solve for the transfer function

$$G(s) = V_0(s)/V_i(s)$$

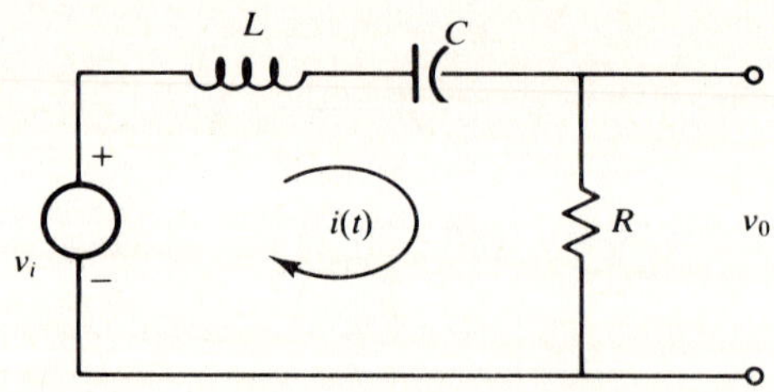

4-8. In the study of aircraft control system dynamics at certain air speeds, an impulse function with a weight of eight is input to the system and the response is measured and found to be approximately as indicated below. Assume the initial conditions to be zero and find:
(a) The system transfer function.
(b) The differential equation governing the system.

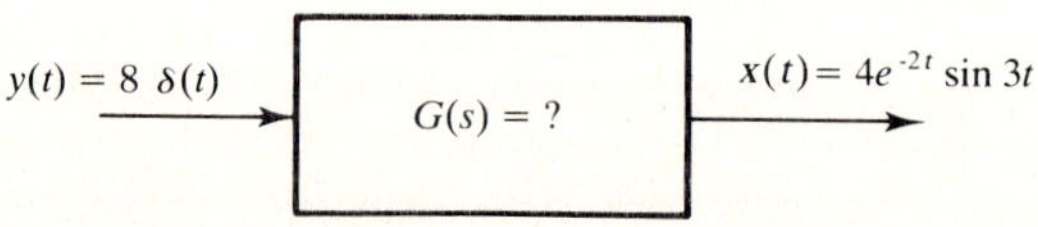

4-9. Plot the pole-zero diagram for the transfer functions given below

(a) $G(s) = \dfrac{s+2}{(s+1)(s+3)}$

(b) $G(s) = \dfrac{s+3}{s(s^2+6s+25)}$

(c) $G(s) = \dfrac{s^2-s-2}{(s+2)(s^2+2s+13)}$

4-10. For the two-mass system shown, find the transfer function $G(s) = Y(s)/F(s)$.

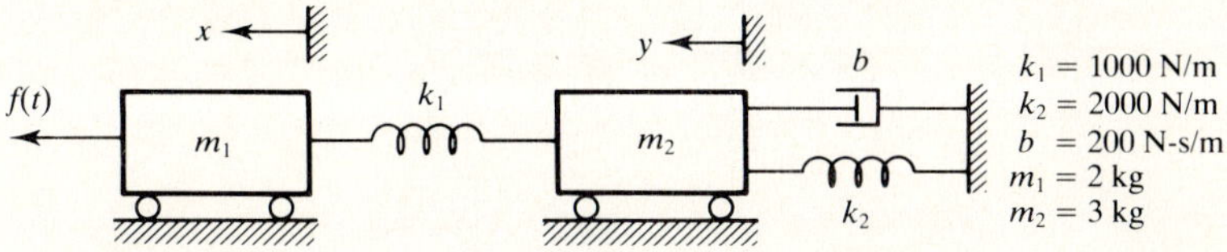

4-11. **(a)** Find the transfer function $G(s) = X(s)/Y(s)$ for the mechanical system.
(b) For the parameters given, evaluate the undamped natural frequency ω_n, the damping ratio ζ, and the damped natural frequency ω_d.

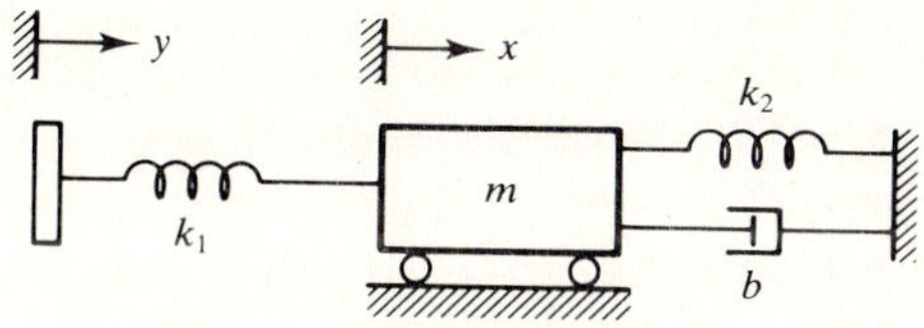

$k_1 = 100,\ k_2 = 200,\ b_1 = 30,\ m = 3$

4-12. **(a)** For the rack and pinion driven through an elastic shaft by an electric motor, find the transfer function $G(s) = X(s)/\Phi(s)$ for the system parameters given

$$k = 100\ \text{N/m} \qquad J = 0.04\ \text{kg}\cdot\text{m}^2 \qquad r = 0.2\ \text{m}$$

$$k_t = 2\ \text{N}\cdot\text{m/r} \qquad m = 4\ \text{kg}$$

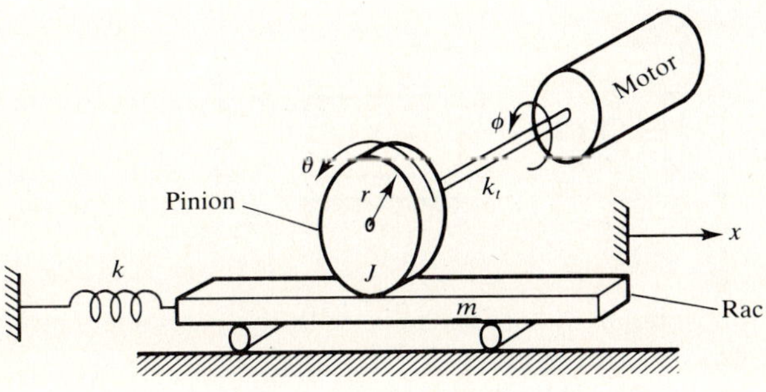

(b) Find the response $x(t)$ for an input $\phi(t) = \omega t \cdot u(t)$ (a constant speed input at the motor end of the shaft).

4-13. For the quadratic characteristic equations given below, find the undamped natural frequency ω_n, the damping ratio ζ, and the damped natural frequency ω_d.

(a) $s^2 + 50s + 1000 = 0$

(b) $s^2 + 40s + 400 = 0$

(c) $s^2 + 80s + 600 = 0$

4-14. **(a)** For the network shown, find the transfer function $V_0(s)/V_i(s)$.
(b) Find the steady-state sinusoidal output voltage if the source voltage is

$$v_i(t) = 10\cos(4 \times 10^6 t)$$

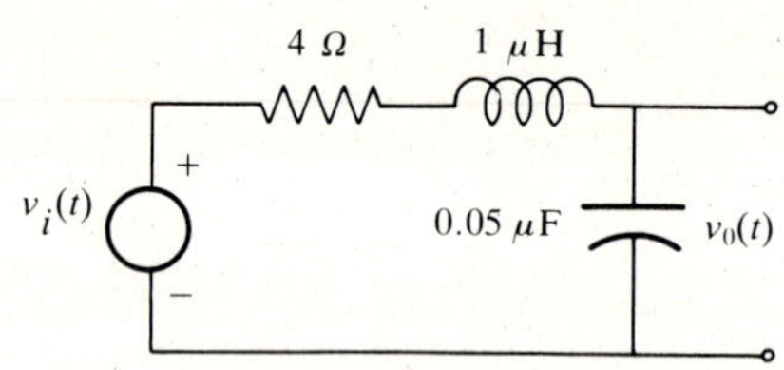

4-15. The mechanical system shown is governed by the differential equation

$$4\frac{d^2x}{dt^2} + 64\frac{dx}{dt} + 400x = 200y$$

(a) Find the transfer function $G(s) = X(s)/Y(s)$.
(b) Find the steady-state response $x(t)$ if the motion $y(t)$ is $y(t) = 3\cos 8t$ meters.

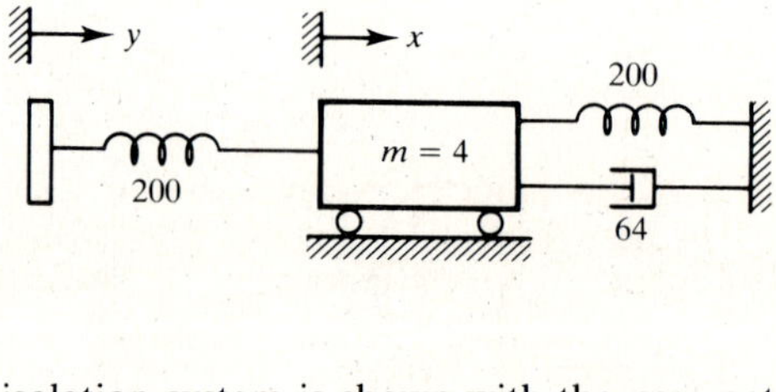

4-16. A vibration isolation system is shown with the parameters given. The governing differential equation is

$$m\frac{d^2x}{dt^2} + b\frac{dx}{dt} + kx = ky + b\frac{dy}{dt}$$

$m = 2\ \text{lb}\cdot\text{s}^2/\text{in.} \qquad b = 6\ \text{lb}\cdot\text{s/in.} \qquad k = 20\ \text{lb/in.}$

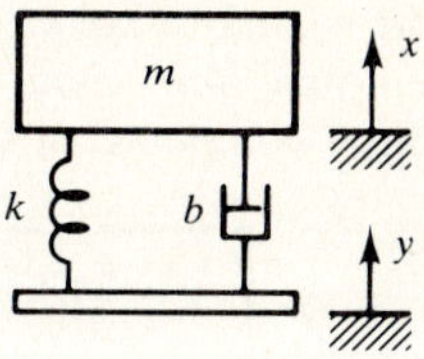

(a) Find the transfer function $G(s) = X(s)/Y(s)$.
(b) Find the steady-state response $x(t)$ if the base motion is $y(t) = 2\cos 5t$.

4-17. Sketch both the magnitude and phase functions as functions of radian frequency ω for the low-pass filter shown.

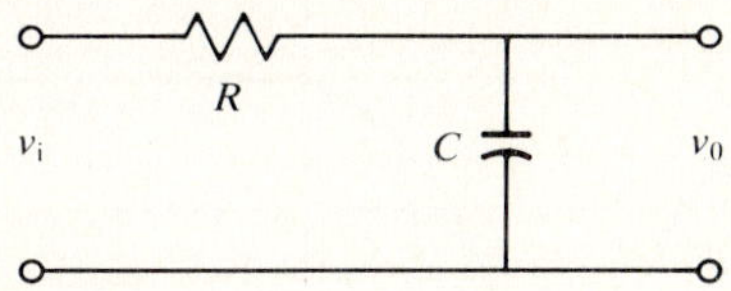

$R = 1000\ \Omega \qquad C = 1\ \mu F$

4-18. For the transfer function shown below, evaluate the steady-state response if the driving function is the sum of two sinusoids or $y(t) = 3\cos 6t + 2\cos 8t$.

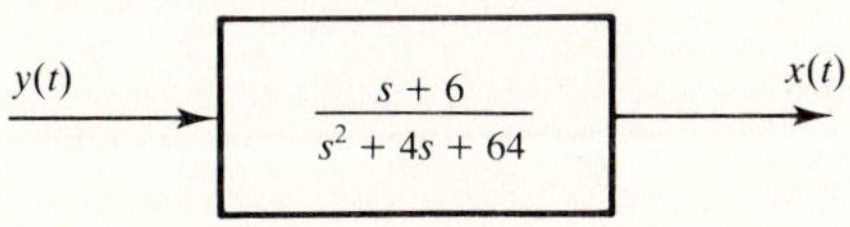

4-19. In the figure below are shown the magnitude and phase characteristics of the frequency response for a stable dynamic system. The input to the system is $y(t) = 5\cos[150t + (\pi/6)]$
(a) Find the amplitude of the steady-state sinusoidal system output.
(b) Find the phase shift contributed by the system.
(c) Write the time-domain expression for the system response $x(t)$.
(d) Repeat the above problem for a driving frequency of $\omega = 1000$ r/s.

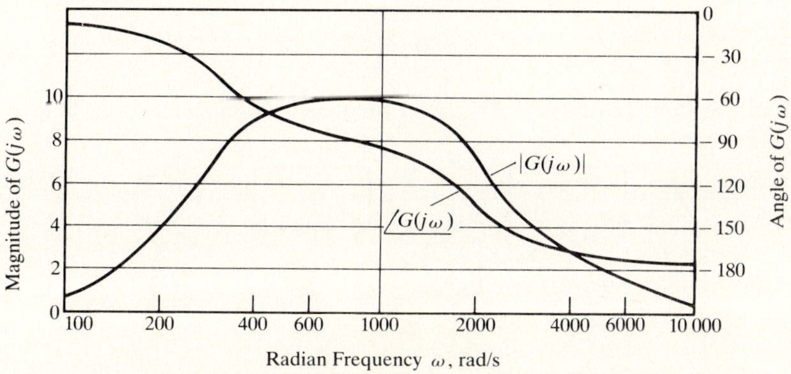

4-20. In an effort to improve the signal-to-noise ratio in the presence of 60 Hz (377 r/s) additive noise, a notch filter may be employed. The situation is illustrated below with the input composed of both signal and sinusoidal noise.

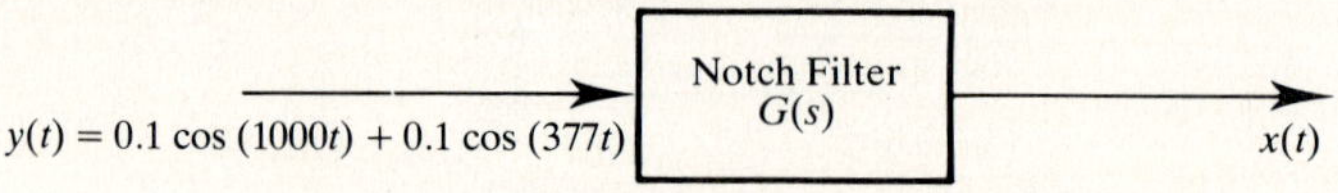

The filter has frequency response magnitude $|G(j\omega)|$ shown below. Comment quantitatively on the improvement of the signal-noise relationship based on the relative amplitudes of each.

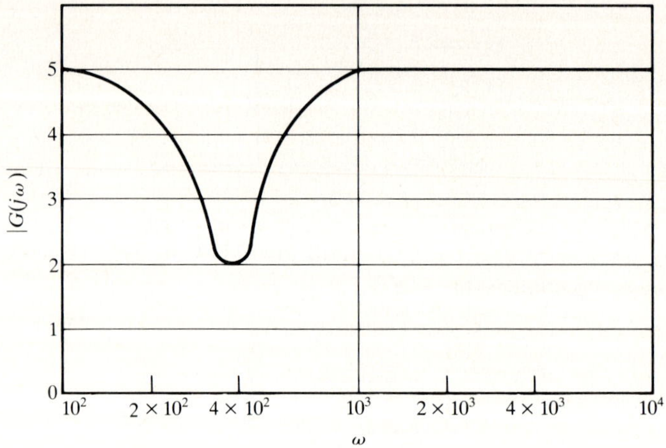

4-21. A first-order system shown below is subjected to a ramp input $y(t) = tu(t)$.

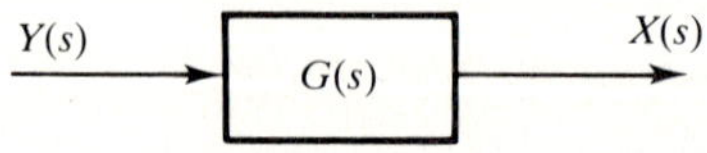

(a) Calculate the system reponse $x(t)$ using the Laplace transform method.
(b) Employing the convolution integral, evaluate the system response noting that the impulse response function is

$$g(t) = \begin{cases} 0 & t < 0 \\ e^{-4t} & t \geq 0 \end{cases}$$

4-22. Find the transfer function $X(s)/R(s)$ for the block diagram given below.

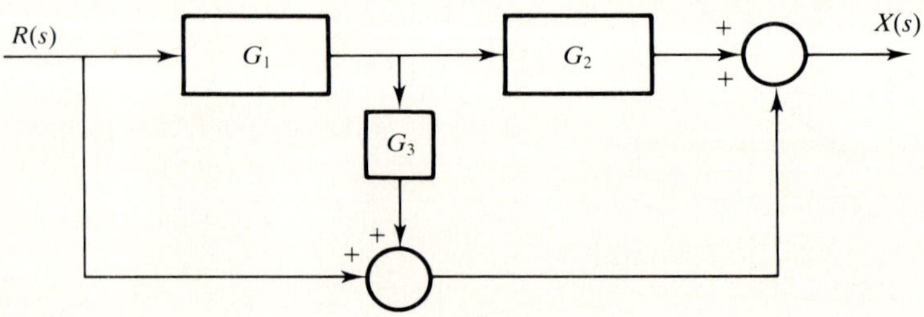

4-23. **(a)** For the block diagram find the transfer function $X(s)/Y(s)$.
(b) Find the differential equation relating $x(t)$ to $y(t)$.

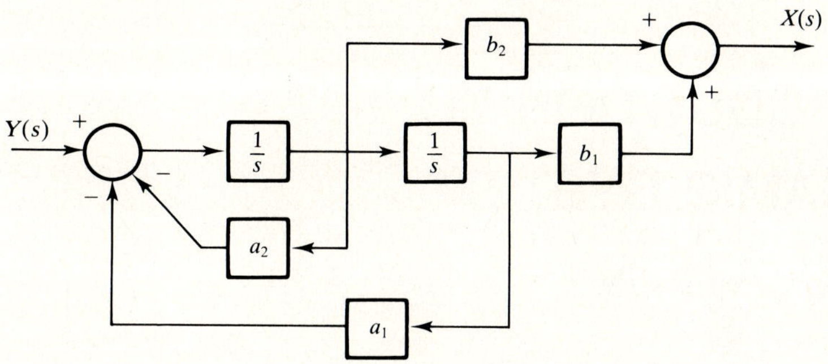

Note that the blocks with transfer functions $1/s$ are integrators, so this diagram represents an analog computer network without the sign inversions.

4-24. Find the transfer function $X(s)/Y(s)$ for the block diagram given below.

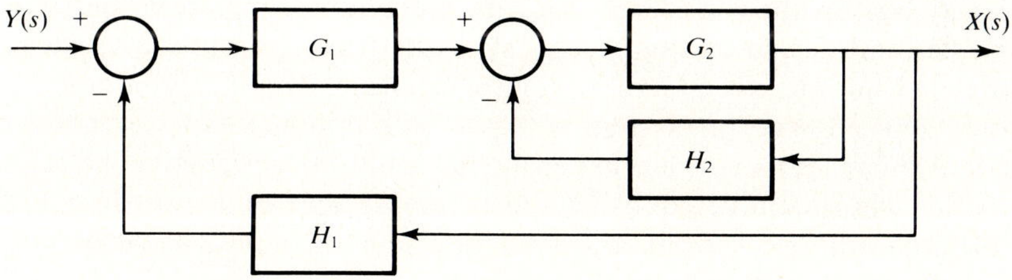

5

INTRODUCTION TO DYNAMIC SYSTEM SIMULATION

5.1 INTRODUCTION

Simulation has, in the past three decades, become a very important tool in the analysis and design of engineering systems. It has also found application in economics and the social and biological sciences. The utility of being able to simulate an engineering system is that we can conduct experiments in the simulation environment. Such experiments could be destructive or at least very costly when conducted using the actual system or a prototype. This kind of situation arises when the system is, for example, an airplane, an automobile, or a costly chemical reactor.

To simulate a system with reliable results we must have a mathematical model of the system, which means we must have a firm grasp on the physical principles which govern its operation. The model commonly takes the form of a set of simultaneous ordinary differential equations, although it could involve either algebraic or partial differential equations also. We shall restrict our discussion to the case of ordinary differential equations only. The idea of simulation is to solve the differential equations under a variety of forcing functions and/or initial conditions to ensure that the system will behave in a prescribed manner or meet a given set of performance specifications. We have seen in previous chapters that, as long as the system is not excessively complex, the response character can be evaluated analytically but, when the system is described by a large system of differential equations, the analytical approach will not be feasible. In addition, nonlinearities present in many systems may also make an analytical approach impossible and, hence, the need for simulation.

Engineering systems can be simulated in a number of ways, but the two most popular involve either analog computers or digital computers. An analog computer is a collection or network of electronic circuitry which, when wired together, implements the differential equation to be solved in the simulation process; the variables of interest are represented as voltages in the network. The process of digital simulation is accomplished on a general-purpose digital computer by solving the governing differential equations by numerical methods. It is common for engineers doing simulations to write programs to solve the differential equations numerically in a high-level computer language such as FORTRAN. On the other hand, there exist digital simulation packages which numerically solve the differential equations. These packages are programmed in the fashion of patching together the elements of an analog computer. Two digital simulation routines in wide use today are CSMP (Continuous System Modeling Program) and ISL (Interactive Simulation Language).

In this chapter we shall explore both methods of simulation. Although analog computers have to a wide extent been replaced with digital simulators, the methodology employed in analog simulation is popular and useful when employing digital simulators. Also, analog computers are useful when the portion of the system being simulated is to be coupled to some existing laboratory hardware. This would be the case in the design of an aircraft autopilot wherein the hardware built could be coupled to airframe dynamics represented in the analog computer. This situation is illustrated in Fig. 5.1.

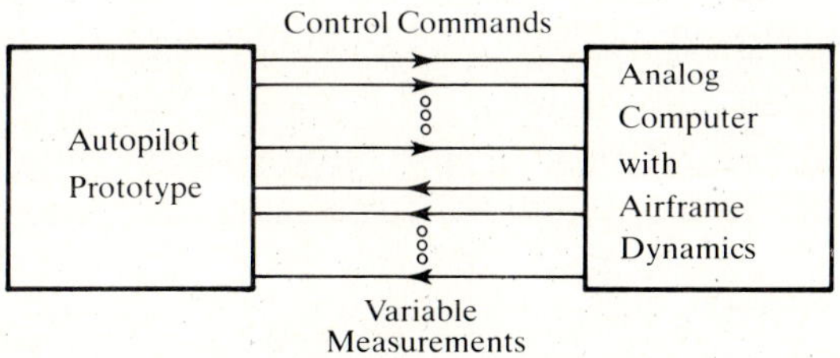

FIGURE 5.1 Interaction of an analog computer with prototype hardware in an autopilot design situation.

5.2 THE OPERATIONAL AMPLIFIER

Originally the term "operational amplifier" was used in the field of analog computation to describe amplifiers which performed various mathematical operations. It was found that the application of negative feedback around a dc amplifier would produce a circuit with a precise gain characteristic which depends on the feedback elements only. With proper selection of feedback components, operational amplifiers can be made to add, subtract, average, integrate, and multiply signals.

The modern operational amplifier comes in integrated circuit form. These integrated circuit chips have very nearly ideal properties at a cost of a fraction of a dollar per amplifier. The properties of this device are high voltage gain, high input resistance, low output resistance, low internal noise

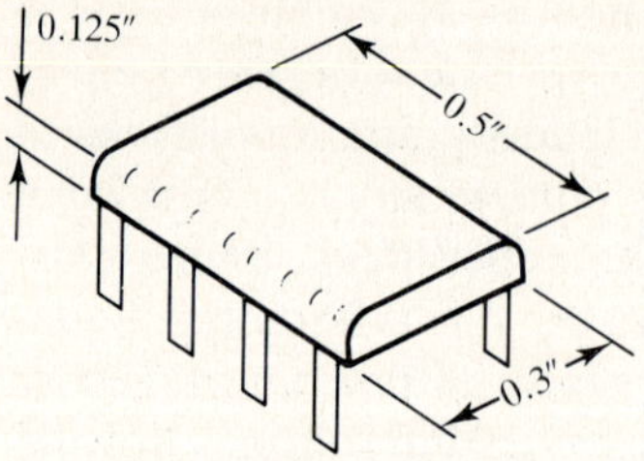

FIGURE 5.2 Typical eight-pin integrated circuit operational amplifier.

generation, and low power dissipation. A typical integrated circuit operational amplifier package is illustrated in Fig. 5.2, which shows its compact size. The actual integrated circuit inside the package is about an eighth of an inch square, with the package providing for ease of attachments. Typical operational amplifiers require a dual (±) power supply of from 1 to 20 volts.

Lest this device be thought of as simple, inside its neat little package, the op amp, as it is commonly called, is a regular maze of resistors, capacitors, and transistors, all in integrated circuit form. A typical circuit diagram for an op amp is shown in Fig. 5.3. The design and fabrication of integrated circuits is a complete technology in itself and we shall be concerned only with the terminal properties and application of the device. There are many possible ways of representing an operational amplifier schematically but the most common ways are shown in Fig. 5.4. Terminal

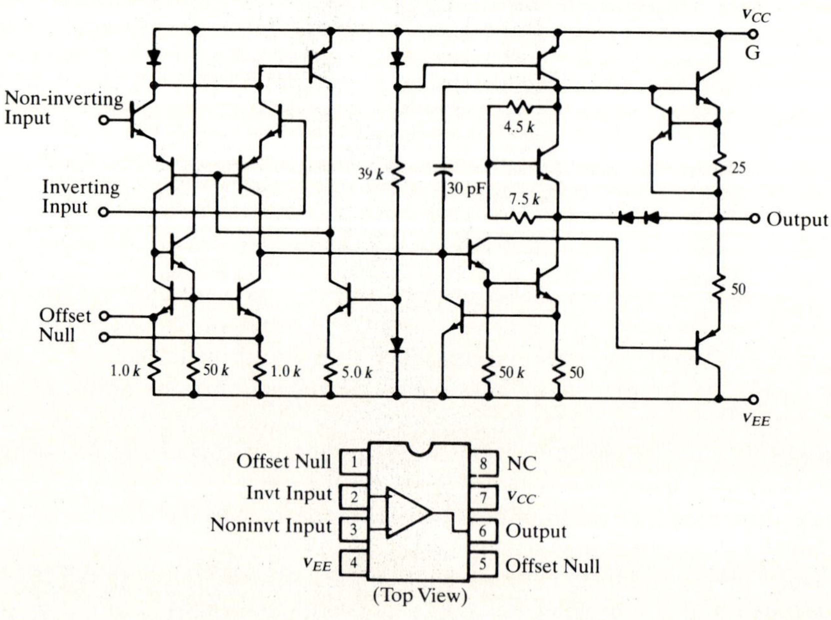

FIGURE 5.3 Circuit schematic for a typical operational amplifier, the μ741. (Courtesy of Motorola, Inc.)

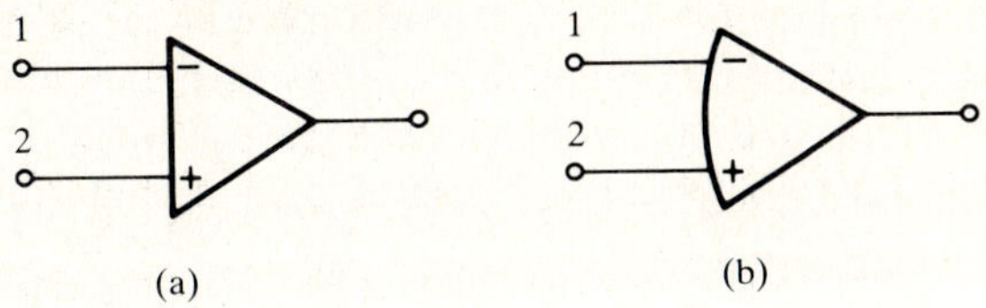

FIGURE 5.4 Common op amp circuit symbols.

one in the diagram is commonly referred to as the "inverting input" because, when the potential at terminal 1 goes positive, the output is driven negative; likewise terminal 2 is the "noninverting input." These designations are commonly indicated by the minus and plus signs, respectively.

Consider the highly idealized model of an op amp illustrated in Fig. 5.5. The input-output relation for this device is

$$v_0 = A(v_2 - v_1) \tag{5.2.1}$$

so long as v_0 does not exceed either the positive or negative supply voltages (typically ± 20 volts). For a typical operational amplifier, the gain A is roughly 10^5. It is clear that, if the gain is this large and the output cannot exceed 20 volts in magnitude, the input voltage difference must be in the microvolt range. It doesn't take a very discerning student to ask the question, "What can we do with a device which requires such a small input to avoid output saturation?" The answer to this question lies in the way we apply the op amp, that is, the way we use other circuit elements and the very large gain to build very useful devices for instrumentation and analog computation.

The ideal operational amplifier would have the following properties:

1. The input resistances R_1 and R_2 of Fig. 5.5 are both infinite.
2. The output resistance R_0 is zero.
3. The gain is very large.

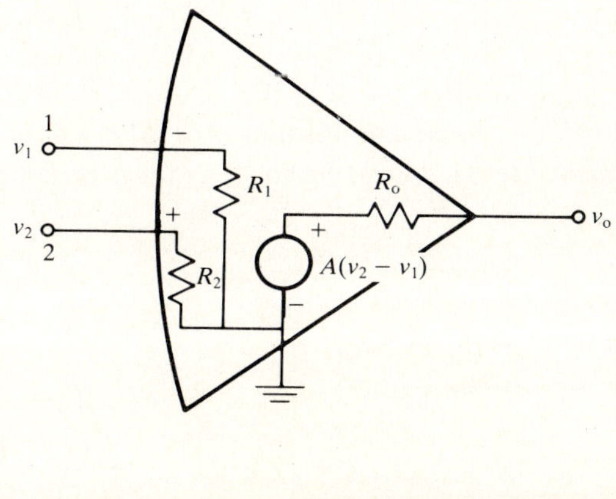

FIGURE 5.5 Ideal operational amplifier model.

In truth, no operational amplifier made exactly meets these specifications, but for many commercially available units these assumptions are very nearly met. We shall use these properties along with the input-output relation (Eq. (5.2.1)) to derive some useful operational amplifier circuits.

5.3 USEFUL OPERATIONAL AMPLIFIER CIRCUITS

We shall now use the properties of the operational amplifier just discussed to derive some electronic circuits which are useful in both analog computation and instrumentation. The two properties which we shall find most useful are those of high input impedance and high gain.

5.3.1 The Voltage Follower

Consider the simple circuit of Fig. 5.6, wherein the output is connected directly to the inverting input and the input signal is connected to the noninverting input terminal. If we note that the input to the inverting terminal is the output voltage v_0, we can write the expression for the output as

$$v_0 = -A(v_0 - v_{in}) \tag{5.3.1}$$

We may solve for v_0 to give

$$v_0 = \frac{A}{1 + A} v_{in} \tag{5.3.2}$$

If we now take the gain A to be a large positive number, then the ratio on the right side becomes essentially unity so that

$$v_0 = v_{in} \tag{5.3.3}$$

At first this doesn't appear to be a very interesting circuit but if we note that the input impedance at the v_{in} terminal is high, then this device is useful as a buffer in instrumentation systems to isolate a transducer or signal source from further elements of a signal processing system. It is also of interest to

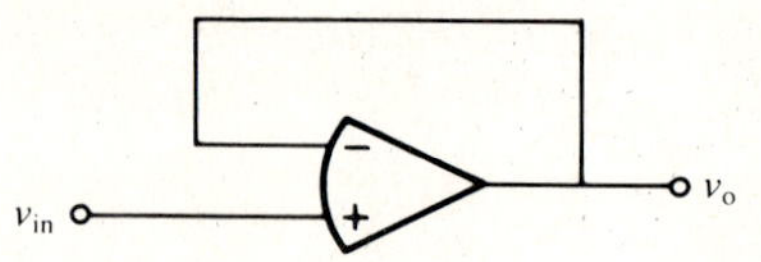

FIGURE 5.6 Voltage follower circuit.

note that the output resistance is very low, so this considerable impedance change can also be useful.

5.3.2 Noninverting Amplifier

Consider now the circuit configuration of Fig. 5.7, which involves the operational amplifier and two resistors R_1 and R_2. The voltage v_1 at the inverting terminal is now a voltage division of v_0 or

$$v_1 = \frac{R_1}{R_2 + R_2} v_0 \tag{5.3.4}$$

Then we may write the output voltage as

$$v_0 = -A(v_1 - v_{in}) \tag{5.3.5}$$

or

$$\frac{v_0}{A} = v_{in} - v_1 \tag{5.3.6}$$

If we now substitute Eq. (5.3.4) for v_1, the result is

$$\frac{v_0}{A} = v_{in} - \frac{R_2}{R_2 + R_1} v_0 \tag{5.3.7}$$

If the gain A is very large, the left side of Eq. (5.3.7) will be very small compared to the terms on the right side and may be taken to be zero, so the input-output relation becomes

$$v_0 = \frac{R_1 + R_2}{R_2} v_{in} \tag{5.3.8}$$

Here the gain is greater than unity and the input and output voltages are in phase with each other. It is a simple matter to choose the resistors to give any desired gain in an instrumentation application.

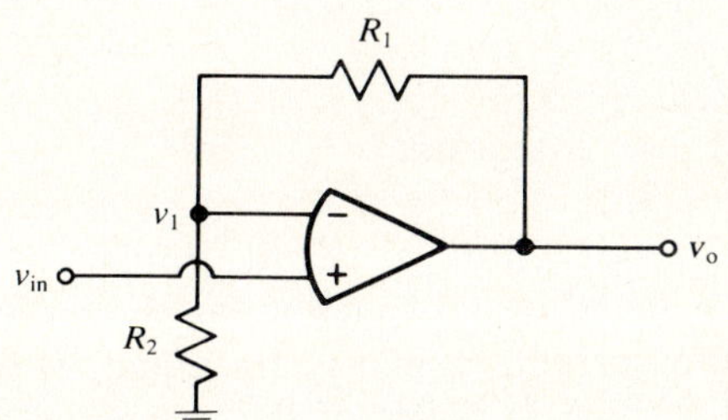

FIGURE 5.7 Noninverting amplifier.

5.3.3 Inverting Amplifier

Consider another configuration of the operational amplifier with two resistors in the circuit of Fig. 5.8.

Since the input impedance of the operational amplifier is high, essentially no current is directed into the amplifier. We can now apply Kirchhoff's current law at the node labeled v to give

$$i_1 = i_2 \tag{5.3.9}$$

Let the currents i_1 and i_2 be written in terms of the voltages shown, giving

$$\frac{v_{\text{in}} - v}{R_1} = \frac{v - v_0}{R_2} \tag{5.3.10}$$

Now let us note the amplifier relation that

$$v_0 = -A(v - 0) = -Av \tag{5.3.11}$$

or

$$v = -\frac{v_o}{A} \tag{5.3.12}$$

If v_0 is in the neighborhood of 10 volts and the gain is greater than 10^5, then v_1 will be near zero and thus neglected in relation (5.3.10) so the input output relation is

$$v_0 = -\frac{R_2}{R_1} v_{\text{in}} \tag{5.3.13}$$

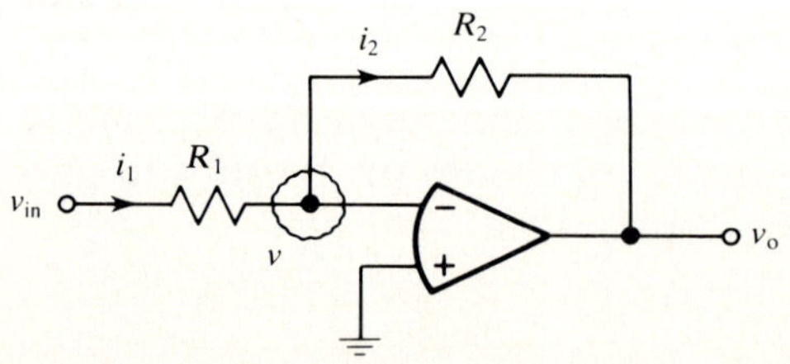

FIGURE 5.8 Inverting amplifier.

5.3.4 Integrator

Consider now a circuit similar to that just considered except with R_2 replaced by a capacitor, as shown in Fig. 5.9. As in the previous example, the currents i_1 and i_2 are equal. If we equate these currents and substitute for them in terms of the respective voltages,

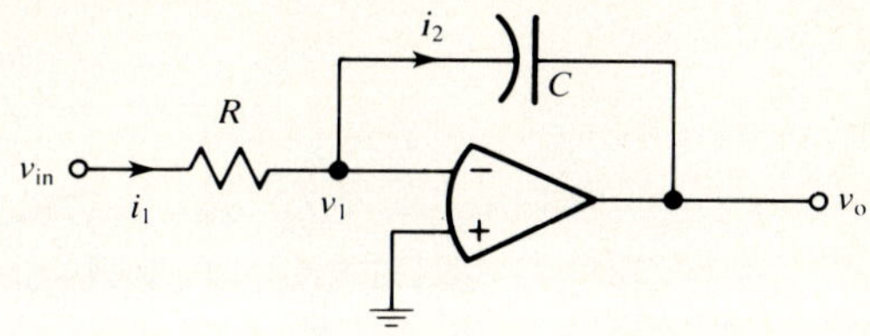

FIGURE 5.9 Operational amplifier integrator.

$$\frac{v_{in} - v}{R} = C \frac{d}{dt} (v - v_0) \tag{5.3.14}$$

Again the voltage v will be small because of the high gain. The input in terms of the output is

$$v_{in} = -RC \frac{dv_0}{dt} \tag{5.3.15}$$

If we separate the variables in this expression, the result is

$$dv_0 = -\frac{1}{RC} v_{in}\, dt$$

Integrating over the time interval 0 to t,

$$\int_{v_0(0)}^{v_0(t)} dv_0 = -\frac{1}{RC} \int_0^t v_{in}(t)\, dt \tag{5.3.16}$$

from which the output is

$$v_0(t) = -\frac{1}{RC} \int_0^t v_{in}(t)\, dt + v_0(0) \tag{5.3.17}$$

The resistor and capacitor values can be chosen by the designer to give an appropriate output, which is the integral of the input voltage plus the initial voltage. This circuit is useful for averaging signals and, as we shall show, it is useful for the solution of ordinary differential equations.

5.3.5 Summing Inverter

Consider now the network illustrated in Fig. 5.10, which is the same as the inverting amplifier except that it has two inputs $v_1(t)$ and $v_2(t)$. Application of Kirchhoff's current law at node v yields

$$i_1 + i_2 = i_3 \tag{5.3.18}$$

The respective currents may be evaluated in terms of the voltages given:

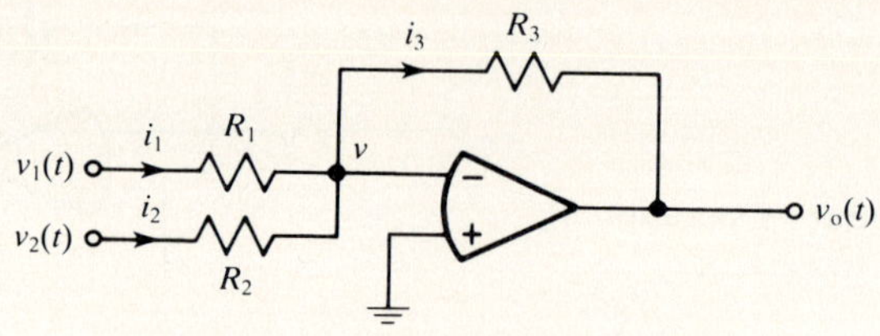

FIGURE 5.10 Summing inverter.

$$\frac{v_1 - v}{R_1} + \frac{v_2 - v}{R_2} = \frac{v - v_0}{R_3} \tag{5.3.19}$$

As in the previous work, the high gain will dictate that the potential v be much smaller than the other voltages so

$$\frac{v_1}{R_1} + \frac{v_2}{R_2} = -\frac{v_0}{R_3} \tag{5.3.20}$$

If we solve for the output voltage the relation is

$$v_0 = -\left[\frac{R_3}{R_1} v_1 + \frac{R_3}{R_2} v_2\right] \tag{5.3.21}$$

Thus the output is the weighted sum of the inputs; we refer to this as a summing amplifier. This idea can be extended to n inputs, in which case the output is the negative of the weighted sum of the inputs or

$$v_0 = -\left[\frac{R_f}{R_1} v_1 + \frac{R_f}{R_2} v_2 + \cdots + \frac{R_f}{R_n} v_n\right] \tag{5.3.22}$$

where R_f is the resistor connected between the output and input.

5.3.6 Summing Integrator

Consider the network of Fig. 5.11, which is identical to that of Fig. 5.9 except now there are two inputs. Application of Kirchhoff's current law to node v gives

$$i_1 + i_2 = i_3 \tag{5.3.23}$$

Then these currents are evaluated in terms of the respective voltages

$$\frac{v_1 - v}{R_1} + \frac{v_2 - v}{R_2} = C \frac{d}{dt}(v - v_0) \tag{5.3.24}$$

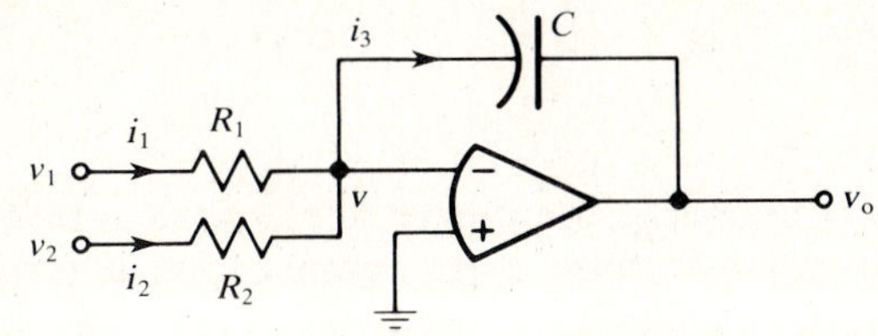

FIGURE 5.11 Summing integrator circuit.

Since the gain of the amplifier is high, the voltage v must be small, so the input-output relation is

$$v_0(t) = -\int_0^t \left[\frac{1}{R_1C} v_1(t) + \frac{1}{R_2C} v_2(t)\right] dt + v_0(0) \qquad \textbf{(5.3.25)}$$

and thus we see that the output voltage is the initial output voltage less the integral of the weighted sum of the input voltages. This concept can easily be extended in the case of more input voltages.

We have seen in this section that we have the ability to amplify, sum, integrate, and invert time-varying signals. We shall put this ability to good use in the next section.

Before we move on to the concept of the analog computer we shall first consider what is perhaps the simplest element of the analog computer system, the potentiometer, which is illustrated in Fig. 5.12. We shall assume that whatever is attached to the output terminal is of sufficiently high input impedance that no current exists in the output; so the device output voltage v_o is simply a fraction of the input voltage v_i:

$$v_0 = \frac{x}{l} v_i \qquad \textbf{(5.3.26)}$$

or

$$v_0 = Kv_i \qquad 0 \le K \le 1 \qquad \textbf{(5.3.27)}$$

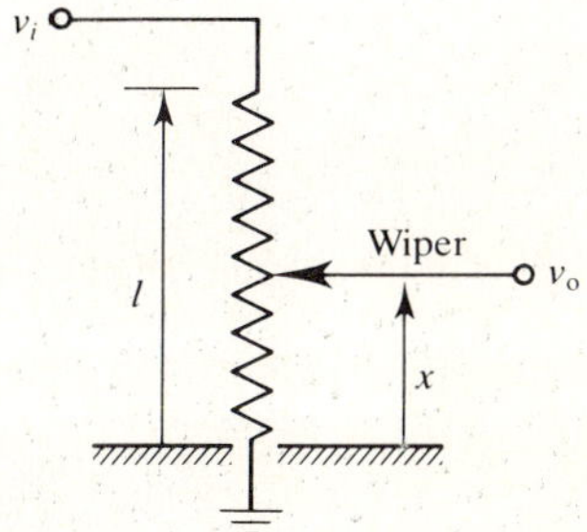

FIGURE 5.12 Simple potentiometer.

5.4 THE ANALOG COMPUTER

In the previous section we developed the electronic networks which make up an analog computer. These devices and their commonly employed analog computer symbols are summarized in Table 5-1. Instead of drawing all the resistors and capacitors for the devices, the analog computer symbols are used to represent the whole device from a terminal view to yield simulation diagrams which are easily readable. Precision resistors and capacitors, chosen by patching wires to the selected capacitors and resistors, are used for summers and integrators. These components for modern analog computers are chosen so the available values of K_i are 0.1, 1, and 10. A small analog computer and its patch board are illustrated in Fig. 5.13(a) and (b), respectively. The voltage range for most modern analog computers is ± 10 volts; however, the advent of the 100-volt transistor may yield a new generation of analog computers with greater dynamic range.

As the beginner might suspect, the analog computer solves a differential equation by the process of integration. Suppose we have a second-order differential equation of the form

$$\ddot{x} + a_1\dot{x} + a_2x = y(t) \tag{5.4.1}$$

with $y(t)$ known and $x(0) = \dot{x}(0) = 0$. It is clear that if $\ddot{x}$ were known it could be integrated twice to yield the solution $x(t)$; however, it is not known. The key to the puzzle lies in the differential equation (5.4.1) which can be solved for $\ddot{x}$ or

$$\ddot{x} = -a_1\dot{x} - a_2x + y(t) \tag{5.4.2}$$

Suppose $\ddot{x}$ has been integrated twice to give $\dot{x}$ and x. Then $\ddot{x}$ can be constructed by the operation of summing $\dot{x}$, x, and y according to Eq. (5.4.2). This is a chicken and egg argument which is best illustrated with an example.

Example 5.1

Consider the second-order constant coefficient differential equation

$$\ddot{x} + 5\dot{x} + 3x = 2 \tag{a}$$

with initial conditions

$$\dot{x}(0) = 1 \qquad x(0) = 4$$

We can solve the differential equation for the highest order derivative or

$$\ddot{x} = -5\dot{x} - 3x + 2 \tag{b}$$

TABLE 5-1 Analog Computer Circuit Elements and Symbols

Device	Circuit	Symbol	Input-Output Relation
Summer	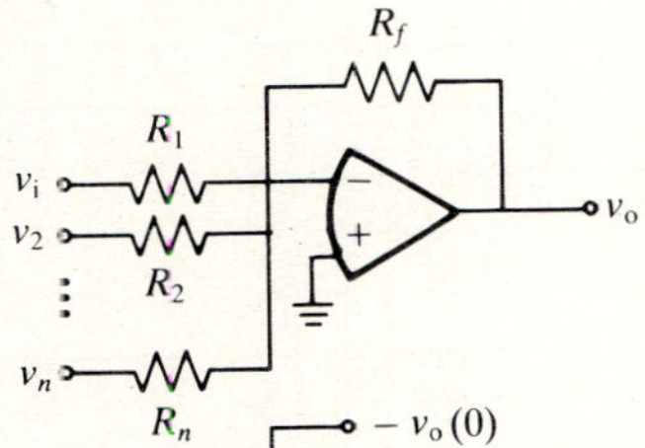	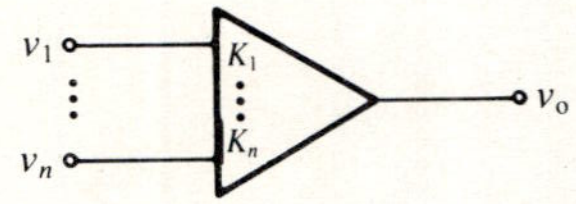	$v_o = -[K_1v_1 + \cdots + K_nv_n]$ $K_i = \frac{R_f}{R_i}$
Integrator	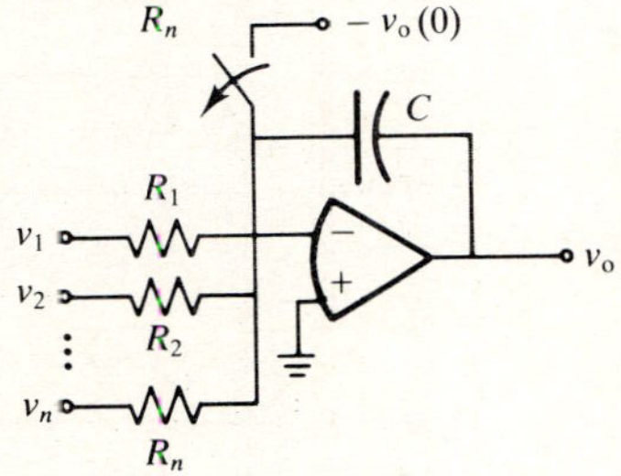	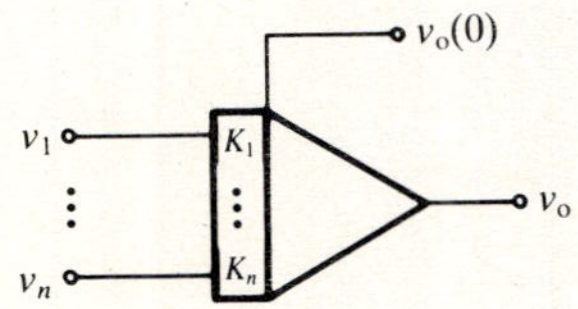	$v_o(t) = -\int_0^t K_1v_1(t) + \cdots + K_nv_n(t)\,dt$ $+ v_o(0)$ $K_i = \frac{1}{R_iC}$
Potentiometer	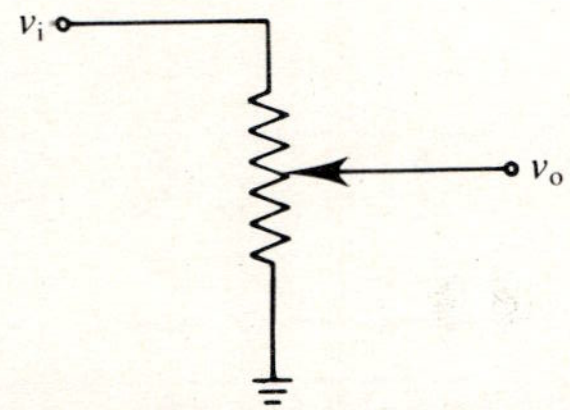	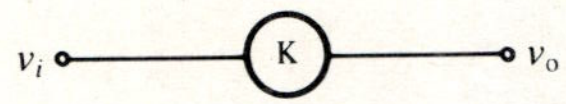	$v_o = Kv_i$ $0 \leq K \leq 1$

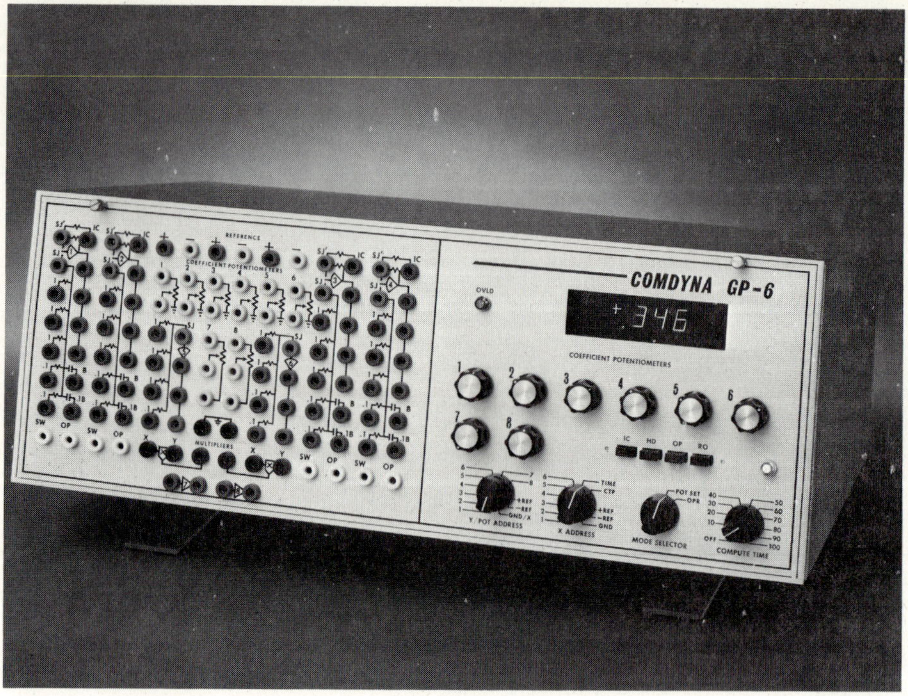

(a)

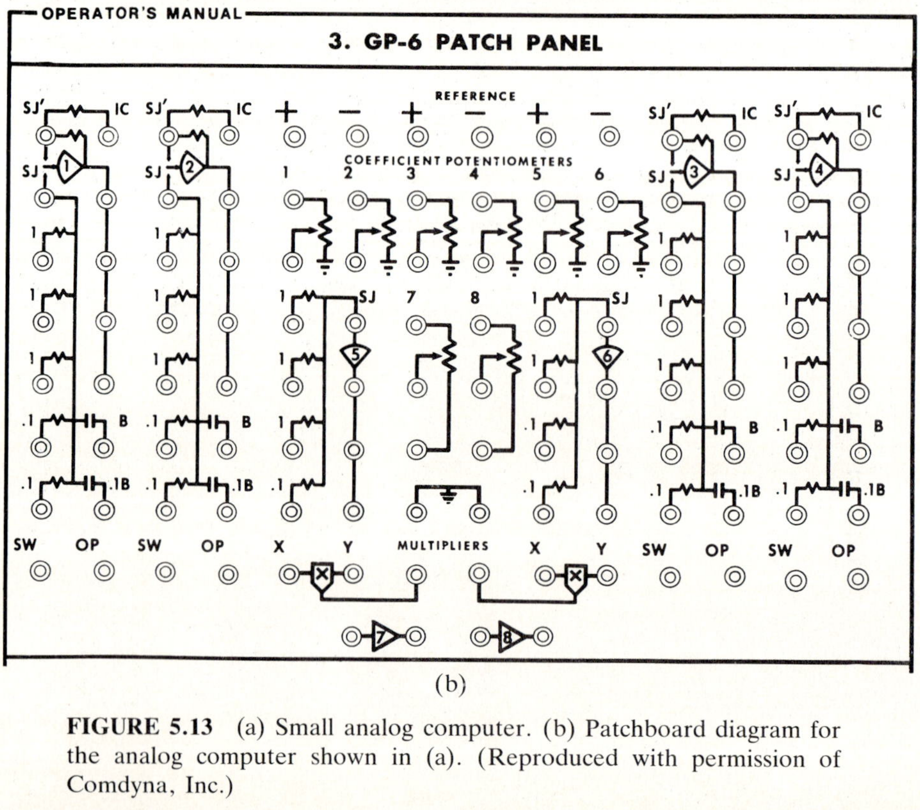

(b)

FIGURE 5.13 (a) Small analog computer. (b) Patchboard diagram for the analog computer shown in (a). (Reproduced with permission of Comdyna, Inc.)

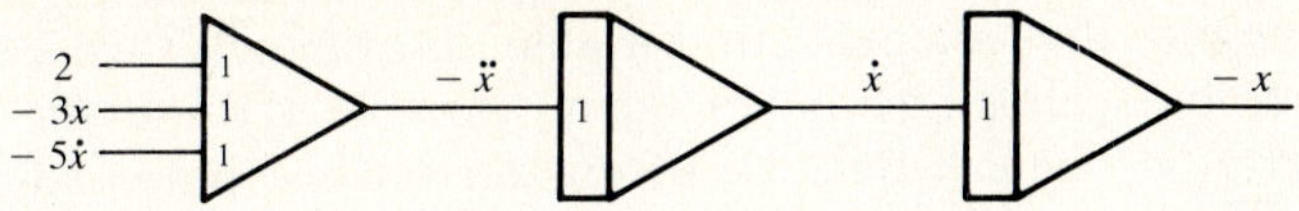

FIGURE 5.14 Summation and successive integration to generate the terms of a differential equation.

Let us assume that $-\ddot{x}$ is the output of the summer. Then, according to Eq. **(b)**, the inputs to the summer must sum to $\ddot{x}$, which is equivalent to the right side, as illustrated in Fig. 5.14. Further, if we successively integrate the output of the summer, we generate first $\dot{x}$ and then $-x(t)$, as illustrated in Fig. 5.14. We can complete the analog computer solution to the problem by feeding back $x(t)$ and $\dot{x}(t)$ with the correct weights, as illustrated in Fig. 5.15. One must be particularly careful with the algebraic signs as one or more errors will change the complete nature of the solution.

Particular attention must be directed to the initial conditions input to the integrators. The key to this problem is illustrated in the integrators in Table 5.1. Note that the initial condition input must be the negative of the desired initial output voltage. For the left integrator, the desired initial output voltage is 1 volt so the input should be -1 volt. For the right integrator, the desired initial output voltage is -4 volts so the input should be $+4$ volts.

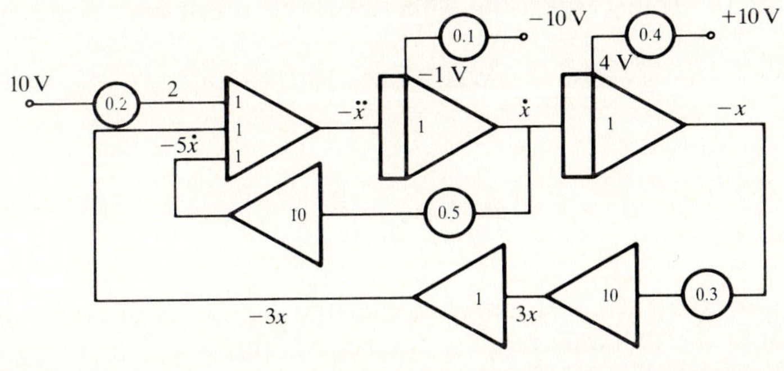

FIGURE 5.15 Complete analog computer simulation diagram for Eq. **(b)**.

5.5 TIME AND MAGNITUDE SCALING

Modern analog computers have a dynamic range of ± 10 volts, requiring that maximum magnitude of the variables be kept within this range but also still be large enough to be measured with ordinary recording equipment. For some systems we wish to simulate, the system dynamics may be too fast for modern analog signal recording equipment (strip chart recorders or x-y plotters) and, as a result, it is difficult to obtain hard copy of the responses without changing the independent variable. Fortunately, we can alleviate this problem by scaling the independent variable (t) and the dependent variables [$x(t)$, $y(t)$, etc.]. Because of the derivatives, time scaling will also affect the magnitude of some variables. Often a combination of time and magnitude scaling will be required.

5.5.1 Time Scaling

Let us consider the difficulties encountered when solving the following differential equation on the analog computer:

$$\frac{d^2x}{dt^2} + 100x = 0 \qquad x(0) = 2, \frac{dx(0)}{dt} = 0 \tag{5.5.1}$$

A simulation diagram for the solution of this problem is given in Fig. 5.16. Let us solve this simple problem analytically by the Laplace transform method to give

$$X(s) = \frac{2s}{s^2 + 100} \tag{5.5.2}$$

If we invert the transform using Table A-2 the result is

$$x(t) = 2 \cos 10t \tag{5.5.3}$$

The first derivative is

$$\frac{dx}{dt} = -20 \sin 10t \tag{5.5.4}$$

Examination of Eq. (5.5.3) indicates that the output of the right integrator will be within the dynamic range of that integrator and that 10 r/s is not an extreme frequency for modern recording equipment. If, on the other hand, we examine Eq. (5.5.4), we note that the output of the left integrator ($-dx/dt$) will range from $+20$ volts to -20 volts, clearly out of the dynamic

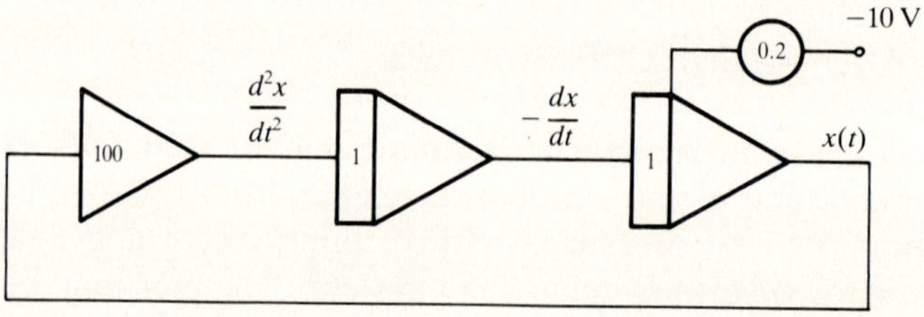

FIGURE 5.16 Analog simulation for the solution of Eq. (5.5.1).

range of a contemporary analog computer. If we examine the second derivative, which is the output of initial inverter, we find that it is

$$\frac{d^2x}{dt^2} = -200 \cos 10t \tag{5.5.5}$$

which is even further out of the dynamic range of the computer. (We might have expected these problems when we had the high gain in the initial inverter.)

The cure for these problems is to scale the temporal variable (t) such that all the coefficients on the left side of the differential equation are as near unity as possible. Let us continue to use this example to illustrate the simple technique known as *time scaling*.

Define a new independent variable τ, which will be a simple multiple of t or

$$t = \frac{\tau}{\alpha} \tag{5.5.6}$$

where α is some yet-unknown positive constant. Let us differentiate this expression implicitly to give

$$dt = \frac{d\tau}{\alpha} \tag{5.5.7}$$

Now note that the denominators of the higher order derivatives are respectively the larger powers (Eq. (5.5.7)) or

$$dt^2 = \frac{d\tau^2}{\alpha^2} \tag{5.5.8}$$

and in general

$$dt^n = \frac{d\tau^n}{\alpha^n} \tag{5.5.9}$$

Let us now return to the differential equation (Eq. (5.5.1)) and substitute relation (5.5.8), which gives

$$\alpha^2 \frac{d^2x}{d\tau^2} + 100x = 0 \tag{5.5.10}$$

Division by α^2 gives

$$\frac{d^2x}{d\tau^2} + \frac{100}{\alpha^2}\, x = 0 \tag{5.5.11}$$

If the goal is to reduce the coefficients on the left side of the differential equation to near unity, a choice of $\alpha = 10$ will yield

$$\frac{d^2x}{d\tau^2} + x = 0 \tag{5.5.12}$$

The solution of this differential equation and the respective derivatives can be given by substitution of Eq. (5.5.6) into Eqs. (5.5.3), (5.5.4), and (5.5.5) to give

$$x(\tau) = 2 \cos \tau$$

$$\frac{dx}{d\tau} = -2 \sin \tau \tag{5.5.13}$$

and

$$\frac{d^2x}{d\tau^2} = -2 \cos \tau$$

All these values will be well within the dynamic range of the 10-volt computer. The analog computer diagram for the solution of Eq. (5.5.12) is given in Fig. 5.17. Note that the undesirable high gain of the initial inverter has been eliminated by time scaling. The only problem which remains is the interpretation of the recorder outputs of $x(\tau)$ and $dx/d\tau$, as illustrated in Fig. 5.18. The temporal axis must be descaled and relation (5.5.6) tells us how this must be done; the descaled temporal axis is labeled in Fig. 5.18. As far as $x(t)$ is concerned, we now have the complete story but since we have also recorded $-(dx/d\tau)$, we need to descale the vertical axis on this plot. This can be done by noting that

$$\frac{dx}{dt} = \alpha \frac{dx}{d\tau} \tag{5.5.14}$$

All of the higher order derivatives must also be descaled by a factor of α to the appropriate power. We shall demonstrate this concept further with an example problem after our discussion of magnitude scaling.

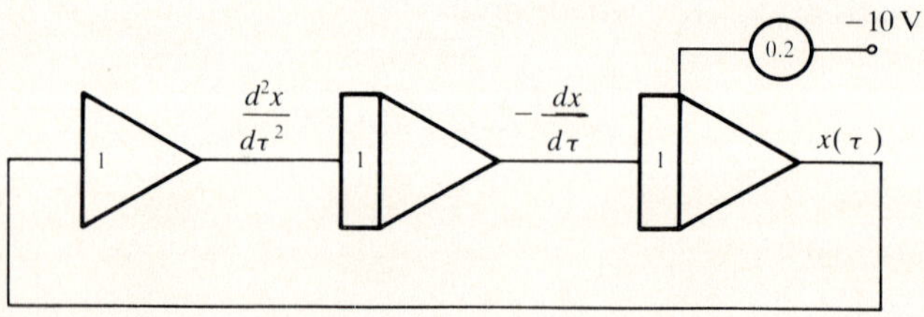

FIGURE 5.17 Analog simulation diagram for Eq. (5.5.12).

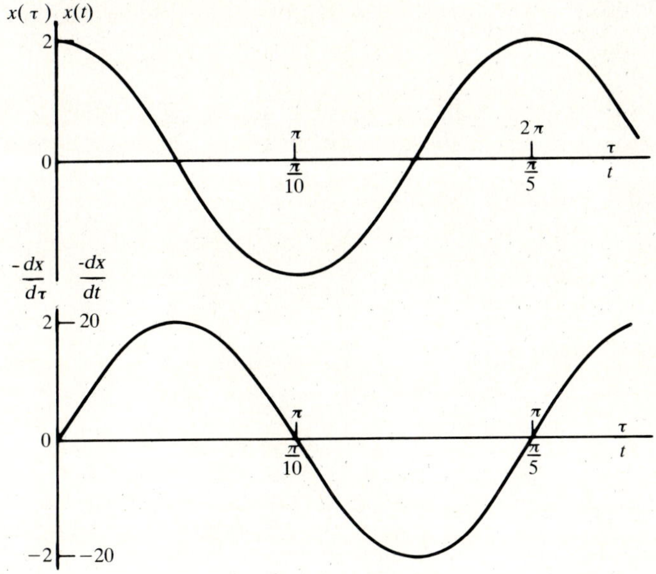

FIGURE 5.18 Typical recorded outputs for the simulation of Eq. (5.5.12).

5.5.2 Magnitude Scaling

In some cases problems are posed where the variables are either too small or too large for present analog computers. In the case where the variables are too small, they become buried in the semiconductor noise which is present to a very small degree in all analog computers. Of course, the signals being too large will saturate the amplifiers of the computer. This is best discussed in the context of a simple example. Consider the differential equation

$$\frac{d^2x}{dt^2} + 2\frac{dx}{dt} + 2x = 50, \; x(0) = \frac{dx(0)}{dt} = 0 \tag{5.5.15}$$

where the left side has been chosen so that no time scaling will be required. If we discard the derivatives to seek the constant steady-state solution, the result is

$$x(\infty) = 25 \tag{5.5.16}$$

which is, of course, far too large for modern analog computers. To magnitude scale this expression, we simply define a new dependent variable $y(t)$ related to $x(t)$ by

$$x = \beta y \tag{5.5.17}$$

If we substitute this expression and its derivatives into relation (5.5.15) we get

$$\beta \frac{d^2y}{dt^2} + 2\beta \frac{dy}{dt} + 2\beta y = 50 \tag{5.5.18}$$

If we now divide through by β we get

$$\frac{d^2y}{dt^2} + 2 \frac{dy}{dt} + 2y = \frac{50}{\beta} \tag{5.5.19}$$

A choice of $\beta = 25$ yields

$$\frac{d^2y}{dt^2} + 2 \frac{dy}{dt} + 2y = 2 \tag{5.5.20}$$

Note that this choice of $\beta = 25$ was arbitrary and many other choices such as $\beta = 20$ or 30 would also work. The problem can now be solved on the analog computer and the solution can be descaled by relation (5.5.17).

In the case where there is more than one dependent variable in simultaneous equations, each may be scaled by a different factor but the independent variable t in all equations must always be scaled by the same factor. A detailed example will now illustrate both time and magnitude scaling.

Example 5.2

Consider the problem of simultaneous differential equations posed in Example 3.4 which involved the coupled equations

$$\frac{d^2\theta}{dt^2} + 125\theta - 25x = 0$$

$$\frac{d^2x}{dt^2} + 40x - 40\theta = 1$$

with all initial conditions equal to zero. The coefficients on the left side of both equations are large enough to require time scaling, so let $t = \tau/\alpha$ in both equations which gives

$$\frac{d^2\theta}{d\tau^2} + \frac{125}{\alpha^2}\theta - \frac{25}{\alpha^2}x = 0$$

$$\frac{d^2x}{d\tau^2} + \frac{40}{\alpha^2}x - \frac{40}{\alpha^2}\theta = \frac{1}{\alpha^2}$$

A good value of α^2 will be 25, although there is range of acceptable values near 25. The time-scaled differential equations are now

$$\frac{d^2\theta}{d\tau^2} + 5\theta - x = 0$$
$$\frac{d^2x}{d\tau^2} + 1.6x - 1.6\theta = 0.04 \tag{a}$$

We see that time scaling has handled the problems on the left side of the equation but the forcing function is now quite small (40 millivolts) and, as a result, the response will be of the same order of magnitude.

Let us now change dependent variables by letting

$$\theta = \beta_1 y_1$$

and

$$x = \beta_2 y_2$$

Substitution of these relations into Eq. **(a)** gives

$$\frac{d^2y_1}{d\tau^2} + 5y_1 - \frac{\beta_2}{\beta_1} y_2 = 0$$

and

$$\frac{d^2y_2}{d\tau^2} + 1.6y_2 - 1.6\frac{\beta_1}{\beta_2} y_1 = \frac{0.04}{\beta_2}$$

In this particular problem it is convenient to let $\beta_1 = \beta_2 = 0.2$, although in general the two β_i could be different. The resulting scaled differential equations are

$$\frac{d^2y_1}{d\tau^2} + 5y_1 - y_2 = 0$$
$$\frac{d^2y_2}{d\tau^2} + 1.6y_2 - 1.6y_1 = 0.2$$

with zero initial conditions. Solving these equations for the highest order derivatives gives

$$\frac{d^2y_1}{d\tau^2} = -5y_1 + y_2$$
$$\frac{d^2y_2}{d\tau^2} = -1.6y_2 + 1.6y_1 + 0.2$$

These two expressions tell us how to patch the analog computer to give the solution. One possible simulation diagram is illustrated in Fig. 5.19. Of course, the recorded outputs must be descaled in both time and magnitude.

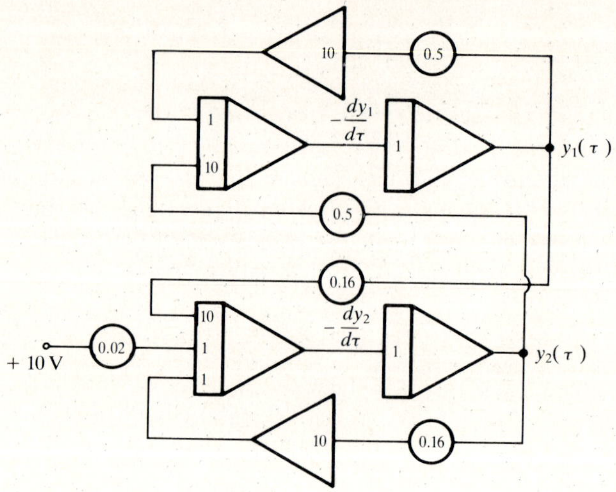

FIGURE 5.19 Analog computer simulation diagram for simultaneous differential equations.

5.6 SIMULATION OF TRANSFER FUNCTIONS

In this section we shall be concerned with the simulation of a transfer function of the form

$$G(s) = \frac{X(s)}{Y(s)} = \frac{b_m s^m + \cdots + b_1 s + b_0}{s^n + a_{n-1}s^{n-1} + \cdots + a_1 s + a_0} \tag{5.6.1}$$

If we cross-multiply this expression and invert the indicated transform relations, the resulting differential equation is

$$x^n + a_{n-1}x^{n-1} + \cdots + a_1\dot{x} + a_0 x = b_m y^m + \cdots + b_1\dot{y} + b_0 y \tag{5.6.2}$$

If we were to take a conventional approach to the solution of this differential equation, we would have to generate the derivatives of $y(t)$ which appear on the right side of Eq. (5.6.2).

We shall now explore a technique which avoids the need to differentiate signals. Differentiation can be particularly difficult if low levels of high-frequency noise are present in $y(t)$. Consider a signal $x(t)$ which is composed of a useful signal $s(t)$ and a small unwanted sinusoidal noise component $A \sin \omega t$ or

$$y(t) = s(t) + A \sin \omega t \tag{5.6.3}$$

The derivative is

$$\dot{y} = \dot{s} + A\omega \cos \omega t \tag{5.6.4}$$

Note that noise amplitude in the first derivative is $A\omega$ and, if ω is greater than unity, the noise amplitude grows every time it is differentiated. Of course the larger ω is, the worse the problem. Differentiation emphasizes high-frequency content; it has the characteristics of a highpass filter. The important idea here is that, if we differentiate repeatedly, the high-frequency noise, which is always present, will eventually dominate the desired signal.

There is, however, a simple method to solve the differential equation (Eq. (5.6.2)), or to simulate the transfer function (Eq. (5.6.1)), which does not require differentiation of the input function $y(t)$.

Let us return to relation (5.6.1) and multiply the ratio $X(s)/Y(s)$ by unity, expressed as $W(s)/W(s)$ to give

$$\frac{X(s)W(s)}{W(s)Y(s)} = \frac{b_m s^m + \cdots + b_1 s + b_0}{s^n + a_{n-1}s^{n-1} + \cdots + a_1 s + a_0} \qquad n > m \tag{5.6.5}$$

Let us now define the ratio $W(s)/Y(s)$:

$$\frac{W(s)}{Y(s)} = \frac{1}{s^n + a_{n-1}s^{n-1} + \cdots + a_1 s + a_0} \tag{5.6.6}$$

which is equivalent to the auxiliary differential equation

$$w^n + a_{n-1}w^{n-1} + \cdots + a_1\dot{w} + a_0 w = y(t) \tag{5.6.7}$$

Since we have made this definition it, follows from Eq. (5.6.6) that

$$\frac{X(s)}{W(s)} = b_m s^m + \cdots + b_1 s + b_0 \tag{5.6.8}$$

which tells us that the time-domain output is

$$x(t) = b_m w^m + \cdots + b_1\dot{w} + b_0 w \tag{5.6.9}$$

To simulate the transfer function, it is now simply a matter of solving differential equation (5.6.7) and by summation constructing the output $x(t)$ from the outputs of the respective integrators according to Eq. (5.6.9). An example illustrates how simple this technique really is.

Example 5.3

Sketch the analog simulation diagram to simulate the transfer function

$$G(s) = \frac{4s + 3}{s^2 + 2s + 5} = \frac{X(s)}{Y(s)}$$

The auxiliary differential equation which arises from the denominator is

$$\ddot{w} + 2\dot{w} + 5w = y(t) \tag{a}$$

and the output which is dictated by the numerator is

$$x(t) = 4\dot{w} + 3w \tag{b}$$

This is illustrated in Fig. 5.20.

The dashed line separates the simulation diagram into two distinct parts associated respectively with expressions **(a)** and **(b)**. The structure to solve Eq. **(a)** is a feedback structure with forward integration, while the structure to create the output is a feed-forward structure. This will always be the case for transfer functions with denominators of higher order than numerators.

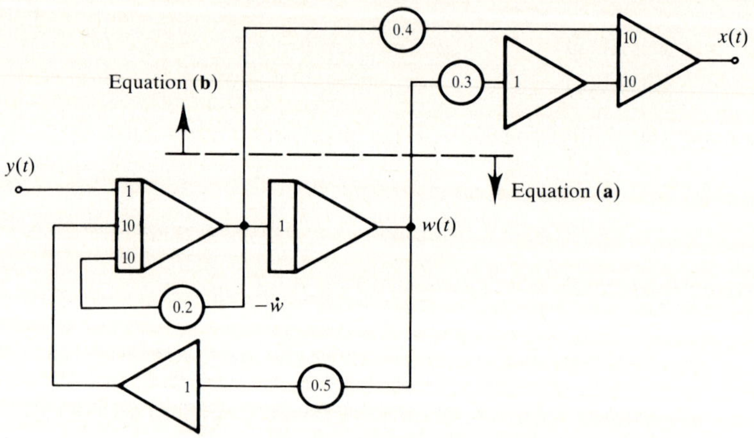

FIGURE 5.20 Simulation diagram for a transfer function.

If the denominator and numerator are of equal order (n), then long division must be used to put the transfer function in the form of a constant plus a lower order numerator divided by the same denominator. This will add a feed-forward path directly from the input to the output. This is illustrated in the following example.

Example 5.4

Consider the simulation of the transfer function

$$G(s) = \frac{s + 0.5}{s + 1} = \frac{X(s)}{Y(s)}$$

Perform long division to yield

$$G(s) = 1 - \frac{0.5}{s+1}$$

Now the simulation can be split into two parts:

$$\dot{w} + w = y(t)$$

and

$$x(t) = y(t) - 0.5w(t)$$

The simulation diagram is given in Fig. 5.21.

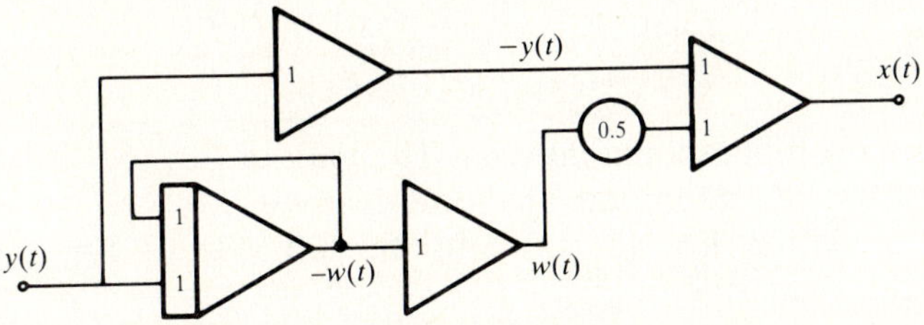

FIGURE 5.21 Simulation diagram for transfer function with first-order numerator and denominator.

5.7 INTRODUCTION TO DIGITAL SIMULATION

So far in this chapter we have concentrated on simulation using analog computers. For several centuries it has been known that differential equations could be solved numerically. However, only in the past two decades has the advent of the digital computer made numerical solutions attractive for system simulation. In fact, it was the need for solving a large system of differential equations associated with problems of fluid dynamics that spurred the development of the electronic digital computer by John von Neumann.

We shall introduce the topic of digital simulation in this chapter because this is a topic of much current interest. In-depth coverage of the topic is the subject of several textbooks. We shall limit our discussion to the solution of first-order differential equations of the form

$$\frac{dx}{dt} = f[x(t), t] \qquad x(0) = x_0 \tag{5.7.1}$$

However, the method is easily applicable to higher order differential equations by decomposing an equation of order n into n first-order differential equations to be solved simultaneously. All methods of numerical solution are accomplished by dividing the temporal (t) axis into increments of width T seconds and then calculating the solution to the differential equation at these points in time. This interval T is commonly called the computational interval.

5.7.1 Euler's Method

Euler's method is sometimes referred to as the forward difference or point-slope method and is given by simply approximating the derivative by the difference of the function between two successive sample times divided by the elapsed-time interval T. Thus

$$\left.\frac{dx}{dt}\right|_{t=nT} \simeq \frac{x[(n+1)T]-x(nT)}{T} \tag{5.7.2}$$

where T is the computational interval. The limit of Eq. (5.7.2) as T becomes small is, in fact, the definition of the derivative.

If we substitute Eq. (5.7.2) into the left side of Eq. (5.7.1), while evaluating the right side of (5.7.2) at $t = nT$, we get

$$\frac{x[(n+1)T]-x(nT)}{T} = f[x(nT), nT] \tag{5.7.3}$$

We may solve this expression for $x[(n+1)T]$ to get

$$x[(n+1)T] = x(nT) + f[x(nT)]T \tag{5.7.4}$$

Note from the original differential equation (Eq. (5.7.1)) that $f[x(nT), nT]$ is the slope of the $x(t)$, t trajectory. Relation (5.7.4) states that, if we know $x(nT)$ and the time dependence of $f(x, t)$, we may use this slope to extrapolate to the next point on the solution trajectory. This situation is illustrated in Fig. 5.22. This method works well so long as T is small enough that the curvature of the trajectory at time nT can be neglected. Euler's method is not the best numerical method for solving differential equations but it is the simplest and the easiest to program.

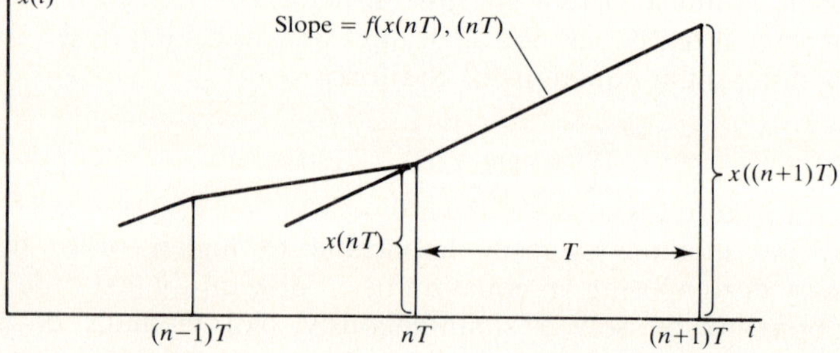

FIGURE 5.22 Solution projection using Euler's method.

Example 5.5

Solve the linear first-order differential equation by Euler's method.

$$\frac{dx}{dt} = -x + 1 \qquad x(0) = 0$$

From the result of Section 4.5.1, the time constant of this system is one second, so a good choice of the computation interval is $T = 0.1$ s. Approximation of the derivative by the forward difference gives

$$\frac{x(n+1)T - x(nT)}{0.1} = -x(nT) + 1$$

$$x(n+1)T = (1 - 0.1)x(nT) + 0.1 \qquad \textbf{(a)}$$

If $x(0) = 0$, we can calculate, for $n = 0$, $n + 1 = 1$, and $T = 0.1$,

$$x(0.1) = (0.9)(0) + 0.1 = 0.1$$

and similarly

$$x(0.2) = (0.9)(0.1) + 0.1 = 0.19$$

and so on. It is a simple matter to program expression **(a)** on a digital computer to give the solution trajectory illustrated in Fig. 5.23, which is compared to the analytical solution of the equation which can be given by Laplace transforms to be

$$x(t) = 1 - e^{-t}$$

which is also illustrated in Fig. 5.23. Note that there is some error between the numerical solution and the analytical solution. If T had been selected to be smaller, the errors would be smaller.

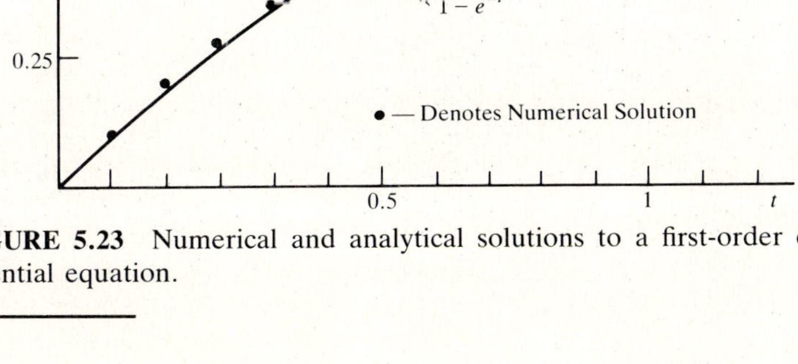

FIGURE 5.23 Numerical and analytical solutions to a first-order differential equation.

5.7.2 The Runge-Kutta Method

Another method which is popular for the numerical solution of ordinary differential equations is the Runge-Kutta method. In fact, if a standard method exists, it is this one. Although this method is of a higher degree of computational complexity than Euler's method, it is not excessively complex. We shall present the method without derivation and the interested reader may consult the excellent text by Hornbeck.[2] Again, we shall give only the result for first-order differential equations of the type of Eq. (5.7.1), repeated here for convenience.

$$\frac{dx}{dt} = f[x(t), t] \qquad x(0) = x_0 \tag{5.7.1}$$

Given the value of the function at time $t = nT$, the algorithm for calculating the next solution point is

$$x[(n+1)T] = x(nT) + \frac{1}{6}(k_1 + 2k_2 + 2k_3 + k_4) \qquad n = 0, 1, 2, \ldots \tag{5.7.5}$$

where the k_is are given by

$$k_1 = f[x(nT), nT] \cdot T \tag{5.7.6}$$

$$k_2 = f\left[x(nT) + \frac{k_1}{2}, \left(n + \frac{1}{2}\right)T\right] \cdot T \tag{5.7.7}$$

$$k_3 = f\left[x(nT) + \frac{k_2}{2}, \left(n + \frac{1}{2}\right)T\right] \cdot T \tag{5.7.8}$$

and

$$k_4 = f[x(nT) + k_3, (n+1)T] \cdot T \tag{5.7.9}$$

Given $x(0)$, one can evaluate k_1 and then use k_1 to evaluate k_2 and so on until k_4 has been evaluated. Now that all the k_i are known, we may evaluate Eq. (5.7.5) to get $x(T)$ and then return to calculation of the k_i for the next step, and so it goes. If we were to evaluate a pair of coupled first-order equations, then a separate set of k_i would be needed for each equation. The programmer must be careful to somehow distinguish between the k_i for each equation.

Example 5.6

Let us consider the hand computation of the Runge-Kutta method for the problem posed in Example 5.5 for the same time interval $T = 0.1$.

$$\frac{dx}{dt} = -x + 1 \qquad x(0) = 0$$

In this problem

$$f(x, t) = -x + 1,\ f[x(nT), nT] = -x + 1$$

Now begin the computation of the k_is, noting that, for $n = 0$, $x(nT) = 0$.

$$k_1 = (-0 + 1)(0.1) = 0.1$$

and

$$k_2 = \left(-\frac{0.1}{2} + 1\right)(0.1) = 0.095$$

$$k_3 = \left(-\frac{0.095}{2} + 1\right)(0.1) = 0.09525$$

$$k_4 = (-0.09525 + 1)(0.1) = 0.090475$$

Now use relation (5.7.5) to calculate the answer:

$$x(0.1) = 0 + \frac{1}{6}\,[0.1 + (2)(0.095) + (2)(0.09525) + 0.09475] = 0.0951625$$

It is interesting to compare this result to the exact solution which, at $t = 0.1$ s, is 0.09516. We could repeat the hand calculations for $t = 0.2$ s. But it is very easy to program a computer or hand calculator to do these operations. In Table 5-2 we have given the numerical results for the exact solution, the Euler method, and the Runge-Kutta solution and it is clear that, although somewhat more complex, the Runge-Kutta method gives superior results. In this example, the Runge-Kutta method provides results equivalent to the analytical solution to at least three decimal places for the time range given.

TABLE 5-2 Comparison of Various Numerical Solution Methods

t	Exact Solution	Euler Solution	Runge-Kutta Solution
0	0	0	0
0.1	0.095	0.100	0.095
0.2	0.181	0.190	0.181
0.3	0.259	0.271	0.259
0.4	0.330	0.344	0.330
0.5	0.393	0.409	0.393
0.6	0.451	0.469	0.451
0.7	0.503	0.522	0.503
0.8	0.551	0.570	0.551
0.9	0.593	0.613	0.593
1.0	0.632	0.651	0.632
1.1	0.667	0.686	0.667
1.2	0.699	0.718	0.699

5.7.3 Applications to Higher Order Differential Equations

Thus far in this section we have discussed only the numerical solution of first-order differential equations. However, all engineers must be equipped to solve differential equations which are of higher than first order. In this section we shall discuss the extension of the techniques just explored to the solution of second-order differential equations. This technique is then easily extended to systems of higher order.

Consider here the differential equation

$$\ddot{x} + b\dot{x} + cx = f(t) \tag{5.7.10}$$

Let us now define a new dependent-variable $v(t)$, which is the derivative of $x(t)$ or

$$v(t) = \dot{x}(t) \tag{5.7.11}$$

and thus expression (5.7.10) may be rewritten as

$$\dot{v} = -bv - cx + f(t) \tag{5.7.12}$$

and relation (5.7.11) is

$$\dot{x} = v \tag{5.7.13}$$

Relations (5.7.12) and (5.7.13) are a set of coupled first-order differential equations. We have just shown how a second-order differential equation can be reduced to two first-order equations. It should be clear to the reader how this technique can be used to reduce an nth-order differential equation to n simultaneous first-order equations. For the remainder of the development we shall assume the existence of two coupled first-order differential equations of a form for which Eqs. (5.7.12) and (5.7.13) are a special case.

$$\dot{x} = f[x(t), v(t), t] \tag{5.7.14}$$

and

$$\dot{y} = g[x(t), v(t), t] \tag{5.7.15}$$

These two equations are more general than the specific pair of Eqs. (5.7.12) and (5.7.13) because the right sides of these equations can be linear or nonlinear. Let us now examine the Euler method of solution by making the forward difference approximation for the respective derivatives or

$$\frac{x[(n+1)T - x(nT)]}{T} = f[x(nT), v(nT), nT] \tag{5.7.16}$$

and

$$\frac{v[(n+1)T - v(nT)]}{T} = g[x(nT), v(nT), nT] \tag{5.7.17}$$

If we now solve for $x[(n+1)T]$ and $v[(n+1)T]$, we get

$$x[(n+1)T] = x(nT) + Tf[x(nT), v(nT), nT] \tag{5.7.18}$$

and

$$v[(n+1)T] = v(nT) + Tg[x(nT), v(nT), nT] \tag{5.7.19}$$

If the initial conditions $x(0)$ and $v(0)$ are known and the samples of the forcing function are known then relations (5.7.18) and (5.7.19) can be stepped forward in time to give the approximate solution to the problem, namely $x(nT)$ and $v(nT)$ versus time $t = nT$.

Now let us consider how the Runge-Kutta method given in Section 5.7.2 might be extended to handle the solution of second-order systems. Let us again consider the coupled first-order differential equations (Eqs. (5.7.14) and (5.7.15)). Recall that for a single differential equation we defined a set of k_i in relations (5.7.6) to (5.7.10) to be computed in the course of one step of the solution. In a similar manner let us define a set of k_i associated with relation (5.7.14) and a similar set of l_i associated with Eq. (5.7.15). The k_i and l_i are defined as follows:

$$\begin{aligned}
k_1 &= f[x(nT), v(nT), nT] \cdot T \\
l_1 &= g[x(nT), v(nT), nT] \cdot T \\
k_2 &= f\left[x(nT) + \frac{k_1}{2}, v(nT) + \frac{l_1}{2}, \left(n + \frac{1}{2}\right)T\right] \cdot T \\
l_2 &= g\left[x(nT) + \frac{k_1}{2}, v(nT) + \frac{l_1}{2}, \left(n + \frac{1}{2}\right)T\right] \cdot T \\
k_3 &= f\left[x(nT) + \frac{k_2}{2}, v(nT) + \frac{l_2}{2}, \left(n + \frac{1}{2}\right)T\right] \cdot T \\
l_3 &= g\left[x(nT) + \frac{k_2}{2}, v(nT) + \frac{l_2}{2}, \left(n + \frac{1}{2}\right)T\right] \cdot T \\
k_4 &= f[x(nT) + k_3, v(nT) + l_3, (n+1)T] \cdot T \\
l_4 &= g[x(nT) + k_3, v(nT) + l_3, (n+1)T] \cdot T
\end{aligned} \tag{5.7.20}$$

The expressions for stepping the two solutions forward in time are then similar to relation (5.7.5):

$$x[(n+1)T] = x(nT) + \frac{1}{6}(k_1 + 2k_2 + 2k_3 + k_4) \qquad \textbf{(5.7.21)}$$

$$v[(n+1)T] = v(nT) + \frac{1}{6}(l_1 + 2l_2 + 2l_3 + l_4) \qquad \textbf{(5.7.22)}$$

Example 5.7

Consider the mass-spring-damper system under the action of a constant force, as illustrated in Fig. 5.24. Let the weight of the mass be 40 lb so the mass is $m = 40/32.2 = 1.242\ \text{lb}\cdot\text{s}^2/\text{ft}$. The damping coefficient is $b = 25\ \text{lb}\cdot\text{s/ft}$ and the spring stiffness is $k = 3000\ \text{lb/ft}$ while the suddenly applied force is $p_0 = 500$ lb. Solve the governing second-order differential equation for the displacement and velocity of the mass by the Runge-Kutta method for zero initial conditions using a time interval of $T = 0.01$ s.

The governing differential equation is

$$\ddot{x} + \frac{b}{m}\dot{x} + \frac{k}{m}x = \frac{p_0}{m}u(t)$$

The two coupled differential equations are

$$\dot{x} = v$$

and

$$\dot{v} = -\frac{b}{m}v - \frac{k}{m}x + \frac{p_0}{m}u(t)$$

The k_i and l_i from relations (5.7.20), after substitution of the numerical parameters from the problem statement, are

$$k_1 = [v(nT)]T$$

$$l_1 = [-20.13v(nT) - 2415x(nT) + 402.57]T$$

$$k_2 = \left[v(nT) + \frac{l_1}{2}\right]T$$

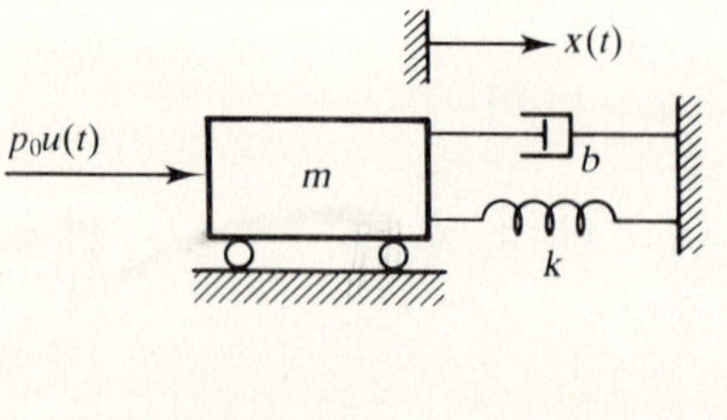

FIGURE 5.24 Mass-spring-damper system under the action of a constant force.

$$l_2 = \left\{-20.13\left[v(nT) + \frac{l_1}{2}\right] - 2415\left[x(nT) + \frac{k_1}{2}\right] + 402.57\right\}T$$

$$k_3 = \left[v(nT) + \frac{l_2}{2}\right]T$$

$$l_3 = \left\{-20.13\left[v(nT) + \frac{l_2}{2}\right] - 2415\left[x(nT) + \frac{k_2}{2}\right] + 402.57\right\}T \qquad \textbf{(a)}$$

$$k_4 = [v(nT) + l_3]T$$

$$l_4 = \{-20.13[v(nT) + l_3] - 2415[x(nT) + k_3] + 402.57\}T$$

and the expressions for stepping the solution forward are

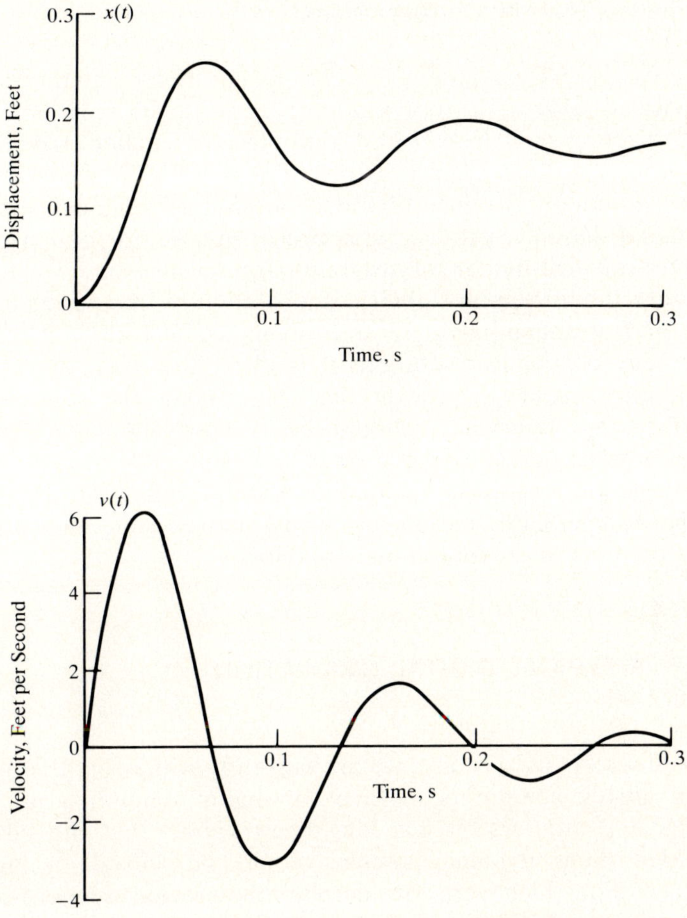

FIGURE 5.25 Runge-Kutta solutions for displacement and velocity for the sprung mass of Example 5.7.

$$x[(n+1)T] = x(nT) + \frac{1}{6}(k_1 + 2k_2 + 2k_3 + k_4)$$

and **(b)**

$$v[(n+1)T] = v(nT) + \frac{1}{6}(l_1 + 2l_2 + 2l_3 + l_4)$$

Expressions **(a)** and **(b)** have been programmed in the BASIC language and the results for $x(t)$ and $v(t)$ are illustrated in Figure 5.25. The BASIC program for this solution is given in Appendix D.6. This program may be easily modified to solve any other second order-system problem.

5.7.4 Numerical Stability Considerations

When we solved a differential equation employing Euler's method we noted that there was a systematic (not random) error in the solution. We also stated that selecting smaller values of T would give smaller errors; likewise larger values of T will give larger errors. It is possible, in some cases, that systems which actually produce bounded outputs for bounded inputs can, when simulated digitally, produce unbounded outputs for bounded inputs. This property is called numerical instability and results when T is too large. One can check the numerical stability of a solution by computing a solution for a value of T and then halving T and solving again. If there is reasonable agreement between the two solutions it is likely that there is no stability problem in either solution. If, on the other hand, the solutions differ radically, the time interval T should be successively halved until two successive solutions agree to the desired precision. Almost all numerical differential equation solution techniques exhibit this difficulty to some degree. The Runge-Kutta method is always more stable than the Euler method, at the cost of computational complexity.

5.8 CSMP: A TYPICAL DIGITAL SIMULATION PROGRAM

Although still widely used today, we are currently witnessing the decline in numbers of analog computers in favor of digital simulation on general-purpose digital computers. It has long been known that the differential equations describing dynamic systems could be solved by numerical methods, several of which were considered in the previous section. The wide availability of general-purpose digital computers with sufficient speed and accuracy has made these techniques feasible without the exhaustive hand-calculator work which was used for simulations in the 1940s.

The convenience of digital simulation is that there is generally no need for magnitude or time scaling because of the wide range of numbers which a digital computer is capable of handling. Even for current eight-bit personal computers, the largest number cannot exceed 10^{+38} in absolute value and the smallest number representable is 10^{-38}. This is a vast improvement over the measurement accuracy, given a dynamic range of ± 10 volts of most contemporary analog computers.

In the early 1960s, several analog computer-like simulation programs were written for the Department of Defense, the first of which was MIMIC, followed soon after by MIDAS. These were large FORTRAN programs with many subroutines which implemented a wide range of functional operations previously thought of as devices or blocks on an analog computer. These devices included summers, potentiometers, integrators, function generators, and a wide range of nonlinear operations such as hysteresis, limiters, relays, and so on. The problem was "patched" to the main program by supplying a deck of data cards which essentially indicated what "devices" were present and the origin of the input signals to each device. Thus all the blocks of a functional block diagram were "tied together" topologically. These simulators were capable of producing numerical response data and, in some cases, line-printer plots. The convenience of not having to scale a problem, yet still employing the functional block diagram methodology of an analog computer, has made digital simulation a powerful tool for contemporary practicing engineers.

More recently there have been other popular simulation programs developed, in particular, ISL (Interactive Simulation Language), ACSL (Advanced Continuous Simulation Language), and CSMP (Continuous System Modeling Program), which are similar in structure to those previously developed.

Here we shall discuss CSMP, a simulation program developed by the IBM Company which is now in its third edition, with CSMP-I being commercially available in the public domain. This discussion is meant to be indicative of only some of the capabilities of CSMP as it is a typical and widely used digital simulation language. For a detailed description of CSMP, the interested reader should consult the excellent text of Speckhart and Green.[3]

CSMP is a large FORTRAN program which is "patched" by user-supplied data cards, although the program will also operate from a terminal. One convenience is that FORTRAN statements, functions, and subroutines may be appended to the data cards so all of the power and intrinsic functions of FORTRAN (sine, cosine, etc.) are available to the CSMP user.

Numerical Data Input System parameters and initial conditions can be input with several types of data statements, namely the CONST, PARAM, and INCON statements.

These statements are followed by a list of parameters set equal to numerical values. For example, in defining some constants $k = 400.$ and $m = 50.$, use the constant (CONST) statement

CONST K = 400., M = 50.

Equivalently, this could have been done with a parameter (PARAM) statement:

PARAM K = 400., M = 50.

To assign initial conditions to integrator outputs, it is appropriate to use the INCON statement:

INCON ICX1 = 5., ICX2 = 50.

When the integration operations are specified, the parameters ICX1 and ICX2 will then appear as the initial condition parameters in the integrator statements for integrators 1 and 2.

System Control There are several control statements which provide control of the simulation and hence should be discussed before continuing.

One important statement is the TIMER statement which controls the length of the simulation t, the printer output interval, the printer plot output interval (these need not be the same), and the minimum integration interval. The variables in this statement are

FINTIM = length of simulation in seconds (final time)

OUTDEL = numerical data output interval (output delta time)

PRDEL = printer plot output interval (printer output delta)

DELT = integration interval (fixed integration interval)

DELMIN = minimum variable integration interval (minimum delta time)

If PRDEL is not specified, it is taken to be FINTIM/100. A typical timer statement would be

TIMER FINTIM = 10., OUTDEL = 0.1, PRDEL = 0.1, DELT = 0.05

Since most simulations hinge on the integration of differential equations, it is necessary either to make the operation of integration transparent to the user or to give the user some control over the method. In CSMP the user has the choice of several variable or fixed interval integration schemes, which are shown in Table 5.3.

TABLE 5-3 Integration Methods Available in CSMP

Variable-Step Integration Methods
- RKS—Variable step Runge-Kutta method
- MILNE—Milne method

Fixed-Step Integration Methods
- RSKFX—Fourth-order Runge-Kutta with fixed interval
- SIMP—Simpson's rule integration
- TRAPZ—Trapezoidal rule integration
- ADAMS—Second-order Adams method
- RECT—Rectangular integration

The particular method is selected with the METHOD statement such as

METHOD RSKFX

which selects the fixed-interval Runge-Kutta integration scheme to be used with an integration interval DELT specified in the TIMER statement.

One statement which must be present in all simulations is the ENDJOB statement which must be the final statement in any CSMP program.

Numerical and Graphical Output To get data out from program variables we use a PRINT statement of the form

PRINT X1, X2

This statement will print three columns, the first of which is time (which is called TIME in the program) and then X1 and X2 at a time interval OUTDEL specified in the timer statement. To get printer plots of data generated during a simulation the PRTPLT statement is of the form

PRTPLT X1, X2

which will give a line-printer plot of X1 versus time and another X2 versus time with an interval specified by PRDEL in the TIMER statement with each plot being of duration FINTIM.

The Integrator Block As we have seen, the operation of solving ordinary differential equations on the analog computer involves the operation of integration. The operation of CSMP is similar except for the important fact that there is no sign inversion in the operation of integration. The integration operation we wish to accomplish is

$$x(t) = \int_0^t y(t)dt + x(0) \tag{5.8.1}$$

where $y(t)$ is the input to the integrator and $x(0)$ is the initial integrator

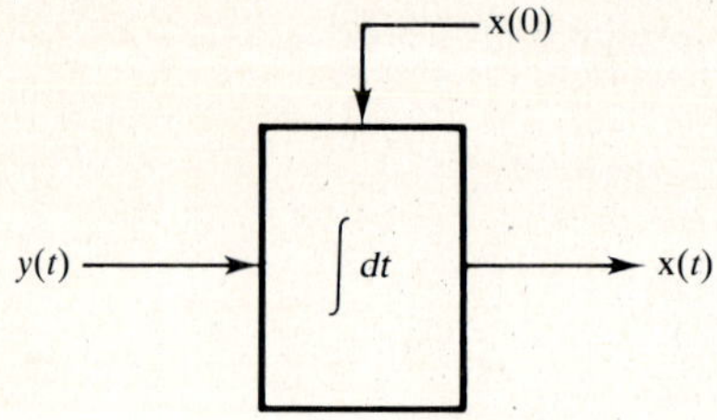

FIGURE 5.26 CSMP integrator block.

output. This operation is represented with the block of Fig. 5.26. The statement to implement the action of Eq. (5.8.1) is

$$X = \text{INTGRL (X0, Y)}$$

where X is the integrator output, Y the integrator input, and X0 the initial condition of the integrator. We shall now present several examples which will illustrate the ease of application of CSMP to simulation of physical systems.

Example 5.8

Let us consider the simulation of a spring-mass-damper system under the action of a suddenly applied constant force, as illustrated in Fig. 5.27. The parameter values are

$$m = \frac{40}{32.2} = 1.242 \text{ lb} \cdot \text{s}^2/\text{ft}$$

$$b = 25 \text{ lb} \cdot \text{s}/\text{ft}$$

$$k = 3000 \text{ lb}/\text{ft}$$

$$p_0 = 500 \text{ lb}$$

The governing differential equation is

$$m\ddot{x} + b\dot{x} + kx = p_0 u(t)$$

As in the analog computer solution, it is best to solve for the highest order derivative or

$$\ddot{x} = (-kx - b\dot{x} + p_0)/m \qquad \textbf{(a)}$$

Clearly, $\ddot{x}$ may be created by the summing process indicated in expression **(a)**. Then, if $\ddot{x}$ is known, $\dot{x}$ and x can be given by successive integration. These ideas are reflected in the block diagram of Fig. 5.28, which is labeled with the CSMP variable names in capital letters. The program for the solution of this

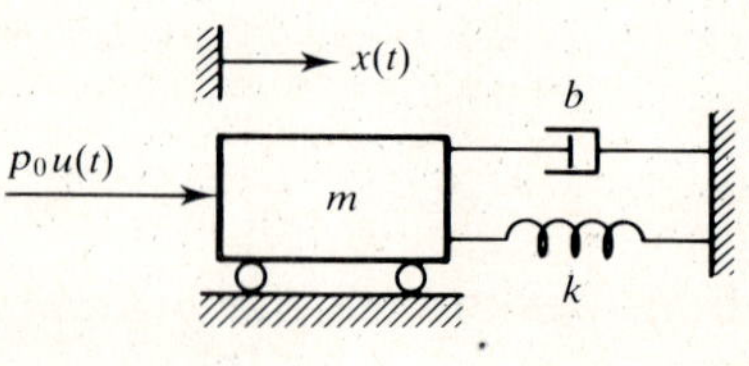

FIGURE 5.27 Spring mass under the action of a constant force.

problem is shown in Table 5-4†. It should be noted here that the statements could occur in any order as the CSMP processor decides the order of execution. If statements need to be evaluated in a particular order, there are special procedures built into CSMP to allow this.

Response data for this problem are displayed in Table 5-5. Line-printer plots for $x(t)$ and $\dot{x}(t)$ are given in Fig. 5.29.

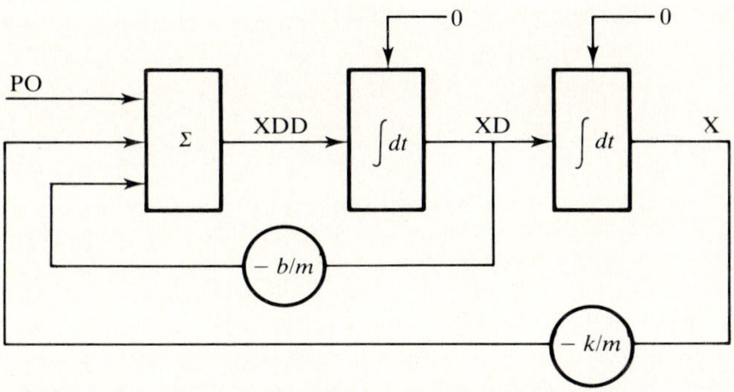

FIGURE 5.28 Block diagram for differential equation solution.

TABLE 5-4 CSMP Program for Mass-Spring-Damper Problem

```
CONST M=1.242,B=25.,K=3000.,PO=500.
XDD=(-K*X-B*XD+PO)/M
X=INTGRL(0.0,XD)
XD=INTGRL(0.0,XDD)
TIMER FINTIM=.25,OUTDEL=.01,PRDEL=.02
PRTPLT X,XD
PRINT X,XD,XDD
LABEL STEP RESPONSE OF MSD SYSTEM
METHOD RKSFX
ENDJOB
```

TABLE 5-5. Numerical Response Data for Mass-Spring-Damper System

TIME	X	XD	XDD
0.00	0.00	0.00	4.0258E 02
2.0000E-02	6.5353E-02	5.6138E 00	1.3172E 02
4.0000E-02	1.8336E-01	5.2493E 00	-1.4599E 02
6.0000E-02	2.5002E-01	1.1551E 00	-2.2458E 02
8.0000E-02	2.3344E-01	-2.4296E 00	-1.1239E 02
1.0000E-01	1.7338E-01	-3.0441E 00	4.5063E 01
1.2000E-01	1.2830E-01	-1.2221E 00	1.1728E 02
1.4000E-01	1.2630E-01	8.9255E-01	7.9536E 01
1.6000E-01	1.5458E-01	1.6517E 00	-4.0406E 00
1.8000E-01	1.8235E-01	9.4768E-01	-5.6956E 01
2.0000E-01	1.8941E-01	-2.1816E-01	-5.0556E 01
2.2000E-01	1.7745E-01	-8.3761E-01	9.1933E 00
2.4000E-01	1.6154E-01	-6.3736E-01	2.5205E 01
2.6000E-01	1.5466E-01	-3.5961E-02	2.9715E 01

†Recall that FORTRAN statements and intrinsic functions may be embedded in a CSMP program.

```
                              MINIMUM                     X       VERSUS TIME                  MAXIMUM
                              0.00                                                             2.5083E-01
TIME          X               I                           RKSFX   INTEGRATION                  I
0.00          0.00            +
1.0000E-02    1.8472E-02      ----+
2.0000E-02    6.5353E-02      -------------+
3.0000E-02    1.2541E-01      ------------------------+
4.0000E-02    1.8336E-01      ------------------------------------+
5.0000E-02    2.2717E-01      --------------------------------------------+
6.0000E-02    2.5002E-01      -------------------------------------------------+
7.0000E-02    2.5083E-01      -------------------------------------------------+
8.0000E-02    2.3344E-01      ----------------------------------------------+
9.0000E-02    2.0491E-01      ----------------------------------------+
1.0000E-01    1.7338E-01      ----------------------------------+
1.1000E-01    1.4616E-01      -----------------------------+
1.2000E-01    1.2830E-01      -------------------------+
1.3000E-01    1.2192E-01      ------------------------+
1.4000E-01    1.2630E-01      -------------------------+
1.5000E-01    1.3854E-01      ---------------------------+
1.6000E-01    1.5458E-01      ------------------------------+
1.7000E-01    1.7028E-01      ---------------------------------+
1.8000E-01    1.8235E-01      ------------------------------------+
1.9000E-01    1.8885E-01      -------------------------------------+
2.0000E-01    1.8941E-01      -------------------------------------+
2.1000E-01    1.8500E-01      ------------------------------------+
2.2000E-01    1.7745E-01      -----------------------------------+
2.3000E-01    1.6897E-01      ---------------------------------+
2.4000E-01    1.6154E-01      --------------------------------+
2.5000E-01    1.5657E-01      -------------------------------+
2.6000E-01    1.5466E-01      ------------------------------+
```

(a)

```
                              MINIMUM                      XD         VERSUS TIME                 MAXIMUM
                              -3.1378E  00                                                        6.1373E  00
TIME          XD               I                           RKSFX      INTEGRATION                    I
0.00          0.00             -------------------+
1.0000E-02    3.5015E  00      ---------------------------------------+
2.0000E-02    5.6138E  00      ---------------------------------------------------+
3.0000E-02    6.1373E  00      -------------------------------------------------------+
4.0000E-02    5.2493E  00      -------------------------------------------------+
5.0000E-02    3.3977E  00      ---------------------------------------+
6.0000E-02    1.1551E  00      --------------------------+
7.0000E-02   -9.2627E-01       --------------+
8.0000E-02   -2.4296E  00      ----+
9.0000E-02   -3.1378E  00      +
1.0000E-01   -3.0441E  00      +
1.1000E-01   -2.3148E  00      ----+
1.2000E-01   -1.2221E  00      -----------+
1.3000E-01   -6.6559E-02       ------------------+
1.4000E-01    8.9255E-01       ------------------------+
1.5000E-01    1.4854E  00      ---------------------------+
1.6000E-01    1.6517E  00      ----------------------------+
1.7000E-01    1.4335E  00      ---------------------------+
1.8000E-01    9.4768E-01       -------------------------+
1.9000E-01    3.4726E-01       --------------------+
2.0000E-01   -2.1816E-01       -----------------+
2.1000E-01   -6.3370E-01       ---------------+
2.2000E-01   -8.3761E-01       --------------+
2.3000E-01   -8.2473E-01       --------------+
2.4000E-01   -6.3736E-01       ---------------+
2.5000E-01   -3.4749E-01       -----------------+
2.6000E-01   -3.5961E-02       ------------------+
```

(b)

FIGURE 5.29 (a) Displacement response versus time. (b) Velocity response versus time.

5.9 SUMMARY

In this chapter we have explored some interesting circuits which have application to instrumentation as well as analog computations. Among the circuits explored were the building blocks associated with a modern analog computer. The use of an analog computer to solve the differential equations associated with dynamic systems was discussed, as well as the technique of employing feedback to generate differential equation dynamics. Time and magnitude scaling were discussed to alleviate problems created by problem variables being too large for allowable voltage limits of modern operational amplifiers or where the actual dynamics were either too fast or too slow for available computers or recording equipment.

Digital simulation was discussed as an alternative to analog simulation and the numerical integration of differential equations was introduced. The CSMP program was presented as a typical digital simulator and several examples of such simulations for linear systems were given.

REFERENCES

1. Bennett, A. W. *Introduction to Computer Simulation*. West Publishing Co., New York, 1974.
2. Hornbeck, R. W. *Numerical Methods*. Quantum Publishers, New York, 1975.
3. Speckhart, F. H., and Green, W. L. *A Guide to Using CSMP—The Continuous System Modeling Program*. Prentice-Hall, Englewood Cliffs, NJ, 1976.

PROBLEMS

5-1. Design an inverting amplifier with a gain of 10. (Pick high values of resistors to give high input impedance and low output impedance). Assume that the open loop op-amp gain is very high.

5-2. Design a noninverting amplifier with a gain of 20.

5-3. Show that the configuration shown below is an integrator. Assume that the operational amplifier has infinite voltage gain. What physical limitations will cause difficulty in building this device?

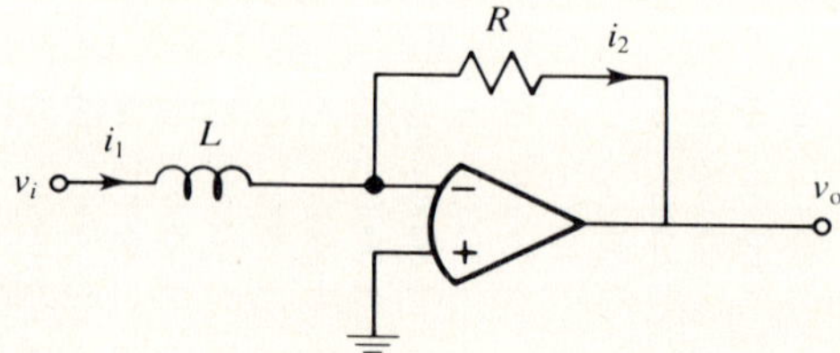

5-4. Assuming that values of R and L are chosen so that i is negligible compared to i_1 and i_2 and v_o/A is small compared to v_o and v_i, write the expression for v_o in terms of v_i and the circuit parameters for the network shown.

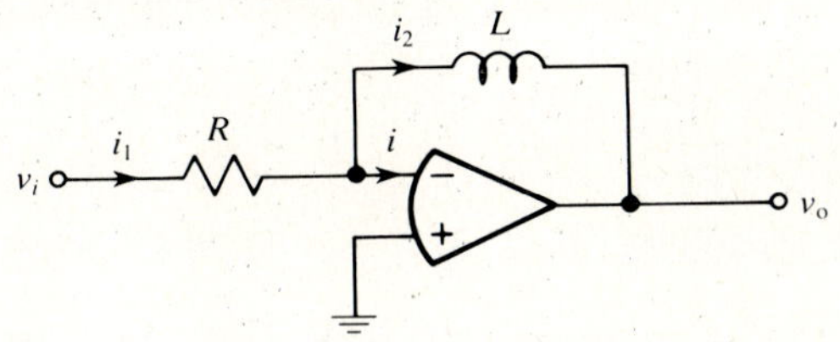

5-5. Sketch the output waveform for the circuit shown below and the input voltage waveform shown.

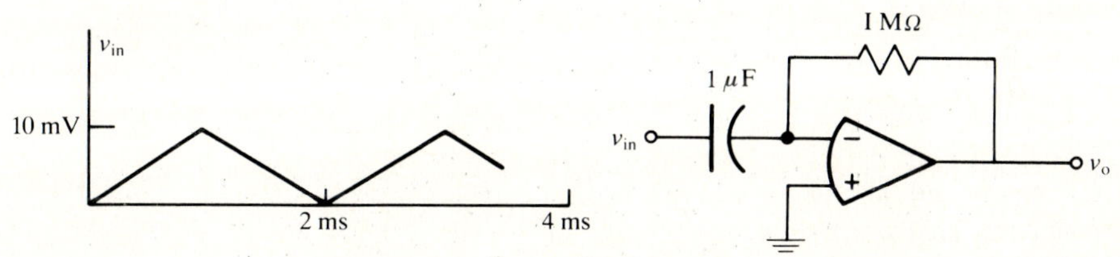

5-6. A square wave with fundamental frequency of 10 Hz and $V_{max} = 10$ volts is to be converted to a triangular waveform with maximum amplitude of 5 volts. Design a circuit to perform this operation.

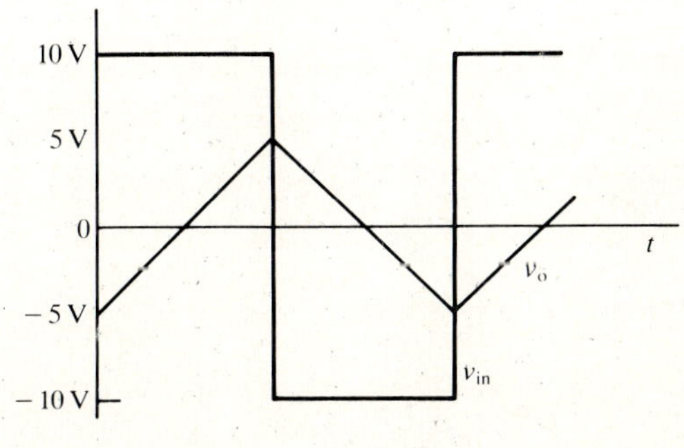

5-7. At $t = 0$, the output of the integrator shown is zero volts and a step function of 5 volts is applied to the input. It is required that the output of the integrator reach negative saturation (-10 volts) at $t = 2$ seconds. Select the value for R to accomplish this.

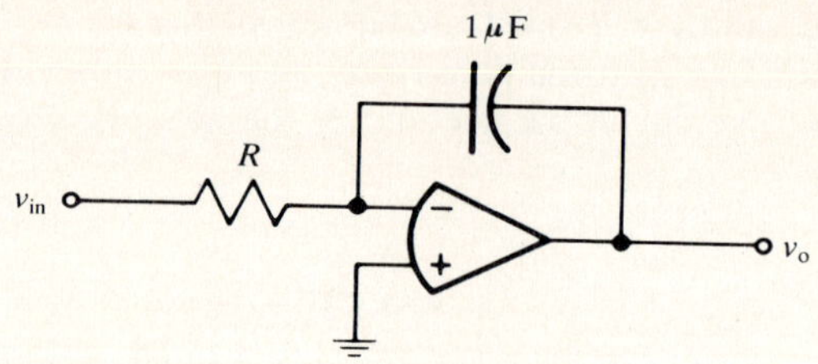

5-8. Prepare an analog simulation of the following differential equation:

$$\frac{d^2x}{dt^2} + 0.7\frac{dx}{dt} + 2x = 20 \quad t \geq 0$$

$$x(0) = 3, \quad \frac{dx(0)}{dt} = -2$$

5-9. Sketch the analog computer network used to solve the differential equation shown below.

$$\frac{d^3v}{dt^3} + 2\frac{d^2v}{dt^2} + 3\frac{dv}{dt} + 5v = 6$$

5-10. Sketch the analog computer diagram to solve the system of differential equations given below.

$$\ddot{x} + 2(\dot{x} - \dot{y}) + x = 2$$

$$\dot{y} + 3(y - x) = 0$$

The initial conditions are

$$x(0) = 0, \dot{x}(0) = 2$$

$$y(0) - 3$$

5-11. Set up the analog computer program for the simultaneous solution of the differential equations shown below.

$$\frac{d^2x}{dt^2} + a_1\frac{d^2y}{dt^2} + a_2\frac{dx}{dt} + a_3x = 0$$

$$\frac{d^3y}{dt^3} + b_1\frac{d^2x}{dt^2} + b_2\frac{dy}{dt} - b_3\frac{dx}{dt} + b_4y = 0$$

5-12. Sketch the analog computer network to solve the following coupled differential equations.

$$\frac{d^2x}{dt^2} + 4\frac{dx}{dt} + 3(x - y) = 0$$

$$\frac{dy}{dt} + y - 2x = 5$$

$$\frac{dx(0)}{dt} = 2,\ x(0) = y(0) = 0$$

5-13. Draw the circuit to solve the following differential equation(s).

(a) $\dfrac{d^3y}{dx^3} + 2y = 4$ **(b)** $\dfrac{d^2x}{dt^2} + 3\dfrac{dx}{dt} + x - y = 2$

$$\frac{dy}{dt} + 2y - x = 0$$

Assume that all initial conditions are zero.

5-14. Find the differential equation solved by the circuit shown:

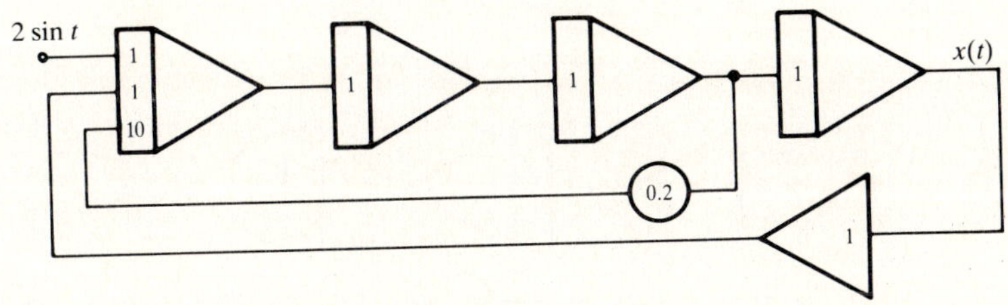

5-15. Write the differential equation solved by the analog computer network given below.

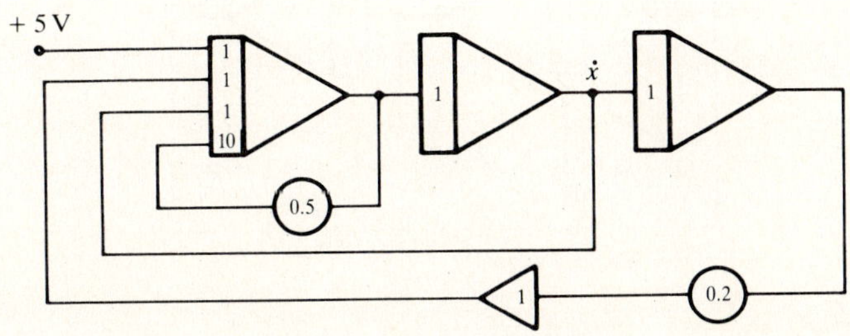

5-16. **(a)** Find the differential equation solved by the analog computer network shown.

(b) Identify correctly all initial conditions.

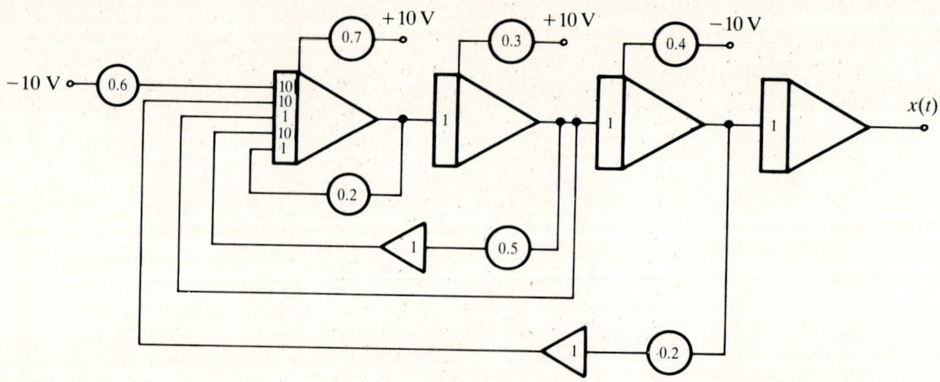

5-17. Write the differential equations solved by the analog computer networks given below.

(a)

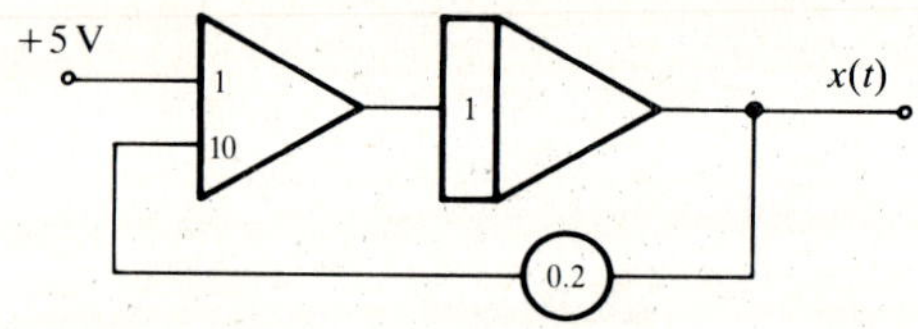

(b)

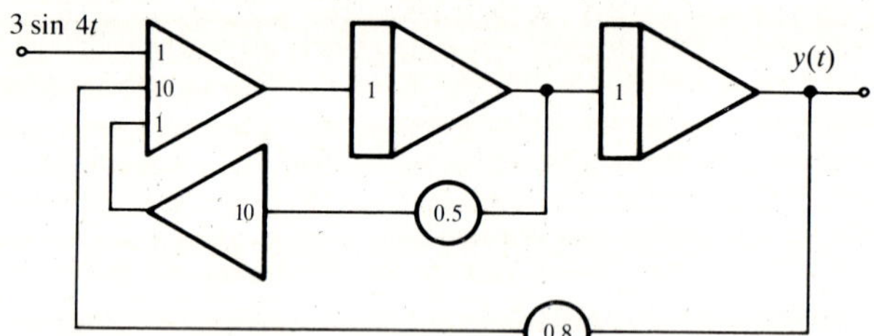

(c)

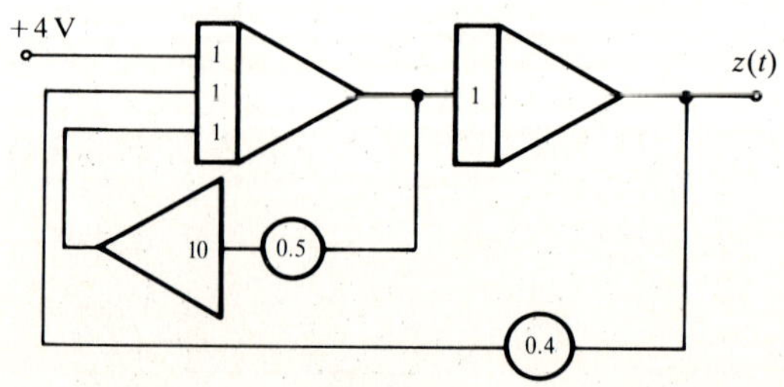

5-18. Design an analog computer network to generate the function $y = e^{-at}$. The way to go about this is to differentiate both sides of the above equation with respect to time, then solve the resulting differential equation using analog computer techniques. Indicate the initial condition at $t = 0$ that must be imposed on the circuit before pushing the "go" button.

5-19. Suppose you need to generate a sinusoidal function $\sin \Omega t$. Note that

$$\sin \Omega t = \Omega \int \cos \Omega t \, dt$$

$$\cos \Omega t = -\Omega \int \sin \Omega t \, dt$$

Set up the analog computer diagram to do this, assuming $\Omega \leq 1$.

5-20. For the differential equation given below, perform the time-scaling operation including the scaling on the initial conditions. Assume that your analog computer is one with a dynamic range of ± 10 volts.

$$\ddot{x} + 30\dot{x} + 1000x = 0$$

$$x(0) = 5$$

$$\dot{x}(0) = -80$$

5-21. For the differential equation given below, the dynamics do not appear to need time scaling. However, the initial conditions indicate that the size of the solution will be outside the range of a ± 10-volt analog computer. Perform the operation of magnitude scaling.

$$\dddot{y} + 2\ddot{y} + 3\dot{y} + y = 0$$

$$y(0) = 50$$

$$\dot{y}(0) = 100$$

$$\ddot{y}(0) = 0$$

5-22. Consider the analog computer solution of the differential equation given below. Perform both time and magnitude scaling in order to make it solvable on an analog computer with a dynamic range of ± 10 volts.

$$\dddot{x} + 30\ddot{x} + 400\dot{x} + 2000x = 0$$

$$x(0) = 30$$

$$\dot{x}(0) = \ddot{x}(0) = 0$$

5-23. A dynamic system is governed by the following pair of differential equations for $t \geq 0$.

$$\ddot{x} + 40\dot{x} + 100(x - y) = 5$$

$$\dot{y} + 50(y - x) = 0$$

$$x(0) = \dot{x}(0) = y(0) = 0$$

Scale the equations so that the resulting scaled system will run on a 10-volt analog computer.

6

APPLICATION OF TRANSFORM METHODS TO SIMPLE THERMAL AND FLUID SYSTEMS

6.1 INTRODUCTION

To this point in this text we have considered electrical and mechanical systems. However, there are a host of other important physical problems which can be solved by the same methods. In this chapter we shall introduce several types of systems for which the associated differential equations can be formulated by application of the principles of conservation of energy, mass, or momentum. In contrast to electrical and mechanical systems, the resulting differential equations are more often nonlinear and thus the Laplace transform method is not applicable to their solution. We shall use this opportunity to discuss and develop some approximate linear models to which the transform method is applicable.

6.2 THERMAL SYSTEMS

To formulate the governing equations for thermal systems we need to employ the principle of conservation of energy, which is better known in the engineering literature as the first law of thermodynamics. There are many detailed mathematical statements of this principle but we shall present it here in its simplest form. For a region of space (a thermodynamic system) we can write this principle for some fixed interval of time as

$$E_{\text{in}} - E_{\text{out}} = \Delta E_{\text{stored}} \tag{6.2.1}$$

Energy may be transported into or out of the system by flow of mass, by heat conduction or by movement of the system boundary. Energy may be stored within the system boundary by the heat capacity of the medium or by increase of potential or kinetic energy inside the boundary.

Equation (6.2.1) may be written in rate form by differentiating Eq. (6.2.1) to give

$$\dot{E}_{\text{in}} - \dot{E}_{\text{out}} = \frac{d}{dt}(E_{\text{stored}}) \tag{6.2.2}$$

This is the most useful form of the energy balance for dynamic system analysis.

6.2.1 Energy Storage

All matter, whether solid, liquid, or gas, stores energy in the thermal agitation of its molecules. This stored energy is commonly referred to in the field of thermodynamics as internal energy. The ability of a material to store this energy is measured by the specific heat c, which is the energy storage per unit mass per unit change in temperature. Then the incremental stored energy for an incremental change in temperature ΔT is

$$\Delta E = \Delta T \int_{\text{Vol}} \rho c \, dVol \tag{6.2.3}$$

If we then perform the indicated integration,

$$\Delta E = \rho c(Vol)\,\Delta T \tag{6.2.4}$$

where the parentheses are used to indicate multiplication. The rate of energy storage (power) is

$$\frac{dE}{dt} = \rho(Vol)c\,\frac{dT}{dt} \tag{6.2.5}$$

Note that the product of density ρ and the volume is the mass m, so the rate of energy storage is

$$\frac{dE}{dt} = mc\,\frac{dT}{dt} \tag{6.2.6}$$

6.3 MODES OF HEAT TRANSFER

Thermal energy is transported by one of three means: conduction, convection, or radiation. Sometimes these effects are compounded such that heat transport occurs simultaneously by several of these modes.

6.3.1 Conduction

This mode of heat transfer governs the transport of heat in solids and stationary liquids when no convective currents exist. Consider the one-dimensional problem illustrated by the wall in Fig. 6.1. Here the temperatures of the left and right walls are T_1 and T_2, respectively, with $T_1 > T_2$. The governing equation for this mode of heat transfer is Fourier's law of conduction, which states that the rate of heat transfer is

$$\dot{q} = \frac{kA}{\Delta x}(T_1 - T_2) \tag{6.3.1}$$

where A is the area normal to the direction of heat transport, Δx is the wall thickness, and k is the thermal conductivity of the wall material. The quantity $\Delta x/kA$ is often given the name of thermal resistance. Values for thermal conductivities k of some typical engineering materials are given in Table 6-1. The dominant mode of heat transfer in fluids (liquids and gases) is generally not conduction unless special precautions are taken to prevent the development of convective currents in the bulk of the fluid. We shall address this problem in the next section.

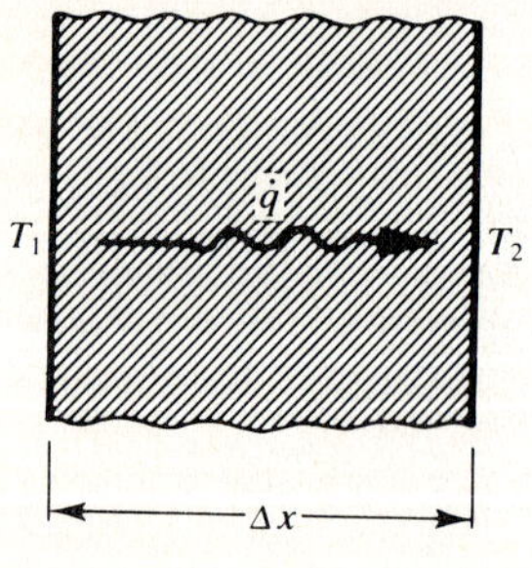

FIGURE 6.1 Conduction heat transfer in a wall.

TABLE 6-1

Material	Thermal Conductivity k W/(m · K)
Concrete	1.36
Water at 20°C	0.597
Fiber insulating board at 20°C	0.0484
Plate glass at 20°C	0.761
Steel at 20°C	44.5
Fiberglass at 20°C	0.039
Air at 20°C	0.056
and one atmosphere	0.0262
Styrofoam	0.0345

6.3.2 Convection

Convection is the transfer of thermal energy (heat) at a solid-fluid interface due to a temperature gradient between a solid surface and the bulk of a fluid, as illustrated in Fig. 6.2. If the fluid is caused to circulate by some external mechanism (fan, pump, etc.), the heat transfer is referred to as forced convection. If the only fluid motion only takes place due to the buoyancy caused by temperature gradients, the mechanism is referred to as free convection. For both types of convection the governing equation is Newton's law of cooling or

$$\dot{q} = hA(T_1 - T_2) \tag{6.3.2}$$

where A is the area of the solid-fluid interface and h is referred to as the film coefficient. Approximate values for film coefficients for various physical situations are given in Table 6-2. The range of values given in Table 6-2 is due to the many variables involved in the free and forced convection problems.

In the forced convection problem, the film coefficient h is a function of free-stream velocity of the fluid $\bar{V}$, the fluid density ρ, the thermal conductivity k, the temperatures T_1 and T_2, the fluid specific heat c, viscosity μ, and the geometry of the situation. For forced convection from single bodies immersed in flowing fluids, the film coefficient is a function of the Reynolds number $N_R = \bar{V}L\rho/\mu$, where L is the characteristic size dimension of the immersed body. For Reynolds numbers greater than 500, an approximate

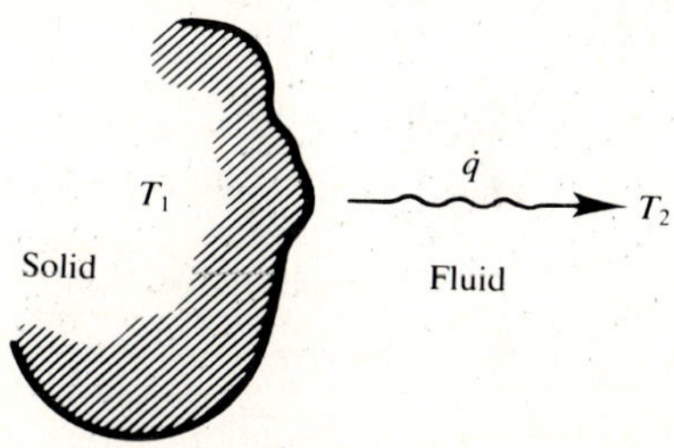

FIGURE 6.2 Convective heat transfer mechanism.

TABLE 6-2 Approximate Values of Convective Film Coefficient

Mechanism	h, (W/(m$^2\cdot$K))
Free convection in air	5–50
Forced convection in air	25–250
Forced convection in water	250–15,000
Boiling water	2,500–25,000
Condensing water vapor	5,000–100,000

relation given from an accumulation of experimental data for the film coefficient is

$$h = \frac{k}{L} 0.31(N_R)^{0.6} \tag{6.3.3}$$

The free convection problem differs from the forced convection problem because the only velocities developed in the fluid are due to bouyant effects. The geometry of the surface plays an important role in the film coefficient. For example, the free convection from upward-facing horizontal surfaces is considerably different from that for a down-facing horizontal surface. If we define ΔT as the positive temperature difference, there exist empirical relations for the film coefficients for air at one 'atmosphere. One popular relation for air at one atmosphere is

$$h = C\left(\frac{\Delta T}{L}\right)^n \tag{6.3.4}$$

where the coefficient C and exponent n are functions of two dimensionless parameters, the Grashof number and the Prandtl number, given respectively by relations (6.3.5) and (6.3.6):

$$N_G = \frac{\beta g \rho^2 L^3 \Delta T}{\mu^2} \tag{6.3.5}$$

and

$$N_P = \frac{\mu c}{k} \tag{6.3.6}$$

where β is the coefficient of thermal expansion, g the acceleration of gravity,

TABLE 6-3 Film Coefficients for Free Convection in Air

Geometry	$N_G N_R$	C	n	L
1. Vertical planes and cylinders	10^4–10^9	1.41	1/4	Height
2. Vertical planes and cylinders	10^9-10^{12}	1.31	1/3	1
3. Horizontal cylinders	10^3–19^9	1.32	1/4	Diameter
Horizontal cylinders	10^9–10^{12}	1.24	1/3	1
4. Horizontal plates, heated facing upward or cooled facing downward	10^5–2×10^7	1.32	1/4	Length of side
	2×10^7–3×10^{10}	1.52	1/3	1
5. Horizontal plates, cooled facing upward or heated facing downward	3×10^5–3×10^{10}	0.585	1/4	Length of side

ρ the density, L the characteristic length, ΔT the temperature difference, μ the fluid viscosity, c the specific heat, and k the fluid thermal conductivity. For surfaces having different horizontal and vertical dimensions, the following relation is suggested for calculation of the length parameter L.

$$\frac{1}{L} = \frac{1}{L_{\text{vert}}} + \frac{1}{L_{\text{horiz}}} \tag{6.3.7}$$

Values of C and n for various Grashof-Prandtl number products are given in Table 6-3.

6.3.3 Radiation

The two bodies illustrated in Fig. 6.3 at respective temperatures T_1 and T_2 can exchange energy by electromagnetic radiation without the need of a transmission medium. This type of energy transmission is governed by the Stefan-Boltzmann law which is

$$\dot{q} = C(T_1^4 - T_2^4) \tag{6.3.8}$$

where the temperatures are absolute temperatures. The coefficient C depends on the geometries, colors, textures, and areas of both surfaces. We should not underemphasize the complexity of the calculation of the coefficient C. Unless one of these temperatures is much greater than the other, the heat transfer will be small compared to that transferred by convection and conduction. For a detailed treatment of radiation it is suggested that the reader consult Ref. 1.

Let us now define the thermal resistance to be the measure of difficulty with which heat is transferred per unit of temperature gradient causing the heat transfer. For conduction the resistance is defined as

$$R = \frac{\Delta x}{kA} \tag{6.3.9}$$

while for convection the thermal resistance is

$$R = \frac{1}{hA} \tag{6.3.10}$$

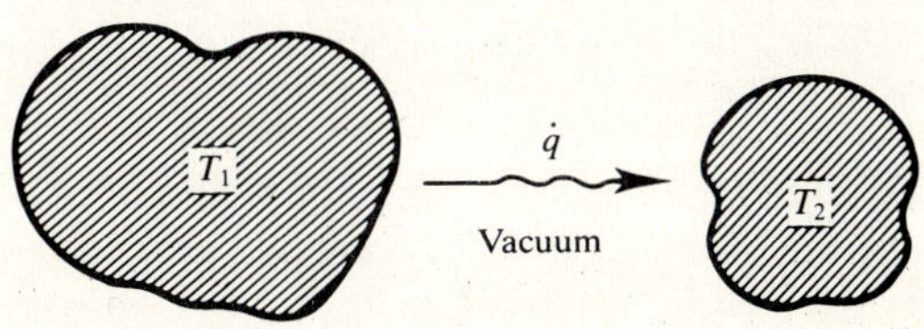

FIGURE 6.3 Transmission of energy by thermal radiation.

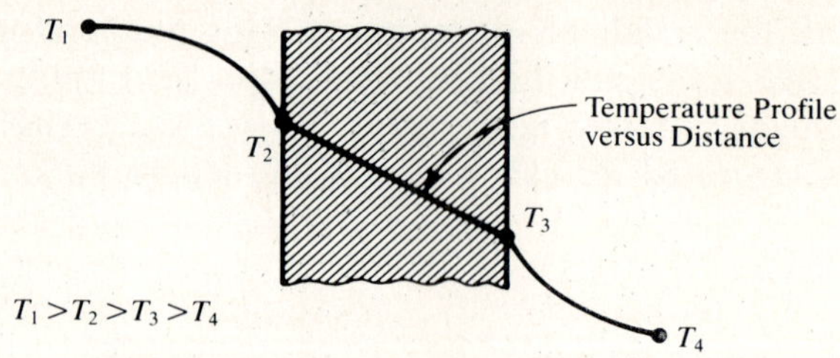

FIGURE 6.4 Heat transfer through a wall.

When there is simultaneous heat transfer through a wall by conduction, with convective boundary layers on either side of the wall, as illustrated in Fig. 6.4, the analysis becomes more complicated as we shall now show. The heat transfer for the left boundary layer is

$$\dot{q}_{12} = \frac{1}{R_1}(T_1 - T_2) \tag{6.3.11}$$

while that for the wall is

$$\dot{q}_{23} = \frac{1}{R_2}(T_2 - T_3) \tag{6.3.12}$$

and that for the right boundary layer is

$$\dot{q}_{34} = \frac{1}{R_3}(T_3 - T_4) \tag{6.3.13}$$

If we equate these rates of heat transfer and eliminate the wall temperatures T_2 and T_3, we get

$$\dot{q}_{14} = \frac{1}{R_{eq}}(T_1 - T_4) \tag{6.3.14}$$

where

$$R_{eq} = R_1 + R_2 + R_3 \tag{6.3.15}$$

We see that the resistances to heat transfer behave as electrical resistors connected in series, while the temperatures behave as voltages and the heat transfer rates behave as currents. This simple electrical analogy is given in Fig. 6.5.

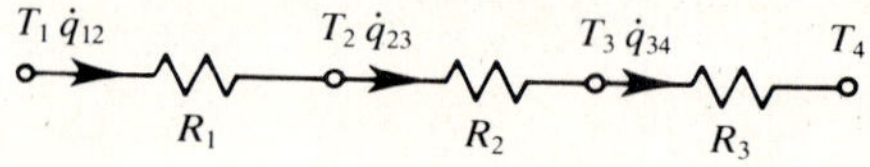

FIGURE 6.5 Electrical analogy for heat flow.

Example 6.1

Calculate the steady-state heat transfer through the one-meter-high vertical wall shown in Fig. 6.6, which has a convective boundary layer on the right surface. The height of the wall is one meter, the internal temperature $T_1 = 20°C$, and the external temperature is $T_3 = 0°C$. Assume that the thermal conductivity of the wall is $k = 0.04 \text{ W/m} \cdot \text{K}$. The temperature T_2 is unknown and must be found in the course of the problem. We know that the steady-state rate of heat transfer from across the wall and the boundary layer must be the same. The conduction heat transfer per unit area is

$$\frac{\dot{q}_{\text{cond}}}{A} = \frac{k}{\Delta x}(T_1 - T_2)$$

while that for convection is

$$\frac{\dot{q}_{\text{conv}}}{A} = h(T_2 - T_3)$$

Clearly the interface temperature T_2 must be between 0° and 20°C, so let us initially assume $T_2 = 10°C$. The film coefficient may be calculated from relation 1 from Table 6-3.

$$h = 1.41(10)^{1/4} = 2.51 \text{ W/(m}^2 \cdot °\text{C)}$$

The convective heat transfer rate is

$$\frac{\dot{q}_{\text{conv}}}{A} = 2.51(10 - 0) = 25.1 \text{ W/m}^2$$

The conduction heat transfer is

$$\frac{\dot{q}_{\text{cond}}}{A} = \frac{0.04}{0.2}(20 - 10) = 2 \text{ W/m}^2$$

which is vastly different from the convection transfer. Clearly our assumption of $T_2 = 10°C$ is incorrect and we must try again. Since the convection heat transfer is much larger, it is clear that $T_2 = 10°C$ is too high. If we assume $T_2 = 5°$ and go through the process once more we get

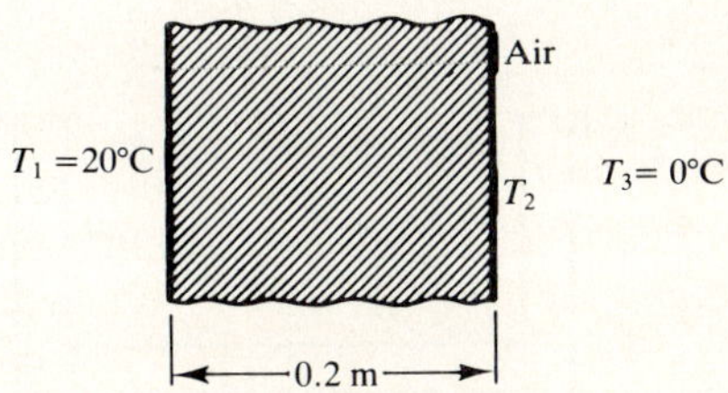

FIGURE 6.6 Combined conduction and convection heat transfer through a wall.

$$\frac{\dot{q}_{\text{conv}}}{A} = 10.5 \text{ W/m}^2$$

and

$$\frac{\dot{q}_{\text{cond}}}{A} = 3 \text{ W/m}^2$$

Clearly T_2 is still slightly too high and adjustments must be made until the rates of heat transfer by both modes are the same. For a trial temperature of $T_2 = 2°\text{C}$, the convective film coefficient is $1.68 \text{ W/(m}^2 \cdot °\text{C)}$ and the rates of heat transfer are, respectively,

$$\frac{\dot{q}_{\text{conv}}}{A} = 3.35 \text{ W/m}^2$$

and

$$\frac{\dot{q}_{\text{cond}}}{A} = 3.6 \text{ W/m}^2$$

This may be sufficiently accurate for much engineering work but, if a better answer is required, one more trial will result in improved accuracy. The total thermal resistance for this wall and the single boundary layer is

$$R = \frac{1}{A}\left(\frac{\Delta x}{k} + \frac{1}{h}\right) = \frac{1}{A}\left(\frac{0.2}{0.04} + \frac{1}{1.68}\right) = \frac{0.595}{A}\ \frac{°\text{C}}{W}$$

Example 6.2

Let us find a dynamic model for a thermal bath commonly used to control the effect of environmental temperature variations on experiments conducted by chemical and mechanical engineers. The system is shown in Fig. 6.7.

We shall provide a stirrer in order to ensure a uniform temperature distribution in the liquid, which will considerably simplify the resulting model. While this is not necessary, we shall assume that $T(t) > T_0(t)$ and we shall ignore the energy contributed by the stirrer; in practice we could measure that contribution.

The rate of energy stored is given by expression (6.2.4) to be

$$\frac{d}{dt}(E_{\text{stored}}) = mc\frac{dT}{dt} \tag{a}$$

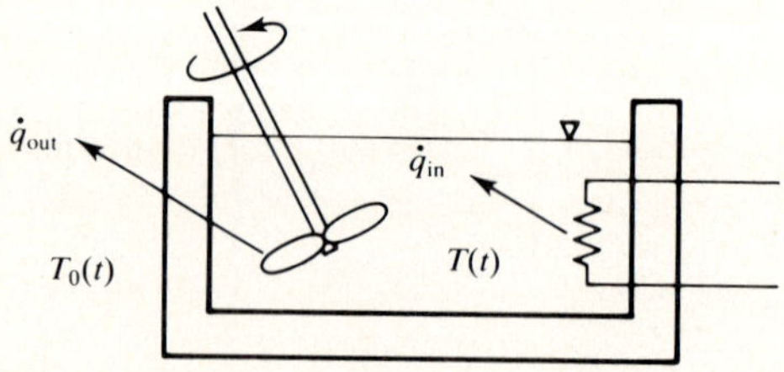

FIGURE 6.7 A temperature bath.

where m and c are, respectively, the mass and specific heat of the liquid. The rate of energy transport out of the system is that transported away from the liquid body by convection, conduction, and radiation. For small temperature differences between the tank and the surroundings, these effects can be treated together and the energy loss from the walls and the free surface can be treated as a linear function of the temperature difference $T - T_0$ which, from relation (6.3.14), is

$$\dot{q}_{out} = \frac{1}{R}[T - T_0] \tag{b}$$

where both T and T_0 are functions of time. The constant R will be a function of the surface area, the wall area and the insulating properties of the walls, and the ambient conditions of the surrounding air.

The rate of energy transfer into the system from the heater will simply be denoted as q_{in}. If we now substitute Eqs. **(a)** and **(b)** into relation (6.2.2), the result is a differential equation governing the system:

$$\dot{q}_{in} - \frac{1}{R}(T - T_0) = mc\frac{dT}{dt} \tag{c}$$

If we rearrange the terms we get

$$mc\frac{dT}{dt} + \frac{1}{R}T = \frac{1}{R}T_0 + \dot{q}_{in} \tag{d}$$

If we examine this expression, we see that this system has two possible stimuli or forcing functions, namely, the environmental temperature $T_0(t)$ and the heat added by the heater $q_{in}(t)$.

If we take the Laplace transform of equation **(d)** and solve for $T(s)$, the result is

$$T(s) = [hAT_0(s) + \dot{q}_{in}(s)]\left[\frac{1}{mcs + 1/R}\right] \tag{e}$$

This expression may be represented by the block diagram of Fig. 6.8.

If the environmental temperature T_0 is a constant, we can define a new

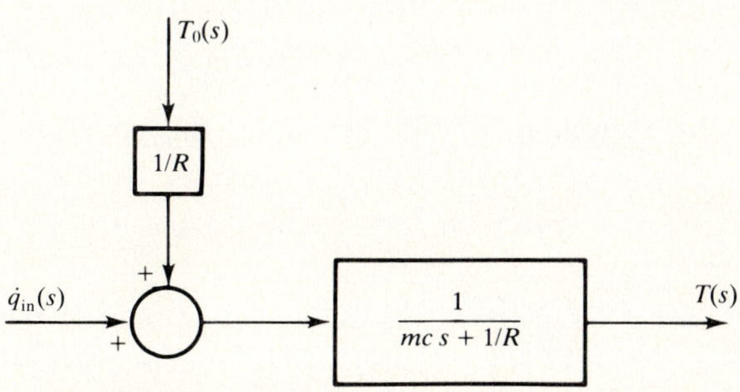

FIGURE 6.8 Thermal system block diagram.

temperature variable to be the difference between the tank temperature $T(t)$ and the environmental temperature $T_0(t)$ or

$$\theta(t) = T(t) - T_0(t) \tag{f}$$

If we differentiate this with respect to time, the result is

$$\frac{d\theta}{dt} = \frac{dT}{dt} \tag{g}$$

If we now substitute these relations into Eq. **(d)**, the governing differential equation for a constant environmental temperature becomes

$$mc\frac{d\theta}{dt} + \frac{1}{R}\theta = \dot{q}_{in} \tag{h}$$

This is a first-order system with a time constant of $\tau = mCR$. This makes intuitive sense since the speed of response should be inversely proportional to the mass and directly proportional to the ease of heat transfer.

6.4 FLUID SYSTEMS

In this section we shall consider only incompressible fluids (liquids) for which the density ρ is constant. There are several fundamental principles which are useful in the formulation of the governing differential equations for fluid systems.

6.4.1 The Principle of Conservation of Mass

The first concept is that of the principle of conservation of mass which, simply stated, for a system boundary over some fixed time interval is

$$m_{in} - m_{out} = \Delta m_{stored} \tag{6.4.1}$$

If we make a rate statement of this principle, the governing expression is

$$\dot{m}_{in} - \dot{m}_{out} = \frac{d}{dt}(m_{stored}) \tag{6.4.2}$$

There are some more sophisticated statements of the mass balance, but that given in Eq. (6.4.2) is perhaps the most useful in the modeling of fluid tank systems.

Example 6.3

Formulate a differential equation for the mass balance of the constant surface area tank shown in Fig. 6.9. The rate of change of mass in the tank is

$$\frac{d}{dt}(m_{\text{stored}}) = \rho A \frac{dh}{dt}$$

Then the mass balance (Eq. (6.3.2)) yields

$$\dot{m}_{\text{in}} - \dot{m}_{\text{out}} = \rho A \frac{dh}{dt} \tag{a}$$

which is the desired differential equation. In this case the outflow $\dot{m}_{\text{out}}$ is a function of the depth of the fluid in the tank. In order for us to describe this dependence we shall have to develop some new ideas in the next section.

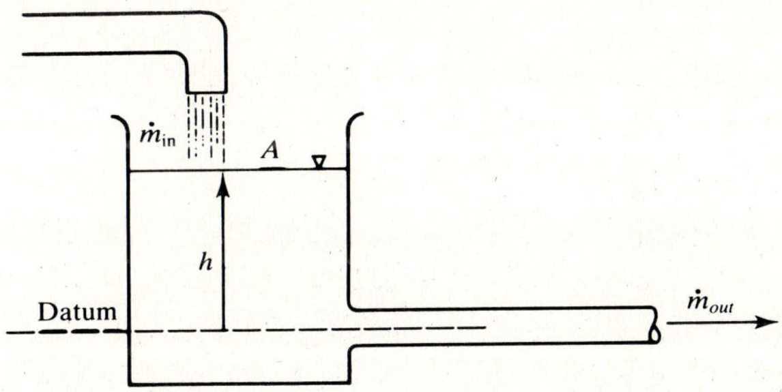

FIGURE 6.9 A prismatic tank with inflow and outflow.

Example 6.4

Consider a simplified analysis of a hydraulic actuator and spool valve illustrated in Fig. 6.10. We shall make the following simplifying assumptions to derive the model for this system.

1. There is no fluid leakage around the spools of the valve and there is no leakage around the piston.
2. Neglect the compressibility of the hydraulic fluid.
3. The flow rate through the spool valve is proportional to the spool displacement $x(t)$.

For an incompressible fluid, the mass balance is equivalent to a volume rate balance. For the actuator cylinder, the rate at which the piston sweeps the cylinder is equal to the flowrate Q_{in} or

$$A \frac{dy}{dt} = Q_{\text{in}}(t) \tag{a}$$

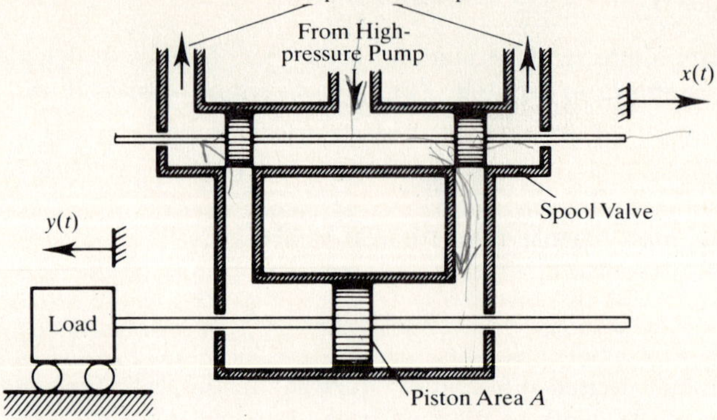

FIGURE 6.10 Simple hydraulic actuator and spool valve.

If we now employ assumption 3, the result is

$$Q_{in}(t) = Kx(t) \tag{b}$$

If we now substitute Eq. **(b)** into Eq. **(a)** for $Q_{in}(t)$, the result is

$$A\frac{dy}{dt} = Kx(t) \tag{c}$$

If we transform this expression and ignore the initial condition $y(0)$ the result is

$$AsY(s) = KX(s)$$

If we solve for the transfer function

$$G(s) = \frac{Y(s)}{X(s)} = \frac{K}{As}$$

which is shown in Fig. 6.11. Now let us investigate the case where $x(t)$ is a sudden constant displacement X_0 occurring at $t = 0$, as illustrated in Fig. 6.12(a). The transform of the spool motion is

$$X(s) = \frac{X_0}{s}$$

The response is then given by multiplying the input by the transfer function

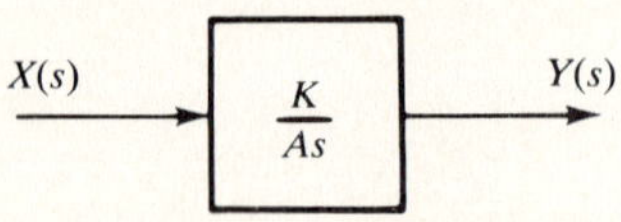

FIGURE 6.11 Block diagram for sample hydraulic actuator system.

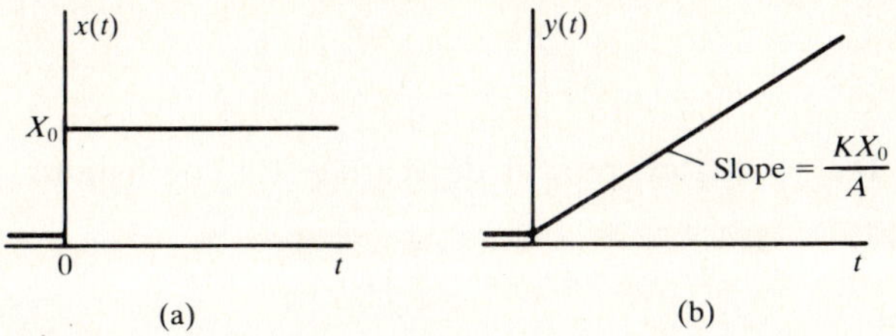

FIGURE 6.12 (a) Sudden constant spool valve displacement. (b) Actuator response.

$$Y(s) = G(s)X(s) = \frac{KX_0}{As^2}$$

After inversion of this simple transform, the result for the displacement of the load is

$$y(t) = \frac{KX_0}{A}\,t$$

which is illustrated in Fig. 6.12(b). Of course this motion cannot be sustained for a long time because the piston will come to rest against the end of the cylinder, but within the range of actuator motion this is a reasonable model for light loads. If the loads require large forces, then fluid leakage and compressibility must be considered and a more complicated model results.

The device is often referred to as a hydraulic integrator because the load displacement is the integral of the spool valve displacement, as can be seen from the governing differential equation (Eq. **(c)**).

6.4.2 A More Thorough Analysis of the Hydraulic Actuator

Let us now consider the more complicated case which arises when the loads on the actuator are high, requiring accompanying high-pressure operation. In this case neither the leakage around the piston nor the compressibility of the liquid can be neglected. The pressure drop across the orifice Δp is a function of the source pressure p_h and the pressure at the piston p_L. We shall assume that high pressure of the pump is fixed; thus the volume flow rate through the value $q(t)$ is a function of the valve position $x(t)$ and pressure $p_L(t)$ on the high-pressure side of the piston or

$$q = f(x, p_L) \qquad \textbf{(6.4.3)}$$

If the flow is zero for $x = 0$ and $p_L = 0$, then the flow rate can be written as

$$q = \left(\frac{\partial q}{\partial x}\right)_0 x + \left(\frac{\partial q}{\partial p_L}\right)_0 p_L \tag{6.4.4}$$

Let us define the evaluated partial derivatives as coefficients C_x and C_p or

$$C_x = \left(\frac{\partial q}{\partial x}\right)_0 \tag{6.4.5}$$

and

$$C_p = -\left(\frac{\partial q}{\partial p_L}\right)_0 \tag{6.4.6}$$

and thus relation (6.4.4) is

$$q = C_x x - C_p p_L \tag{6.4.7}$$

The flow into the cylinder is now comprised of the following components:

$$q = q_0 + q_L + q_c \tag{6.4.8}$$

where q_0 is the component of flow which causes the piston to move, q_L the leakage component around the piston, and q_c the compressed portion. The piston motion flow rate is

$$q_0 = A\,\frac{dy}{dt} \tag{6.4.9}$$

and the compressible component is

$$q_c = \frac{V}{\beta}\,\frac{dp_L}{dt} \tag{6.4.10}$$

where V is the effective volume of the fluid under compression and β is the bulk modulus of the fluid. The leakage component will be assumed proportional to the cylinder pressure p_L, which is essentially the pressure differential across the piston.

$$q_L = K_L p_L \tag{6.4.11}$$

Substituting relations (6.4.7), (6.4.9), (6.4.10), and (6.4.11) into relation (6.4.8), the result is

$$q = C_x x - C_p p_L = A\,\frac{dy}{dt} + \frac{V}{\beta}\,\frac{dp_L}{dt} + K_L p_L \tag{6.4.12}$$

The force developed by the piston is

$$F = Ap_L \tag{6.4.13}$$

Consider now a load which is inertial with mass M in the presence of viscous damping b such that

$$F = M\frac{d^2y}{dt^2} + b\frac{dy}{dt} = p_L A \tag{6.4.14}$$

Solving for p_L gives

$$p_L = \frac{M}{A}\frac{d^2y}{dt^2} + \frac{b}{A}\frac{dy}{dt} \tag{6.4.15}$$

and the derivative is

$$\frac{dp_L}{dt} = \frac{M}{A}\frac{d^3y}{dt^3} + \frac{b}{A}\frac{d^2y}{dt^2} \tag{6.4.16}$$

Substitution of Eqs. (6.4.15) and (6.4.16) into Eq. (6.4.12) gives a governing differential equation of

$$\frac{MV}{A\beta}\frac{d^3y}{dt^3} + \left[(K_L + C_p)\frac{M}{A} + \frac{Vb}{\beta A}\right]\frac{d^2y}{dt^2} + \left[(K_L + C_p)\frac{b}{A} + A\right]\frac{dy}{dt} = C_x x \tag{6.4.17}$$

If we transform this differential equation and solve for a transfer function $Y(s)/X(s)$ we get

$$\frac{Y(s)}{X(s)} = \frac{C_x}{s\left\{\frac{MV}{A\beta}s^2 + \left[(K_L + C_p)\frac{M}{A} + \frac{Vb}{\beta A}\right]s + \left[(K_L + C_p)\frac{b}{A} + A\right]\right\}} \tag{6.4.18}$$

We see that, as before, there is integrator behavior because of the factorable s in the denominator. However, this integrator is softened by the dynamics of the quadratic factor in the denominator.

6.5 THE MOMENTUM BALANCE

Let us consider a three-dimensional steady flowfield, which is illustrated in Fig. 6.13(a). We shall assume that the viscous fluid effects are negligible and there is no work being done in the fluid by external forces. A streamtube will be chosen which is of incremental length dL in the direction of the flow, with a small area A at the inlet and normal to the flow and area $A + dA$ at

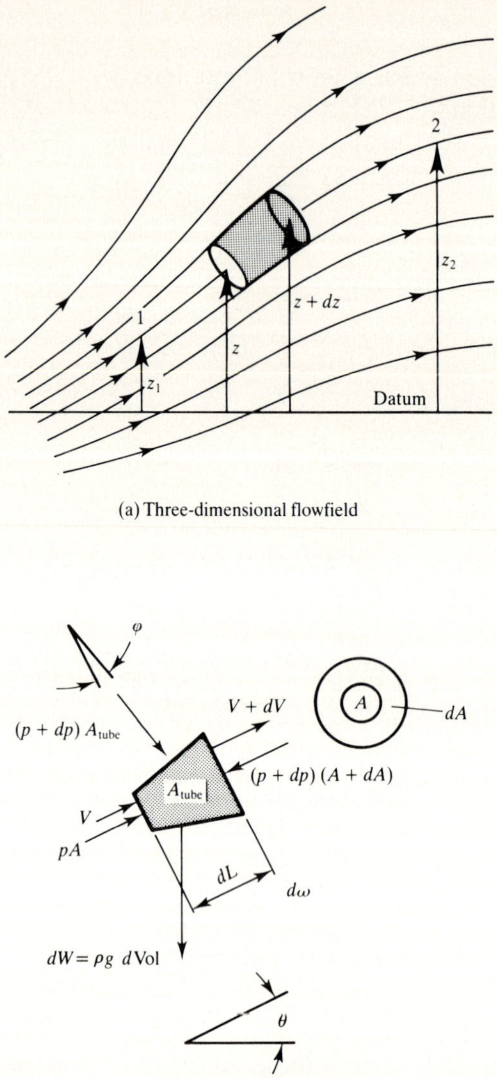

(a) Three-dimensional flowfield

(b) Streamtube freebody diagram

FIGURE 6.13 Three-dimensional flowfield. (b) Streamtube free-body diagram.

the outlet. The area of the lateral surface of the streamtube will be denoted as S_{tube}. The pressure at the entrance will be denoted as p, while the exit end is $p + dp$ and the entering velocity will be denoted as V, while the exit velocity will be termed $V + dV$. A free-body diagram of the streamtube is shown in Fig. 6.13(b).

The volume of the element will be termed *dVol* and is

$$dVol = \left(\frac{2A + dA}{2}\right) dL \tag{6.5.1}$$

assuming a truncated conical section of average area $(A + A + dA)/2$. Let us now apply Newton's second law to this stream tube, which gives

$$pA - (p + dp)(A + dA) - \rho g \sin\theta\left(A + \frac{dA}{2}\right) dL$$
$$+ \left(p + \frac{dp}{2}\right) S_{\text{tube}} \sin\phi = \left(\rho + \frac{d\rho}{2}\right)\left(A + \frac{dA}{2}\right) dL \, \frac{dV}{dt} \tag{6.5.2}$$

Note that the velocity $V = dL/dt$ and that $S_{\text{tube}} \sin\phi = dA$ and $dL \sin\theta = dz$. So after dropping terms which have products of differentials in Eq. (6.5.2), the result is

$$-p\, dA - dp\, A - \rho g \frac{dz}{dL} A\, dL + p\, dA = \rho A V\, dV \tag{6.5.3}$$

If we divide by the area A and eliminate the $p\, dA$ terms, the result is

$$-dp - \rho g\, dz = \rho V\, dV \tag{6.5.4}$$

This expression is known as Euler's equation for the flow of an inviscid fluid in the absence of externally applied forces. This expression is applicable to both compressible and incompressible steady fluid flows.

We shall now assume that the fluid is incompressible such that the density ρ may be assumed constant. With this assumption we may integrate this expression along a streamline between points 1 and 2 in the flowfield of Fig. 6.13(a), or, writing this as a definite integral

$$\rho g \int_{z_1}^{z_2} dz + \int_{p_1}^{p_2} dp + \rho \int_{v_1}^{v_2} V\, dV = 0 \tag{6.5.5}$$

The result is then

$$p_1 + \rho g z_1 + \rho \frac{V_1^2}{2} = p_2 + \rho g z_2 + \rho \frac{V_2^2}{2} \tag{6.5.6}$$

This expression is Bernoulli's equation and is a result of considering a momentum balance for an inviscid, incompressible steady flow. We shall now apply this equation to derive the expression for the tank problem examined in Example 6.3.

Example 6.5

Consider the tank illustrated in Fig. 6.14 and consider the mass balance. The mass balance gave us the following relation

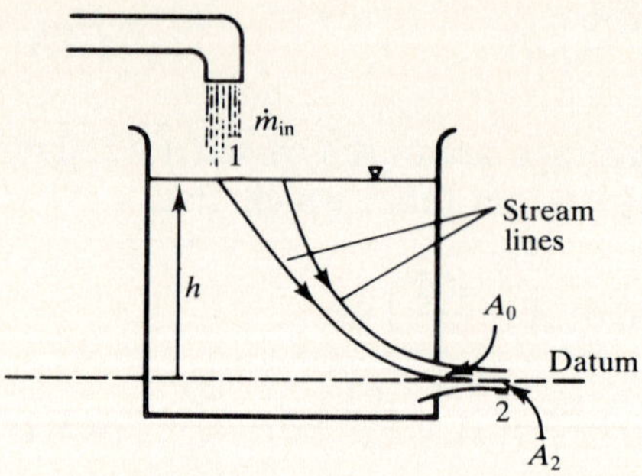

FIGURE 6.14 Fluid tank system.

$$\rho A \frac{dh}{dt} = \dot{m}_{in} - \dot{m}_{out} \qquad \textbf{(a)}$$

We shall now apply Bernoulli's equation (Eq. (6.5.6)) to the streamline illustrated in Fig. 6.14, taking points 1 and 2 as shown and a datum or reference level which coincides with point 2. If the area of the free surface A is larger than that of A_2, we can argue, using the mass balance, that V_1 (the velocity of the surface movement) is smaller than V_2; in fact, we shall assume it to be zero. Since the pressures at points 1 and 2 are atmospheric pressures, we may say $p_1 = p_2$. With our choice of the datum the elevation of point 2 is zero, so $z_2 = 0$ and the result is

$$\rho \frac{V_2^2}{2} = \rho g z_1$$

and, since $z_1 = h$, this reduces to

$$V_2 = \sqrt{2gh} \qquad \textbf{(b)}$$

In the fluid mechanics literature this expression is commonly referred to as Torricelli's equation. (In this analysis we have neglected the effects of the viscosity of the liquid.)

If we examine this problem experimentally, we find that the point at which the pressure in the outlet stream is the same as atmospheric pressure is not in the plane of the orifice but rather downstream a bit where the streamlines are parallel, as illustrated at point 2 in Fig. 6.14. At this point the area A_2 is somewhat smaller than the orifice area A_0. Let us define the contraction coefficient C_c as the ratio of A_2 to A_0 or

$$C_c = \frac{A_2}{A_0} \qquad \textbf{(c)}$$

Experimentally we also observe that, due to viscous effects, V_2 is somewhat less than what would be predicted by the Torricelli equation **(b)**. Let us define the coefficient of velocity C_v as the ratio of actual velocity at point 2 to the ideal velocity predicted by expression **(b)** or

$$C_v = \frac{V_2}{V_{2\,\text{ideal}}} = \frac{V_2}{\sqrt{2gh}}$$

The actual velocity is then

$$V_2 = C_v\sqrt{2gh} \tag{d}$$

The mass flow rate out of the tank is then given as

$$m_{\text{out}} = \rho A_2 V_2 = \rho A_2 C_v \sqrt{2gh} \tag{e}$$

We may write the area A_2 in terms of the contraction coefficient as

$$A_2 = C_c A_0$$

Then the flowrate out of the tank is, from Eq. **(e)**,

$$m_{\text{out}} = \rho A_0 C_c C_v \sqrt{2gh}$$

It is common practice to define the product $C_c C_v$ as the coefficient of discharge, C_d, so the outflow is

$$m_{\text{out}} = \rho A_0 C_d \sqrt{2gh} \tag{f}$$

Discharge coefficients for several shapes of outlet orifices are given in Table 6-4. If expression **(f)** is substituted into the mass balance epression **(a)**, the result is

$$\rho A \frac{dh}{dt} = m_{\text{in}} - \rho A_2 C_d \sqrt{2gh} \tag{g}$$

If we divide by the fluid density we get

$$A \frac{dh}{dt} + A_2 C_d \sqrt{2g} h^{1/2} = Q_{\text{in}}(t) \tag{h}$$

where $Q_{\text{in}}(t)$ is the inlet volume flowrate. Expression **(h)** represents a volume balance which, for an incompressible fluid, is equivalent to a mass balance. The differential equation **(h)** is a nonlinear differential equation because of the $h^{1/2}$ term. Since this differential equation is nonlinear, neither the Laplace transform method nor transfer function concept is applicable. For the special case of no flow-in, $Q_{\text{in}}(t) = 0$, the variables in expression **(h)** may be separated and a closed-form solution to the differential equation given by integration.

TABLE 6-4 Discharge Coefficients for Several Types of Orifices

Sharp-edged	Rounded	Short Tube	Borda
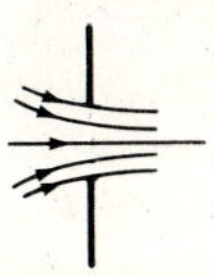		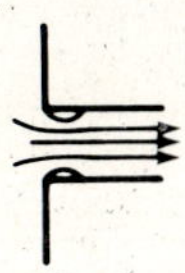	
$C_d = 0.61$	$C_d = 0.98$	$C_d = 0.80$	$C_d = 0.51$

In the next section of this chapter we shall examine how we might handle nonlinear terms for small variations in the variable h about some steady state. The result will be a linear differential equation governing small variations in the head of the tank.

6.6 STEADY FLUID FLOW IN PIPES

Consider the flow in an incompressible viscous fluid in the constant-area pipe illustrated in Fig. 6.15. We shall assume that the velocity profiles at cross sections 1 and 2 are the same and that the average velocity at both cross sections is V. A criterion for determination of whether the flow is laminar or turbulent is the numerical value of the Reynolds number, which is a dimensionless parameter involving the critical physical variables of the problem. The Reynolds number is defined for a circular pipe to be

$$N_R = \frac{V\,d\,\rho}{\mu} \tag{6.6.1}$$

where d is the pipe diameter, ρ the fluid density, and μ the dynamic viscosity of the fluid, all expressed in a consistent set of units such that N_R is truly dimensionless. If the Reynolds number is less than 2300, the flow will be laminar and the relation between the pressure drop $p_2 - p_1$ is a linear function of the velocity V or

$$p_1 - p_2 = \frac{32\mu L}{d^2}\,V \tag{6.6.2}$$

where L is the length of the pipe. Although this is a simple linear relation between pressure drop and velocity, it is often not applicable because the Reynolds number is often greater than 2300 and hence the flow is turbulent. The relation between pressure drop and velocity is for the turbulent case

$$p_1 - p_2 = \frac{fL\rho}{2d}\,V^2 \tag{6.6.3}$$

where f is the friction factor related to Reynolds number by the experimental curves illustrated in Fig. 6.16. The parameter e/d on the curves is commonly referred to as the relative roughness of the inner pipe surface.

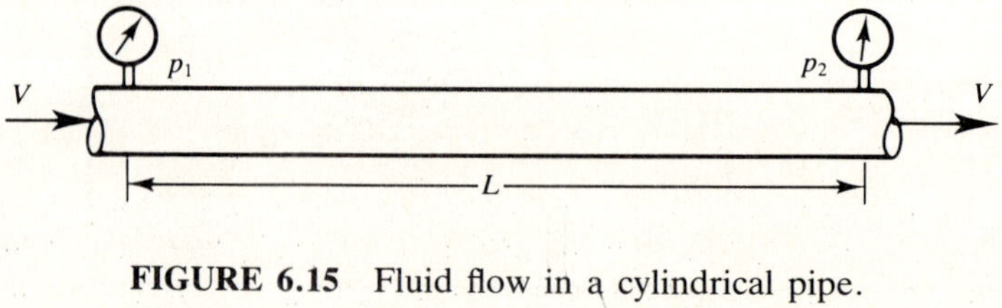

FIGURE 6.15 Fluid flow in a cylindrical pipe.

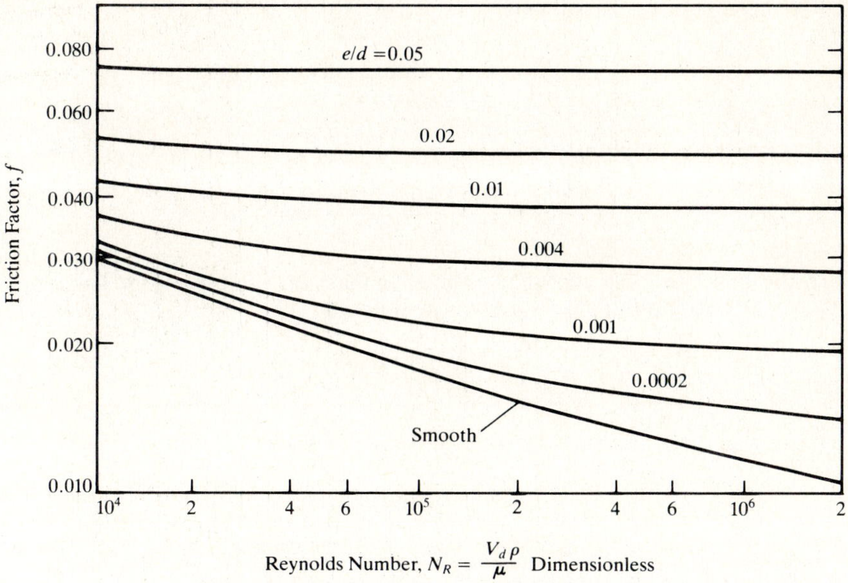

FIGURE 6.16 Dimensionless friction factor for turbulent flow in pipes.

TABLE 6-5 Roughness for Various Types of Commercial Pipe

Pipe Type	e (m)
Uncoated cast iron	2.54×10^{-4}
Galvanized iron	1.52×10^{-4}
Wrought iron	4.32×10^{-5}
Concrete	3×10^{-4}–3×10^{-3}
Riveted steel	9×10^{-4}–9×10^{-3}
Drawn tubing	1.27×10^{-7}
Commercial steel	3.8×10^{-6}

Some typical values of the absolute roughness e in meters for various kinds of pipe are given in Table 6-5.

Example 6.6

Calculate the pressure drop in a 5-cm-diameter uncoated cast iron pipe of length 300 m carrying a mass flowrate of 1 kg/s. The properties of water at standard conditions are

$$\rho = 1000 \text{ kg/m}^3$$

$$\mu = 8 \times 10^{-4} \text{ N} \cdot \text{s/m}^2$$

For this flow rate the velocity is

$$V = \frac{\dot{m}}{\rho A} = \frac{1(4)}{(1000)\pi(0.05)^2} = 0.51 \text{ m/s}$$

The Reynolds number is then

$$N_R = \frac{V\, d\rho}{\mu} = \frac{(0.51)(0.05)(1000)}{8 \times 10^{-4}} = 3.18 \times 10^4$$

so the flow is clearly turbulent. The roughness for the pipe from Table 6-5 is $e = 2.54 \times 10^{-4}$ m and the relative roughness is

$$\frac{e}{d} = \frac{2.54 \times 10^{-4}}{5 \times 10^{-2}} = 5.1 \times 10^{-3}$$

From Fig. 6.16, the friction factor is approximately

$$f = 0.025$$

and from relation (6.6.3) the pressure drop is

$$p_2 - p_1 = (0.025)\,\frac{(300)1000}{2(0.05)}\,(0.51)^2 \text{ N/m}^2\text{/(or Pa)}$$

$$= 1.95 \times 10^4 \text{ N/m}^2 = 2.83 \text{ psi}$$

Example 6.7

Consider the single tank shown in Fig. 6.17 with a long exit tube such that the formulation of Example 6.5 is no longer valid. In this problem we can no longer assume that the pressure at point 2 is zero (atmospheric). If we apply the Bernoulli equation between points 1 and 2 we get, by working in gage pressures,

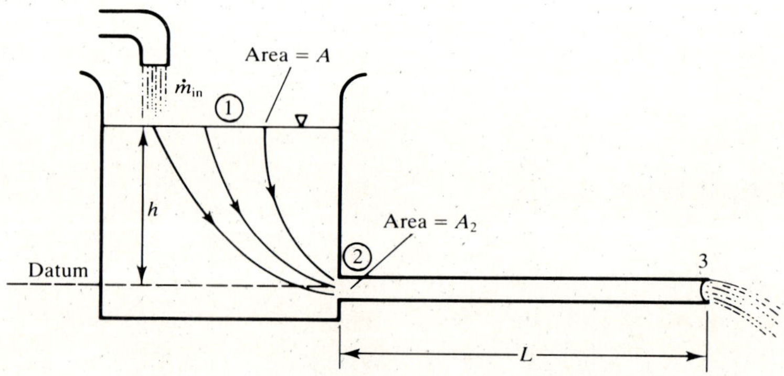

FIGURE 6.17 Tank with long exit tube.

$$\rho g z_1 = p_2 + \rho \frac{V_{2\,\text{ideal}}^2}{2} \qquad \textbf{(a)}$$

If we now assume, because of constant area, that V_2 and V_3 are the same, the inviscid Bernoulli equation would tell us that the pressures at points 1 and 2 are the same. However, this does not allow for the pressure drop due to viscous effects in the pipe. Assume that the actual velocity in the pipe is V_2 and that the pipe flow is turbulent, so the relation for the pipe is

$$p_2 = \frac{fL\rho}{2d} V_2^2 \qquad \textbf{(b)}$$

The coefficient of discharge relates V_2 and $V_{2\,\text{ideal}}$ to give

$$V_2 = C_d V_{2\,\text{ideal}}$$

If we note that $z_1 = h$, relation **(a)** becomes

$$\rho g h = p_2 + \rho \frac{1}{C_d^2} \frac{V_2^2}{2} \qquad \textbf{(c)}$$

If we substitute Eq. **(b)** into Eq. **(c)** we get

$$\rho g h = \frac{fL\rho}{2d} V_2^2 + \frac{\rho}{2C_d^2} V_2^2$$

or

$$gh = \left(\frac{fL}{2d} + \frac{1}{2C_d^2}\right) V_2^2$$

Solving for V_2 gives

$$V_2 = \left(\frac{g}{\dfrac{fL}{2d} + \dfrac{1}{2C_d^2}}\right)^{1/2} \sqrt{h} \qquad \textbf{(d)}$$

Note that the mass flowrate out is given as the product of the density ρ, the area A_2, and the velocity V_2, or

$$\dot{m}_{\text{out}} = \rho A_2 V_2 \qquad \textbf{(e)}$$

If relation **(d)** is substituted into relation **(e)**, which is in turn substituted into the mass balance of equation (Eq. **(a)** from Example 6.3), the result is

$$\rho A \frac{dh}{dt} = \dot{m}_{\text{in}} - \rho A_2 \left(\frac{g}{\dfrac{fL}{2d} + \dfrac{1}{2C_d^2}}\right)^{1/2} \sqrt{h}$$

This nonlinear differential equation can be linearized about a steady-flow condition by the methods given in Section 6.7.

6.7 LINEARIZATION OF NONLINEAR PROBLEMS

In Sec. 6.6 we developed a model of the discharge of a tank which was nonlinear in the height of fluid in the tank. Since this model was not linear the useful tools for examining system response developed earlier in this text were not applicable. In this section we shall explore a technique useful for approximating the solution of nonlinear problems of this type.

First we shall briefly review the concepts of the Taylor series. If we are given a function $y = f(x)$ we could plot this function as illustrated in Fig. 6.18. For the choice of some point b, the function $f(x)$ can be represented everywhere by the Taylor series

$$f(x) = f(b) + \left.\frac{df}{dx}\right|_{x=b}(x-b) + \frac{1}{2}\left.\frac{d^2f}{dx^2}\right|_{x=b}(x-b)^2 + \cdots \qquad \textbf{(6.7.1)}$$

This description becomes cumbersome for any arbitrary x distant from $x = b$ because many derivatives of $f(x)$ must be evaluated to give an accurate representation of $f(x)$.

Let us now consider the case of approximating $f(x)$ in a region near b such that $(x - b)$ is a small number. If $(x - b)$ is small, then the higher powers of $(x - b)$ are even smaller and the first two terms of Eq. (6.7.1) will often give a good approximation of $f(x)$ in the neighborhood of $x = b$ or

$$\hat{f}(x) = f(b) + \left.\frac{df}{dx}\right|_{x=b}(x-b) \qquad (y = mx + b) \qquad \textbf{(6.7.2)}$$

where the circumflex denotes the approximation of $f(x)$. The coordinates $[b, f(b)]$ are commonly referred to as the *operating point* and the approximation given by Eq. (6.7.2) is shown graphically in Fig. 6.18 as a straight-line tangent to $f(x)$ at $x = b$. If we define a new independent variable $x' = x - b$ to be the deviation of x from b, then $f(x' + b)$, the value of the function at the point $x = b$, can be approximated as

$$f(b + x') \cong f(x')\Big|_{x'=0} + \left.\frac{df}{dx'}\right|_{x'=0} x' \qquad \textbf{(6.7.3)}$$

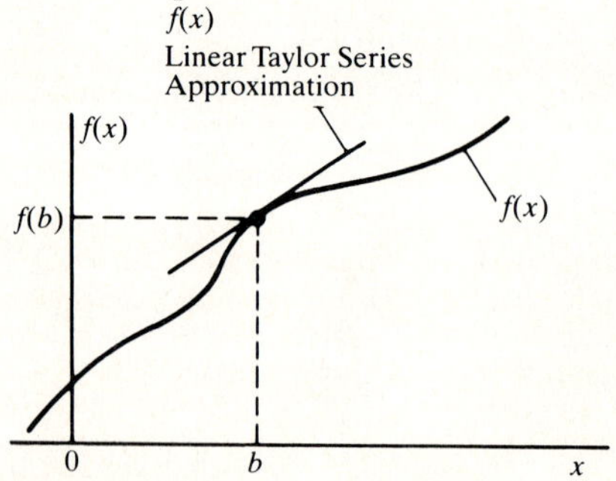

FIGURE 6.18 A nonlinear function to be linearized.

Example 6.8

Let us linearize $y = 1/2x^3$, which is shown in Fig. 6.19 about the point $x = 1$. The function evaluated at $x = 1$ is

$$y(1) = \frac{1}{2}(1)^3 = 0.5$$

exactly and the derivative at $x = 1$ is

$$\left.\frac{dy}{dx}\right|_{x=1} = \left.\frac{3}{2}x^2\right|_{x=1} = \frac{3}{2} = 1.5$$

Thus the approximation is, from Eq. (6.7.2)

$$\hat{f}(x) = 0.5 + 1.5(x - 1) = 1.5x - 1$$

This straight-line approximation is shown as a tangent to the curve at $x = 1$.

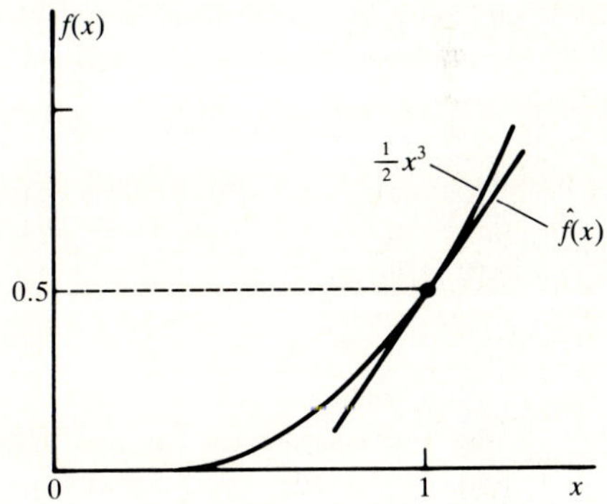

FIGURE 6.19 A cubic and the linear approximation.

The error in a linearizing approximation can be examined several ways but perhaps the easiest way is to calculate the percent error in the approximation by calculating the error and normalizing it with respect to the real value or

$$\%\ \text{error} = \frac{f(x) - \hat{f}(x)}{f(x)} \times 100 \tag{6.7.4}$$

Consider the cubic linearized in Example 6.6. Evaluate the percent error in the approximation at $x = 1.2$. The value of the function is

$$f(1.2) = \frac{(1.2)^3}{2} = 0.864$$

and the value of the approximation is

$$\hat{f}(1.2) = 1.5(1.2) - 1 = 0.80$$

and the percent error is

$$\% \text{ error} = \frac{0.864 - 0.8}{0.864} \times 100 = 7.4\%$$

Generally, the further from the operating point, the larger the error but it is also true that simple examples can be found where this is not the case.

Example 6.9

Let us linearize the dynamic equation of the simple tank system developed in Example 6.5. This equation was

$$A\frac{dh}{dt} + A_2 C_d \sqrt{2g}h^{1/2} = Q_{\text{in}}(t) \qquad \textbf{(a)}$$

Let us assume that a steady-flow condition exists with a height H in the tank and an inflow of $\bar{Q}$. The mass balance for steady flow (constant height in the tank) is, as $dh/dt = 0$,

$$A_2 C_d \sqrt{2gH} = \bar{Q} \qquad \textbf{(b)}$$

We shall make use of this condition later but now let us assume that the flowrate into the tank is given by the steady flow $\bar{Q}$ and a small time-varying part $q(t)$ or

$$Q_{\text{in}}(t) = \bar{Q} + q(t) \qquad \textbf{(c)}$$

Now let us linearize the $h^{1/2}$ term about the point $h = H$. The function to be linearized is $h^{1/2}$, as illustrated in Fig. 6.20 so, if we evaluate the first derivative of $h^{1/2}$, the result is

$$\frac{d}{dh}h^{1/2}\bigg|_{h=H} = \frac{1}{2}h^{-1/2}\bigg|_{h=H} = \frac{1}{2\sqrt{H}}$$

Then the linear approximation is, according to Eq. (6.7.2),

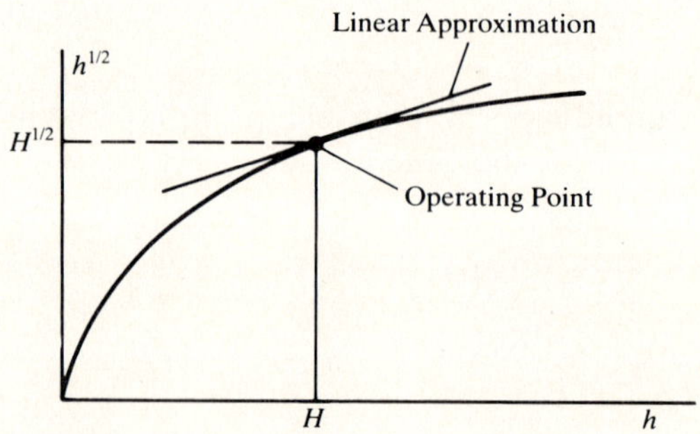

FIGURE 6.20 Nonlinear flow function for tank outflow.

$$h^{1/2} \cong H^{1/2} + \frac{1}{2\sqrt{H}}(h - H)$$

If we substitute this expression and Eq. **(c)** into Eq. **(a)**, the result is

$$A\frac{dh}{dt} + A_2C_d\sqrt{2g}\left[H^{1/2} + \frac{1}{2\sqrt{H}}(h - H)\right] = \bar{Q} + q(t) \qquad \textbf{(d)}$$

Note that the terms of the steady-flow mass balance **(b)** appear on both sides of Eq. **(d)** and therefore cancel. Now let us define a new variable for the change of height from the steady-flow value to be

$$y(t) = h(t) - H$$

Since H is constant, then

$$\frac{dh}{dt} = \frac{dy}{dt}$$

so expression **(d)** reduces to

$$A\frac{dy}{dt} + A_2C_d\sqrt{\frac{g}{2H}}\,y = q(t)$$

which is now a linear differential equation in $y(t)$, the deviation in tank head from the steady-flow value. If we transform this equation, the result is

$$\left(As + A_2C_d\sqrt{\frac{g}{2H}}\right)Y(s) = Q(s)$$

and a transfer function for the tank is

$$G(s) = \frac{Y(s)}{Q(s)} = \frac{1}{As + A_2C_d\sqrt{\frac{g}{2H}}}$$

which is shown in Fig. 6.21. The time constant for this first-order system is

$$\tau = \frac{A}{A_2C_d}\sqrt{\frac{2H}{g}}$$

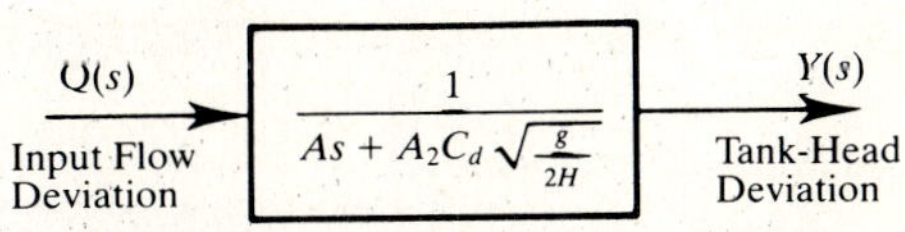

FIGURE 6.21 Transfer function model for tank-head deviation.

6.8 SUMMARY

In this chapter we have explored the fundamental concepts useful in the modeling of thermal and fluid systems. These principles were the laws of conservation of energy, mass, and momentum (Newton's second law). It was found that, in the modeling of fluid systems, the differential equations which result are often nonlinear and as such transform techniques are not applicable. We also considered the linearization of such nonlinear systems for small variations about a steady-state (static) operating point. Transform techniques are then applicable to the resulting linear differential equations.

REFERENCES

1. Jones, J. B., and Hawkins, G. A. *Engineering Thermodynamics*, 2d ed. John Wiley, New York, 1986.
2. Incropera, F. P., and DeWitt, D. P. *Fundamentals of Heat and Mass Transfer*, 2d ed. John Wiley, New York, 1985.
3. Welty, J. R., Wicks, C. E., and Wilson, R. E. *Fundamentals of Momentum, Heat and Mass Transfer*, 3rd ed. John Wiley, New York, 1984.
4. Fox, R. W., and McDonald, A. T. *Introduction to Fluid Mechanics*, 3rd ed. John Wiley, New York, 1985.
5. Gerhart, P. M., and Gross, R. J. *Fundamentals of Fluid Mechanics*. Addison-Wesley, Reading, MA, 1985.

PROBLEMS

6-1. Consider the two-chamber oven shown. The media in both chambers are well stirred such that the bulk temperature of each is constant (respectively, $T_1(t)$ and $T_2(t)$). Assume the environmental temperature to be $T_0(t)$.

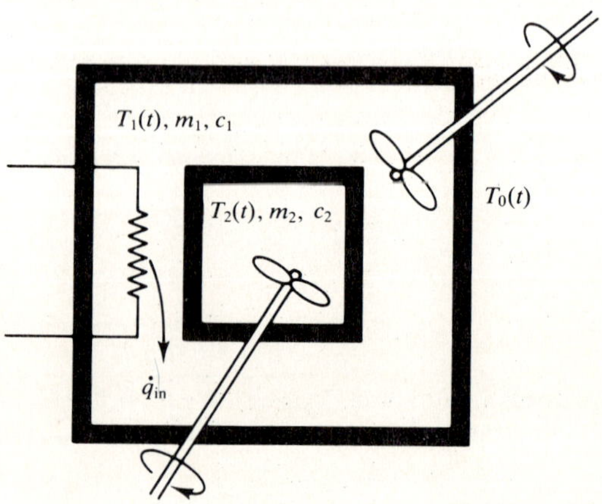

Assume that the heat transfer across all boundaries is proportional to the temperature difference. Write the energy balance equations for each chamber, neglecting the energy contribution of both stirrers.

6-2. Consider the situation illustrated below where a heater is inside an insulated container filled with a liquid of mass m_2 with specific heat c_2. The heater is of mass m_1 with specific heat c_1. The power supplied to the heater is $\dot{q}$. Assume that the rate of energy transfer between the heater body and the liquid is proportional to the temperature difference between the heater and the liquid.

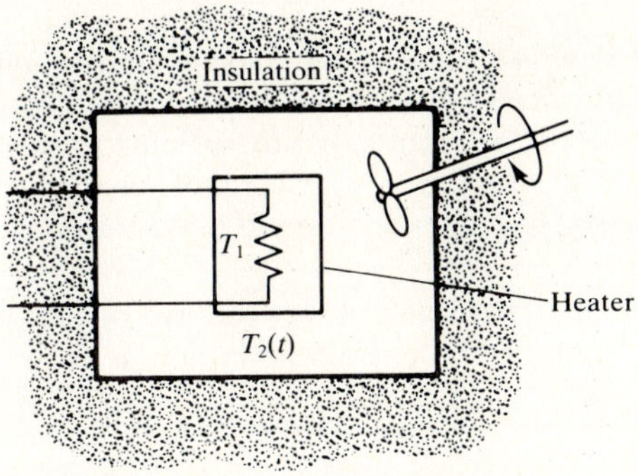

(a) Write the energy balance equations for the heater and the liquid mass.
(b) Derive the transfer function.

$$G(s) = \frac{\mathscr{L}[T_2(t)]}{\mathscr{L}[\dot{q}(t)]}$$

6-3. Consider the insulated chamber shown below which is always full of a liquid of density ρ. The liquid enters with a temperature $T_i(t)$ and leaves at temperature $T_0(t)$. The volume of the tank is V_0 m^3 and the constant flow rate into and out of the tank is Q m^3/s and the specific heat is c J/kg · °C. Write the energy balance for the tank and note that the density will be present in every term.

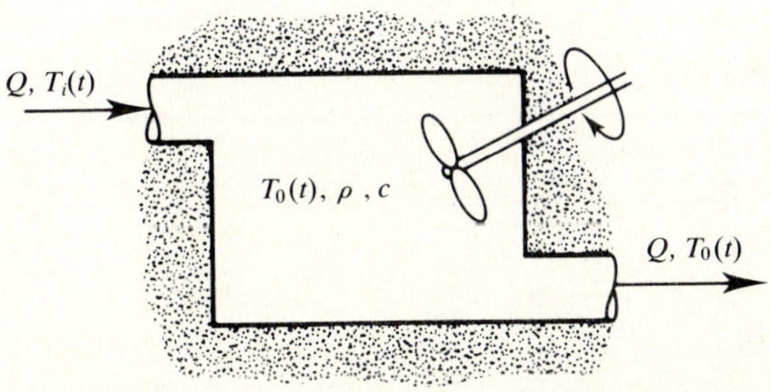

Assume that the tank is well stirred such that the bulk temperature is $T_0(t)$.

6-4. Consider the two-tank system shown in which the respective fluid levels are h_1 and h_2.

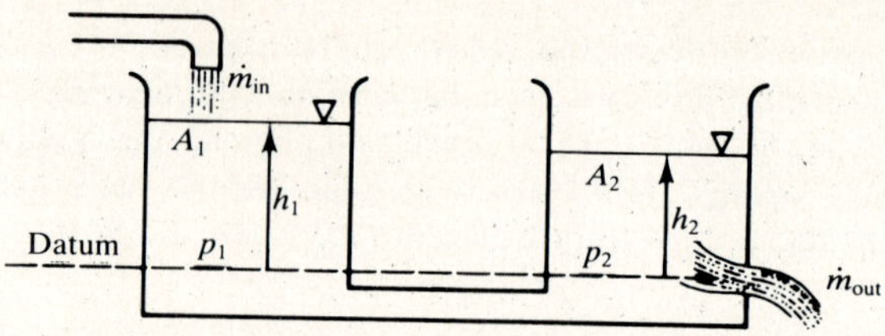

Assume that the flowrate between the tanks is proportional to the square root of the pressure drop $p_1 - p_2$. Also assume that the pressure at the bottom of each tank is hydrostatic or essentially proportional to the height of fluid in that tank. Assume that the outflow $\dot{m}_{out}$ is proportional to the square root of h_2. Employing the principle of conservation of mass derive the governing equations for the two tanks.

6-5. Consider the radiation heat transfer between two bodies in a vacuum. This heat transfer is given by the Stephan-Boltzmann law to be

$$\dot{q}_{1-2} = K(T_1^4 - T_2^4)$$

Consider the case where T_2 is constant and T_1 varies only slightly from T_2 by some variation $\delta T(t)$ or

$$T_1 = T_2 + \delta T(t)$$

Find a linear relationship between $\dot{q}_{1-2}$ and the variation in temperature for $\delta T(t) \ll T_2$.

6-6. Consider the turbulent flow of an incompressible fluid in a pipe of length l. The pressure drop in the pipe is proportional to the square of velocity.

$$p_1 - p_2 = \Delta p = \frac{\rho l f}{2gd} V^2$$

where l is the pipe length, ρ is the fluid density, d is the pipe diameter, g is the acceleration of gravity, and f is the friction factor.

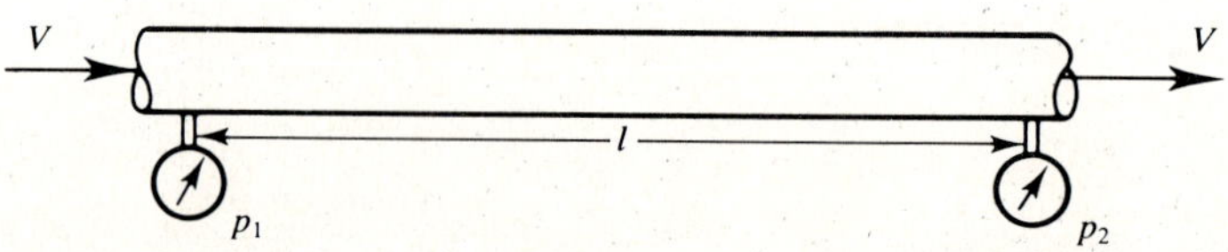

Let us define a steady-flow velocity $\bar{V}$ for which the pressure drop $\overline{\Delta p}$ is given by

$$\overline{\Delta p} = \frac{\rho l f}{2gd} \bar{V}^2$$

Examine the pressure drop $\overline{\Delta p} + \delta p$ which is given for a change in velocity to $\bar{V} + \delta V$ for $\delta V \ll \bar{V}$. Find a linear relation between δp and δV.

6-7. A common way to measure the flowrate of a fluid (in this case air) is by inserting an orifice plate in the line and using a water manometer to measure the change in pressure across the orifice.

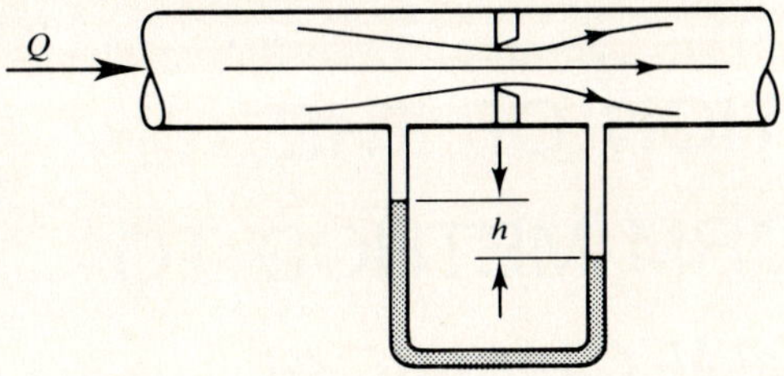

The relation between the manometer column height and the volumetric flowrate is

$$Q = k\sqrt{h}$$

Linearize this relation about a steady flowrate Q_0 and steady height h_0.

7

APPLICATION OF THE TRANSFORM METHOD TO ELECTROMECHANICAL SYSTEMS

7.1 INTRODUCTION

Today there are many devices which depend on the principles of electromechanical energy conversion. These devices include solenoids, relays, motors, generators, loudspeakers, microphones, and others. In this chapter we shall explore the concepts of electromechanical energy conversion and focus on the modeling of two devices which use these concepts to perform their functions. This brief exposure has been highly idealized; for an in-depth treatment the reader is referred to one of several excellent texts on the subject of electromechanical energy conversion (see Sarma and see Fitzgerald, Kingsley, and Umans in the References at the end of this chapter).

7.2 THE LORENTZ FORCE EQUATION

Let us consider an element of a current-carrying conductor of length $d\mathbf{L}$ meters in a constant magnetic field with a magnetic flux density vector $\mathbf{B}$ Wb/m^2, as illustrated in Fig. 7.1. Let us assume at time t that a current $i(t)$ amperes is being carried by the element in the direction specified by the direction of the elemental length vector $d\mathbf{L}$. The magnitude of the $d\mathbf{L}$ represents the length of the element. The elemental force on this element at time t is given by the Lorentz force equation to be

$$d\mathbf{F}(t) = i(t)\, d\mathbf{L} \times \mathbf{B} \tag{7.2.1}$$

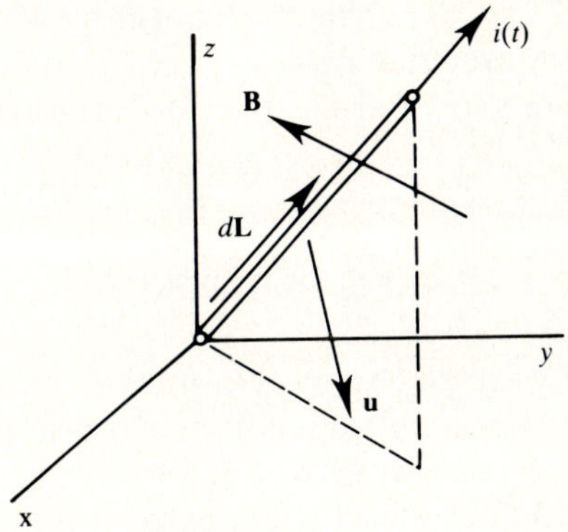

FIGURE 7.1 Element of a current-carrying conductor in a magnetic field.

For a straight conductor of finite length **L** in a uniform field **B** the force developed due to current $i(t)$ is

$$\mathbf{F}(t) = i(t)\mathbf{L} \times \mathbf{B} \qquad (7.2.2)$$

These are the forces which allow the conversion of electrical energy into mechanical energy and thus forces to move loudspeaker cones and torques to rotate motor armatures.

7.3 FARADAY'S LAW

There is another fundamental concept in electromechanical energy conversion which allows us to convert mechanical energy into electrical energy. For the elemental conductor of length $d\mathbf{L}$ meters shown in Fig. 7.1 moving with vector velocity $\mathbf{u}(t)$ m/s in the presence of a fixed magnetic field with flux density $\mathbf{B}$ Wb/m^2, Faraday's law states that the induced electrical potential in the conductor is proportional to the rate of change of flux linkage or in vector form

$$dv(t) = [\mathbf{u}(t) \times \mathbf{B}] \cdot d\mathbf{L} \qquad (7.3.1)$$

For a straight conductor with length vector **L** translating in the presence of field **B** expression (7.3.1) may be integrated to yield

$$v(t) = [\mathbf{u}(t) \times \mathbf{B}] \cdot \mathbf{L} \qquad (7.3.2)$$

Note that in the evaluation of Eq. (7.3.2), the final product is a dot product and hence $v(t)$ has an algebraic sign which may be interpreted as follows. If $v(t)$ is positive, this indicates that the tip of the length vector **L** is at a positive potential relative to the tail of **L**.

This is the mechanism by which rotational mechanical energy is converted to electrical energy, as in the generation of electrical energy with conventional electrical generators.

7.4 THE MOTOR-GENERATOR ARMATURE

Consider the system shown in Fig. 7.2. The current, i_f, called the field current, will be assumed constant so that the flux density **B** it generates in the magnet gap will be constant. Let us assume that the single planar armature loop is at angle θ rotating with angular velocity $\dot{\theta}$ and is carrying a current $i(t)$, as shown. The active elements of the armature, those perpendicular to the field **B**, are located at radius R and are of length L, respectively. Let us first consider the force on the right side of the armature, which is, from relation (7.2.2)

$$F(t) = i(t)LB \tag{7.4.1}$$

in the upward direction (remember the right-hand rule). Similarly, a downward force of the same magnitude exists on the left side of the armature. The lever arm for these forces is $R\cos\theta$ in both cases, so the torque developed on the armature is

$$T(t) = 2i(t)LBR\cos\theta \tag{7.4.2}$$

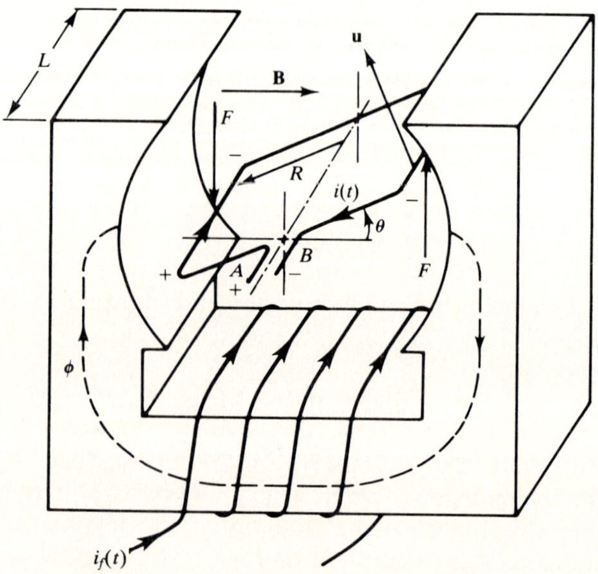

FIGURE 7.2 Simple one-loop armature.

This torque is sinusoidal and thus changes direction every half revolution and would be of little value in an electric motor. However, if we consider the case where the angle θ is small, then the torque is maximum and is

$$T_{max} = 2i(t)LB \tag{7.4.3}$$

If we now use multiple sets of windings and a procedure called commutation such that only the windings with θ near zero, called the active windings, have a current, then the torque caused by these windings will be very nearly constant. Note that the current in such windings is all in the same direction and is called dc. For N turns per active winding, the torque is

$$T_{max} = 2Ni(t)LB \tag{7.4.4}$$

Hence the torque developed in a dc motor is constant and proportional to the product of the armature current and the strength of the field in the magnet gap.

Now let us consider the potential which is simultaneously induced in the same horizontal segments of the armature. It is called the "back emf." It may be shown using relation (7.2.3) that the induced potentials in the end segments of the loop are zero since the $\mathbf{u} \times \mathbf{B}$ vector is always orthogonal to the $d\mathbf{L}$ vector. The potential induced in the left conductor is, from relation (7.3.2),

$$v(t) = uBL \cos \theta(t) \tag{7.4.5}$$

of polarity indicated in Fig. 7.2. The potential induced in the right horizontal conductor is of the same magnitude as that in the left conductor and of a polarity such that it is additive to that of the first conductor. If the velocity u is given by

$$u = R\dot{\theta} \tag{7.4.6}$$

the total induced voltage in the single loop is

$$v(t) = 2R\dot{\theta}BL \cos \theta \tag{7.4.7}$$

Of course, if we rotate the armature at a constant angular velocity, the induced voltage is a sinusoidal waveform of constant peak amplitude and thus we have "invented" the simplest form of an ac generator.

If the armature is made up of separate sets of windings and each is commutated such that the only one connected to external circuitry is that for which θ is small, ideally zero, then the induced voltage is nearly constant which is, for N turns,

$$v(t) = 2RBLN\dot{\theta} \tag{7.4.8}$$

and we have "invented" a unidirectional or dc generator. Note that the induced voltage is proportional to the armature speed and the field strength. The velocity dependence of the induced voltage is useful when the device is used as a tachometer to measure angular speed.

7.5 TERMINAL CHARACTERISTICS OF THE DC MOTOR/GENERATOR

To get reasonable values of torque from a motor or voltage from a generator it is necessary to have a large number of windings on the armature and hence the armature as a circuit will demonstrate the characteristics of electrical resistance and inductance. As a result, the motor may be represented schematically as in Fig. 7.3.

We should note, as in relation (7.4.8), that the induced voltage in the armature (the back emf) $v_b(t)$ is proportional to armature speed ω or

$$v_b(t) = K_b \omega(t) \tag{7.5.1}$$

and that, as in relation (7.4.4), the torque generated is proportional to the armature current or

$$T(t) = K_t i(t) \tag{7.5.2}$$

Let us now formulate the differential equations governing the performance of the motor. For the electrical portion, write a loop equation and substitute Eq. (7.5.1) for the induced voltage $v_b(t)$:

$$v(t) = R_a i + L_a \frac{di}{dt} + K_b \omega \tag{7.5.3}$$

The equation for the mechanical portion of the system is, including the torque from Eq. (7.5.2),

$$J \frac{d\omega}{dt} = -b\omega + K_t i \tag{7.5.4}$$

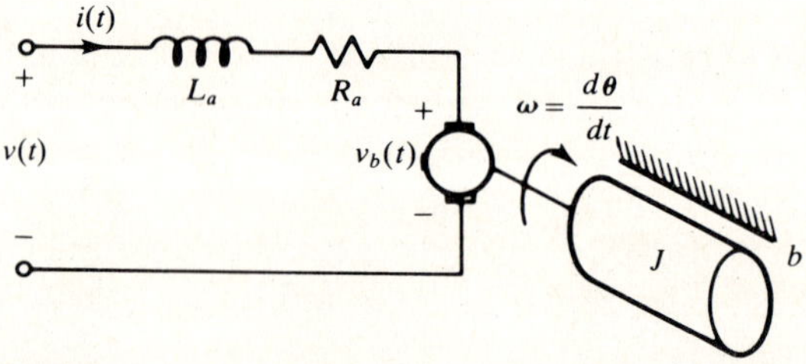

FIGURE 7.3 Schematic of the dc motor/generator.

Let us now examine the s-domain model of the motor by taking the Laplace transform of Eqs. (7.5.3) and (7.5.4) to give

$$V(s) = (R_a + L_a s)I(s) + K_b\Omega(s) \tag{7.5.5}$$

$$(Js + b)\Omega(s) = K_t I(s) \tag{7.5.6}$$

Solution of Eq. (7.5.5) for $I(s)$ gives

$$I(s) = \frac{1}{L_a s + R_a}[V(s) - K_b\Omega(s)] \tag{7.5.7}$$

and solution of Eq. (7.5.6) for $\Omega(s)$ gives

$$\Omega(s) = \frac{K_i}{Js + b} I(s) \tag{7.5.8}$$

These two equations can be interpreted in block-diagram fashion, as illustrated in Fig. 7.4, assuming that the applied voltage $V(s)$ is known and the system output is $\Omega(s)$. It is also interesting to note that the angular displacement of the motor shaft is the integral of the angular velocity or

$$\theta(t) = \int \omega(t)\, dt \tag{7.5.9}$$

and after Laplace transformation

$$\theta(s) = \frac{1}{s}\Omega(s) \tag{7.5.10}$$

This relationship is represented by the right block in Fig. 7.4. Note that there is a feedback path which tends to decrease the potential available to drive current through the armature, that is, the net potential driving the motor is the terminal voltage minus the back emf.

It is interesting to note that there is a relationship between K_b and K_t

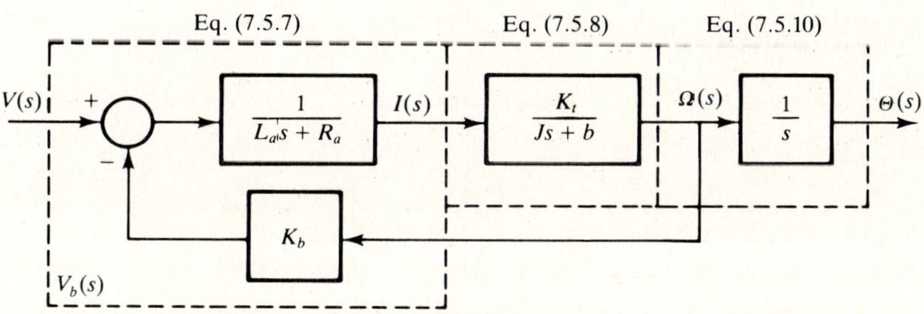

FIGURE 7.4 Block diagram of the dc motor.

for the motor. Consider the mechanical power supplied instantaneously to the armature, the product of torque times speed:

$$p(t) = T(t)\omega(t) \tag{7.5.11}$$

Also, the electrical power taken from the electrical network by rotation of the armature is the back emf times the armature current $i(t)$ or

$$p(t) = v_b(t)i(t) \tag{7.5.12}$$

The back emf is given by relation (7.5.1) and the current can be given by solving relation (7.5.2) for $i(t)$ to give, from relation (7.5.11),

$$p(t) = K_t i(t) \frac{v_b(t)}{K_b} \tag{7.5.13}$$

The principle of conservation of energy says that the electrical and mechanical power must be the same or Eq. (7.5.13) must be equal to Eq. (7.5.12):

$$\frac{K_t}{K_b} = 1 \tag{7.5.14}$$

or the constants K_t and K_b must be numerically the same in the rationalized MKS system.

In Fig. 7.4 the loading property is reflected in the feedback, which effectively decreases the potential available to drive current through the armature.

Consider now what happens if the armature is driven by an externally applied mechanical torque $T(t)$ and the armature circuit is left open such that the armature current is zero. This implies that the voltage drops across the inductance and resistance of the armature are both zero. The equation for the mechanical dynamics is, from Eq. (7.5.4),

$$J\frac{d\omega}{dt} + b\omega = T(t) \tag{7.5.15}$$

and for the electrical dynamics is given by Eq. (7.5.3) to yield

$$v(t) = K_b\omega(t) \tag{7.5.16}$$

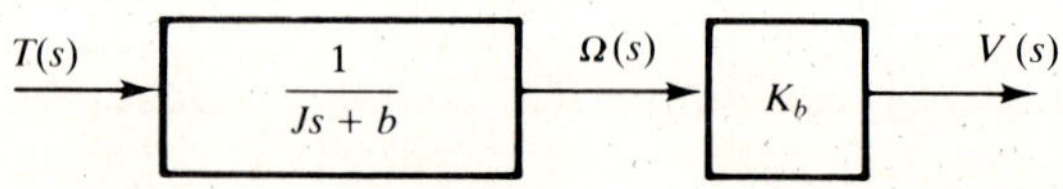

FIGURE 7.5 Transfer functions for the unloaded (open-circuit) dc generator.

Expression (7.5.16) is immediately recognized as the relation describing a tachometer for which the open-circuit terminal voltage is proportional to speed. The transfer function diagram for Eqs. (7.5.15) and (7.5.16) is shown in Fig. 7.5.

7.6 THE ELECTROMECHANICAL LOUDSPEAKER

Another device with which all readers are familiar is the loudspeaker common to most communication and entertainment systems. Consider a simple model of such a speaker wherein only the electrical and mechanical dynamic elements have been represented in Fig. 7.6. The acoustical portions of the system have been omitted for simplicity but could be included if a more complicated electroacoustical model were desired. We shall assume that the coil moves with displacement $x(t)$ along with the base of the speaker cone which has mass, stiffness, and dissipation. The coil will be assumed to carry current $i(t)$ and have respective resistance and inductance R and L. We shall also assume that the force acting on the coil is proportional to the field strength and to the coil current. Also, we shall assume that since the coil moves with velocity dx/dt, induced voltage on the coil is proportional to that velocity. The electrical and mechanical elements have been separated in Fig. 7.7 but the electromechanical coupling has been included in both elements. If we write the loop equation for the electrical elements of Fig. 7.7(a), we get

$$v(t) = L\frac{di}{dt} + Ri + K_b\frac{dx}{dt} \tag{7.6.1}$$

For the mechanical system, Newton's second law gives

$$m\frac{d^2x}{dt^2} = -b\frac{dx}{dt} - kx + K_f i(t) \tag{7.6.2}$$

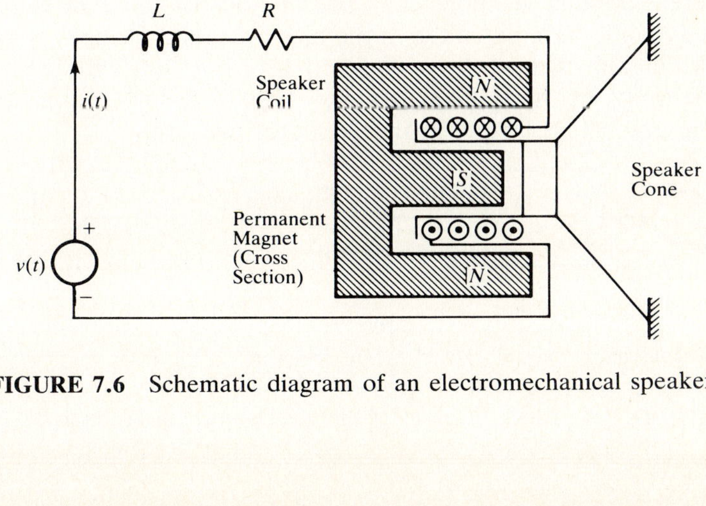

FIGURE 7.6 Schematic diagram of an electromechanical speaker.

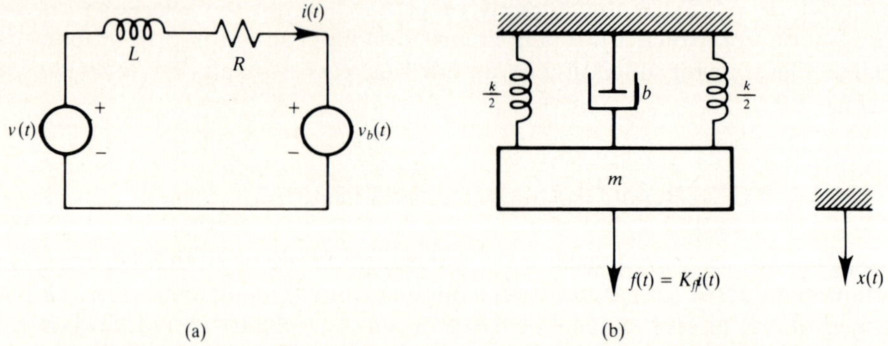

FIGURE 7.7 (a) Electrical elements of the loudspeaker. (b) Mechanical elements of the loudspeaker.

The electromechanical coupling terms in these two equations are now quite obvious. They both must be present in order to have a valid model for electromechanical energy conversion. The term $K_f i(t)$ in the mechanical equation (Eq. (7.6.2)) represents the conversion of electrical energy into mechanical motion, which is the desired effect. The back emf term $K_b(dx/dt)$ in the electrical equation (Eq. (7.6.3)) represents the reflection of the mechanical load into the electrical circuit and hence a load on the fundamental energy source represented by $v(t)$.

Let us now consider the dynamics of this system in the s domain. If we transform Eq. (7.6.1), ignoring the initial conditions, the result is

$$V(s) = (Ls + R)I(s) + K_b sX(s) \tag{7.6.3}$$

and if we also transform Eq. (7.6.2), the result is

$$(mx^2 + bs + k)X(s) = K_f I(s) \tag{7.6.4}$$

From these s-domain relations the interrelations of these mechanical and electrical dynamics are illustrated in the block diagram of Fig. 7.8. Note that this block diagram is much the same as that of Fig. 7.4, with the exception that the stiffness of the speaker cone provides a restoring force for which there is no equivalent torque in the dc motor problem.

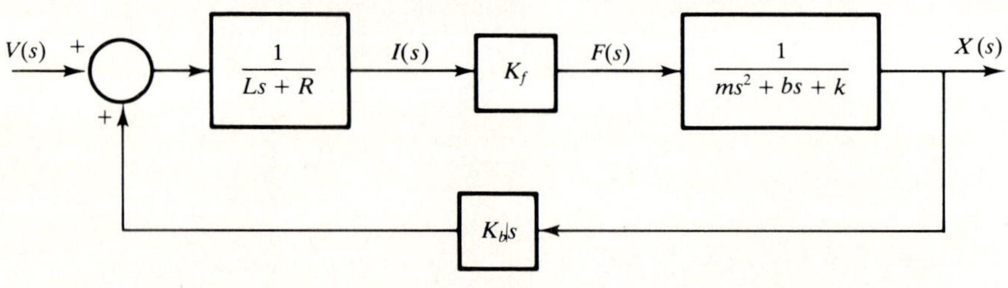

FIGURE 7.8 Block diagram of the electromechanical loudspeaker.

7.7 CAPACITIVE TRANSDUCERS

The devices considered thus far in this chapter operated on magnetic principles reflected in Faraday's Law and the Lorentz force equation. To expand the scope of our thought and the range of applications let us consider a device which operates on electrostatic principles.

Consider the simple system of Fig. 7.9, which is composed of a simple capacitor, a battery, and a switch. Let one plate of the capacitor be movable, with separation from the stationary plate denoted as $x(t)$. The external force on the movable plate will be denoted as $f(t)$. The capacitance of the device is given by expression (3.9.8) to be

$$C(t) = \frac{\varepsilon A}{x(t)} \tag{7.7.1}$$

where ε is the dielectric strength of the gap, in this case air, and A is the area of each plate. Let us initially connect the battery to charge the capacitor, using stationary plates with spacing X_0, and then open the switch. The energy stored on the plates of the capacitor is

$$W_s = \frac{C_0 V_0^2}{2} = \frac{\varepsilon A V_0^2}{2X_0} \tag{7.7.2}$$

or

$$W_s = \frac{Q^2}{2C} = \frac{Q^2 X_0}{2\varepsilon A} \tag{7.7.3}$$

Let us apply a force $f(t)$ to the movable plate and move it some incremental distance dx. The energy stored is now

$$W_s + dW_s = \frac{Q^2(X_0 + dx)}{2\varepsilon A} \tag{7.7.4}$$

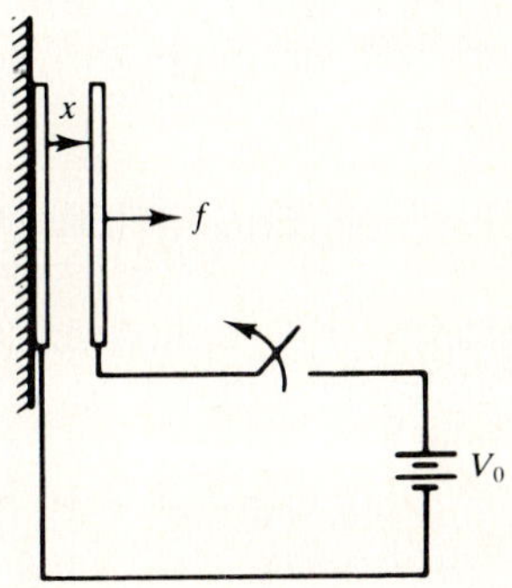

FIGURE 7.9 An electrostatic transducer.

The incremental work done during the incremental movement of the plate is

$$dW_s = \frac{Q^2\,dx}{2\varepsilon A} \tag{7.7.5}$$

and the work done by the external force is

$$dW_s = f\,dx \tag{7.7.6}$$

Using the principle of conservation of energy, we may equate the change in stored energy to the work done:

$$f\,dx = \frac{Q^2\,dx}{2\varepsilon A} \tag{7.7.7}$$

so the force is

$$f = \frac{Q^2}{2\varepsilon A} \tag{7.7.8}$$

If we eliminate the charge, the relation for the force is

$$f = \frac{\varepsilon A V_0^2}{2X_0^2} \tag{7.7.9}$$

The charge and voltage are related by expression (3.9.7) to give

$$Q = C_0 V_0 = \frac{\varepsilon A V_0}{X_0} \tag{7.7.10}$$

Note that, in relation (7.7.8), the force is independent of displacement since the charge is fixed. The capacitor voltage v for any general displacement x is

$$v = \frac{Q}{C} = \frac{Qx}{\varepsilon A} \tag{7.7.11}$$

Note that the capacitor voltage is proportional to displacement x and hence this device might be useful as a displacement transducer.

7.8 AN ELECTROMECHANICAL TRANSDUCER

Let us now complicate the device discussed in the previous section with the addition of mass m, stiffness k, and damping b, along with some circuitry which now remains attached to the capacitor at all times. All this is illustrated in Fig. 7.10(a). This case is somewhat different from the previous one because the charge on the capacitor, the electrostatic force, and the voltage are no longer constant with time.

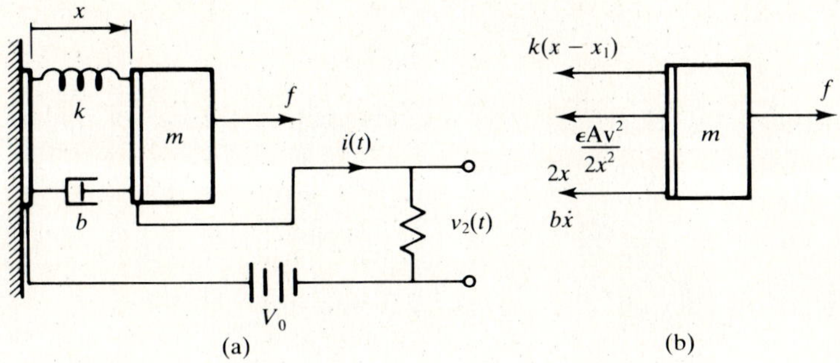

FIGURE 7.10 Capacitative electromechanical transducer. (a) Electrical and mechanical components. (b) Free-body diagram.

We shall denote the equilibrium position of the uncharged plates as x_1. Further, we shall assume that when the electrical circuitry is connected the plates seek a new static equilibrium position because of the attractive force of the charged plates. We shall denote this position as x_0. In this equilibrium condition, the electrostatic force is balanced by a compressive spring force $k(x_1 - x_0)$ to the right, no current flows in the electrical loop, and the total battery voltage V_0 is dropped across the capacitor. The steady-state static equilibrium force balance gives, from relation (7.7.9),

$$\frac{\varepsilon A V_0^2}{2x_0^2} = k(x_1 - x_0) \tag{7.8.1}$$

Now let us consider the case where the mass and circuitry are in dynamic equilibrium. The electrostatic force in this case from relation (7.7.9) is

$$f_e = \frac{\varepsilon A v^2(t)}{2x(t)^2} \tag{7.8.2}$$

where $v(t)$ is the capacitor voltage and $x(t)$ is the displacement between the plates. Consider the free-body diagram of Fig. 7.10(b) and write the equation of motion for the mass to give

$$m\ddot{x} = -b\dot{x} - k(x - x_1) - \frac{\varepsilon A v^2}{2x^2} + f \tag{7.8.3}$$

Now let us define the output voltage to be

$$v_2(t) = Ri(t) = R\,\frac{dq}{dt} \tag{7.8.4}$$

where $q(t)$ is the instantaneous charge on the plates of the capacitor. This charge is related to the terminal voltage by the relation

$$q(t) = R\varepsilon A \frac{d}{dt}\left(\frac{v}{x}\right)$$

Investigation of the electrical loop implies that $v(t) = V_0 - v_2(t)$; then

$$v_2(t) = R\varepsilon A \frac{d}{dt}\left(\frac{V_0 - v_2(t)}{x(t)}\right) \tag{7.8.5}$$

which, after carrying out the indicated differentiation, gives

$$v_2(t) = R\varepsilon A\left[-\frac{\dot{v}_2}{x} - (V_0 - v_2)\frac{\dot{x}}{x^2}\right] \tag{7.8.6}$$

If we also substitute $v(t) = V_0 - v_2(t)$ into relation (7.8.3) we get

$$m\ddot{x} + b\dot{x} + k(x - x_0) = \frac{-\varepsilon A}{2x^2}(V_0 - v_2)^2 + f \tag{7.8.7}$$

Relations (7.8.6) and (7.8.7) are a pair of highly nonlinear differential equations which govern the performance of the sytem. Analytical solution of the equations, if not impossible, is well beyond the scope of the material in this text. As a result we will pursue a path similar to that employed in Section 6.5 when we were dealing with the nonlinearities associated with fluid-flow problems.

We shall define the deviation in the displacement of the movable plate as

$$y = x - x_0 \tag{7.8.8}$$

Substitution of this relation yields governing equations of

$$m\ddot{y} + b\dot{y} + k(y - x_1 + x_0) - \frac{\varepsilon A(V_0 - v_2)^2}{2(x_0 + y)^2} + f \tag{7.8.9}$$

and

$$v_2 = R\varepsilon A\left[\frac{-\dot{v}_2}{x_0 + y} - \frac{(V_0 - v_2)\dot{y}}{(x_0 + y)^2}\right] \tag{7.8.10}$$

In expressions (7.8.9) and (7.8.10) we have nonlinearities of the form $1/(x_0 + y)$ and $1/(x_0 + y_0)^2$ which must be dealt with. If we make two-term Taylor series expansions about $y = 0$ (constant plus linear term) the expansions are

$$\frac{1}{x_0 + y} \simeq \frac{1}{x_0}\left(1 - \frac{y}{x_0}\right) \tag{7.8.11}$$

and

$$\frac{1}{(x_0 + y)^2} \simeq \frac{1}{x_0^2}\left(1 - \frac{2y}{x_0}\right) \tag{7.8.12}$$

If we now substitute these relations into expressions (7.8.9) and (7.8.10), the results are

$$m\ddot{y} + b\dot{y} + ky - k(x_1 - x_0) = -\frac{\varepsilon A}{2x_0^2}(V_0 - v_2)^2\left(1 - \frac{2y}{x_0}\right) + f \tag{7.8.13}$$

and

$$v_2 = \frac{R\varepsilon A}{x_0}\left[-\dot{v}_2\left(1 - \frac{y}{x_0}\right) - (V_0 - v_2)\dot{y}\left(\frac{x_0 - 2y}{x_0^2}\right)\right] \tag{7.8.14}$$

Note that these relations are both still highly nonlinear because of the products of the linearized terms. Also note that the static equilibrium relation (Eq. (7.8.1)) is embedded in relation (7.8.13) and hence some terms will drop out. If we now assume that y is much less than x_0 and v_2 is smaller than V_0, we can retain only linear terms in Eq. (7.8.13) to yield

$$m\ddot{y} + b\dot{y} + ky = \frac{\varepsilon A V_0 v_2}{x_0^2} + \frac{\varepsilon A V_0^2 y}{x_0^3} + f \tag{7.8.15}$$

and if we combine like terms we get

$$m\ddot{y} + b\dot{y} + \left[k - \frac{\varepsilon A V_0^2}{x_0^3}\right]y - \frac{\varepsilon A V_0}{x_0^2}v_2 = f \tag{7.8.16}$$

If we now retain only the linear terms in Eq. (7.8.14) the result is

$$v_2 = \frac{R\varepsilon A}{x_0}\left[-\dot{v}_2 - \frac{V_0}{x_0}\dot{y}\right] \tag{7.8.17}$$

At this point it is clear that Eqs. (7.8.16) and (7.8.17) represent a small-signal, linear model of the capacitive transducer system.

If we transform relations (7.8.16) and (7.8.17) while ignoring any possible initial conditions, the resulting equations are, after denoting that $C_0 = \varepsilon A/x_0$,

$$\left(ms^2 + bs + k - \frac{C_0 V_0^2}{x_0^3}\right)Y(s) - \frac{C_0 V_0}{x_0}V_2(s) = F(s) \tag{7.8.18}$$

and

$$(RC_0 s + 1)V_2(s) + RC_0\frac{V_0}{x_0}sY(s) = 0 \tag{7.8.19}$$

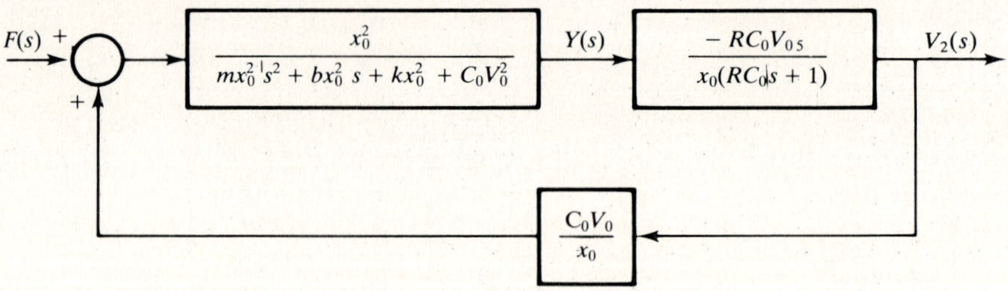

FIGURE 7.11 Block diagram of the capacitative electromechanical transducer.

We can now establish a block diagram relating the applied force $F(s)$ (stimulus) to the output voltage $V_2(s)$ (response). This block diagram is illustrated in Fig. 7.11. Relations (7.8.17) and (7.8.18) can be solved by Cramer's rule to yield a transfer function $G(s) = V_2(s)/F(s)$ for this force transducer, which is

$$G(s) = \frac{V_2(s)}{F(s)} = \frac{-\dfrac{RC_0}{x_0}s}{RC_0ms^3 + (m + RC_0b)s^2 + (b + kRC_0)s + \left(k - \dfrac{C_0V_0^2}{x_0^2}\right)} \qquad \textbf{(7.8.20)}$$

If one were interested in the frequency response of this transducer, it could be evaluated by letting $s = j\omega$.

7.9 SUMMARY

In this chapter we have explored the world of devices in which electromechanical energy conversion takes place. This can occur due to magnetic effects or by electrostatic effects. It was found that electrostatic coupling leads to nonlinear systems which must be linearized about an operating point to pursue the problem of system dynamics.

REFERENCES

Sarma, M. S. *Electric Machines*. Wm. C. Brown, Dubuque, IA, 1985.

Fitzgerald, A. E., Kingsley, Jr., C., and Umans, S. D. *Electric Machinery*, 4th ed. McGraw-Hill, New York, 1983.

Slemon, G. R., and Straughen, A. *Electric Machines*. Addison-Wesley, Reading, MA, 1982.

Chuston, A. N. *Electric Machinery*. Reston, Englewood Cliffs, NJ, 1986.

PROBLEMS

7-1. Calculate the force vector on a straight wire one meter long along the x axis with a current of 5 amperes in the positive x direction. The magnetic field present is not orthogonal to the wire and is given by $\mathbf{B} = 10\hat{\imath} + 12\hat{\jmath}\ \text{Wb/m}^2$ where $\hat{\imath}$ and $\hat{\jmath}$ are the respective unit vectors in the x and y directions.

7-2. Repeat problem 7-1 with the same one-meter wire (oriented vertically) with the current directed in the positive z direction (vertically).

7-3. The earth's magnetic field strength in the Chicago area is roughly $5.8 \times 10^{-5}\ \text{Wb/m}^2$. Calculate the force vector per unit length of wire for a wire carrying a current of 50 amperes when the wire is horizontal and at a right angle to the field. The earth's field has an angle of incidence with the earth of 73°, as illustrated below.

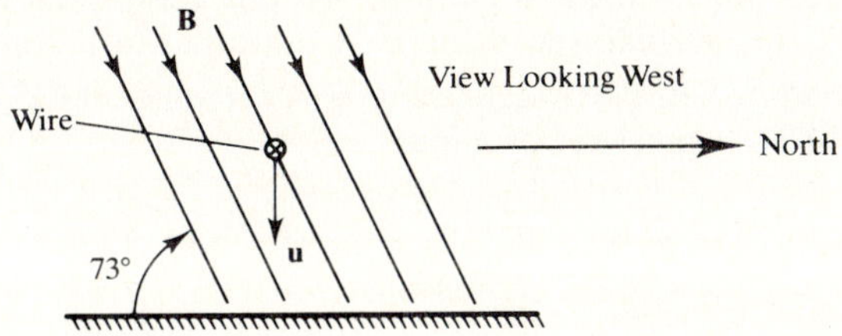

7-4. The wire of problem 7-3 falls under the action of gravity in the presence of the magnetic field of the earth. Calculate the induced voltage per meter of wire length when the velocity $\bar{\mathbf{u}}$ is one meter per second downward.

7-5. Consider the one-loop armature of Fig. 7.2 when the angle of rotation θ is zero. Let the radius R of the loop be 0.5 m, the length L be 2 m, and the field strength is $B = 20\ \text{Wb/m}^2$. Calculate the torque developed in this armature with a current of 10 amperes directed as illustrated in Fig. 7.2.

7-6. Consider the one-loop armature of Fig. 7.2 for a uniform field of $B = 20\ \text{Wb/m}^2$. If the angular velocity $\dot{\theta} = 100$ r/s, find the induced voltage v_{AB} in the loop for angles of $\theta = 0°$, 30°, 60°, and 90°. Assume that $L = 2$ m and $R = 0.5$ m. What will be the induced voltage if the angular velocity is increased to 500 r/s for each of the positions given above?

7-7. The dc motor for which the block diagram is given in Fig. 7.4 is described by the differential equations

$$J\frac{d\omega}{dt} + b\omega - K_t i = 0$$

$$L\frac{di}{dt} + Ri + K_b\omega = v(t)$$

Find the overall transfer function between applied terminal voltage and speed or $G(s) = \Omega(s)/V(s)$.

7-8. For the transfer function found in Problem 7-7, the following parameters are given:

$$J = 0.05 \text{ kg} \cdot \text{m}^2$$

$$b = 0.03 \text{ N} \cdot \text{m} \cdot \text{s/r}$$

$$K_t = 0.4 \text{ N} \cdot \text{m/A}$$

$$L = 20 \text{ mH}$$

$$R = 5\ \Omega$$

$$K_b = 0.4 \text{ V} \cdot \text{s/r}$$

(a) Evaluate the transfer function and substitute numerical values.
(b) Using Laplace transforms, find the speed as a function of time when a step of 20 volts is applied to the armature. Assume zero initial conditions.
(c) Evaluate the steady-state speed for a 20-volt input to the armature.

7-9. Consider the loudspeaker illustrated in Fig. 7.6, for which the block diagram is given in Fig. 7.8. A typical set of parameters is

$$m = 0.2 \text{ kg} \qquad L = 10^{-3} \text{ H} \qquad K_b = 10^2 \text{ V} \cdot \text{s/m}$$

$$k = 10^6 \text{ N/m} \qquad R = 8\ \Omega$$

$$b = 700 \text{ N} \cdot \text{s/m} \qquad K_f = 10^2 \text{ N/A}$$

(a) Find the transfer function $X(s)/V(s)$ and substitute the numerical values of the system parameters.
(b) Evaluate the magnitude and phase of the frequency response function for a range of frequencies from zero to 1000 Hz.

7-10. Consider the capacitive transducer system illustrated below.

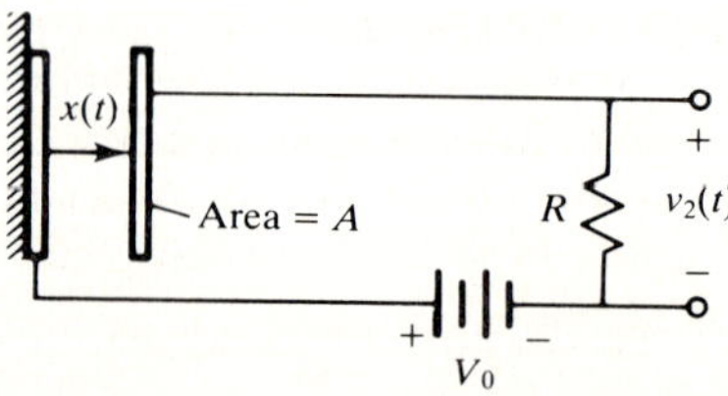

(a) Write the electrical loop equation in terms of the charge on the plates of the capacitor $q(t)$.
(b) Consider the steady-state position X_0 for which the steady capacitor charge is Q_0. Evaluate the steady-state loop equation relating X_0 and Q_0 to V_0.
(c) Linearize the equation from part **(a)** by defining the deviation in displacement to be $x_1(t)$ such that $x(t) = X_0 + x_1(t)$ and the deviation in charge $q_1(t)$ such that $q(t) = Q_0 + q_1(t)$. Drop out the steady-state terms in the resulting loop equation which correspond to the steady state considered in part **(b)**.
(d) Find the transducer transfer function $G(s) = V_2(s)/X_1(s)$.

8

INTRODUCTION TO FOURIER METHODS AND APPLICATIONS

8.1 INTRODUCTION

There are many physical problems in which the signals involved are periodic, but not necessarily sinusoidal. These problems arise in electric circuits, mechanical vibration, acoustics, electronic communication, structural dynamics, turbulent fluid dynamics, and dynamics of machinery. Let us first define periodicity. A function is said to be periodic, with a period T, if

$$f(t+nT)=f(t) \tag{8.1.1}$$

for all integers n. This is shown graphically in Fig. 8.1.

The smallest value of T for which Eq. (8.1.1) holds is called the fundamental period of $f(t)$. Notice that $2T$ or $3T$ or, in general, nT $(n>1)$ are also periods of $f(t)$, but they do not constitute a fundamental period.

Since sinusoids are periodic, we intuitively conclude that it might be possible to represent a periodic function as a linear combination of these sinusoids of different frequencies. In general, infinitely many sinusoids must be used. This method of treating periodic functions is called the Fourier series method. Explicitly we are stating that $f(t)$ can be described in the following form

$$f(t)=\frac{a_0}{2}+\sum_{n=1}^{\infty} a_n \cos n\omega_1 t + b_n \sin n\omega_1 t \tag{8.1.2}$$

where $\omega_1 = 2\pi/T$ is called the fundamental radian frequency. The reason for the constant term being called $a_0/2$ will be discussed in Section 8.3. In our previous work with systems we have seen the ease with which single sinusoidal signals are handled by employing the frequency response function representation of the system. We know that if a signal may be represented as a sum of sinusoids then the total system response to this signal may be

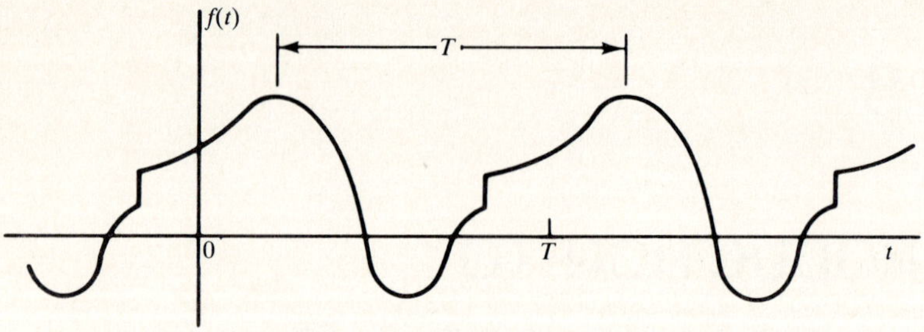

FIGURE 8.1 A typical periodic function.

given by applying the principle of superposition by analyzing the response to each input component separately and simply summing the response to each component to give the total response. Hence we see that representing a function as a linear combination of sinusoids will, indeed, allow us to use superposition as described above. This is our main reason for introducing the Fourier series method.

8.2 INTEGRALS OF COMBINATIONS OF TRIGONOMETRIC FUNCTIONS

Since we are going to work intimately with sinusoids it is important for us to examine the properties of trigonometric functions when integrated over one or more periods. The net area under any single cycle of a sinusoid is zero; hence, the integral over any single cycle is zero for any sinusoid. Since the integral of a single cycle of a sinusoid is zero, the integral over an integer number of cycles is also zero.

It is easy to show with the use of well-known trigonometric identities that the integrals of products of cosines of integer multiples of a fundamental frequency are

$$\int_{t_0}^{t_0+T} \cos n\omega_1 t \cos m\omega_1 t \, dt = \begin{cases} T & m = n = 0 \\ T/2 & m = n \neq 0 \\ 0 & m \neq n \end{cases} \tag{8.2.1}$$

and likewise for sine functions

$$\int_{t_0}^{t_0+T} \sin n\omega_1 t \sin m\omega_1 t \, dt = \begin{cases} 0 & m = n = 0 \\ T/2 & m = n \neq 0 \\ 0 & m \neq n \end{cases} \tag{8.2.2}$$

Similarly, it may be shown that the integrals of sine-cosine products are all zero or

$$\int_{t_0}^{t_0+T} \cos m\omega_1 t \sin n\omega_1 t \, dt = 0 \tag{8.2.3}$$

Relations (8.2.1) to (8.2.3) are commonly referred to as the *orthogonality relations*. These will be useful relationships in the work to follow.

8.3 THE FOURIER EXPANSION

We have hypothesized in Eq. (8.1.2) that we could write any periodic signal or waveform as a sum of sinusoids but we must also include a constant term as in no way can sinusoids or any sum of sinusoids represent a constant or dc component of a waveform. Let us represent an arbitrary periodic waveform $f(t)$ with a *Fourier series* of the form

$$f(t) = \frac{a_0}{2} + \sum_{n=1}^{\infty} a_n \cos n\omega_1 t + b_n \sin n\omega_1 t \tag{8.3.1}$$

The question immediately arises as to what the values of a_0, a_1, $a_2, \ldots, b_1, b_2, \ldots$ should be in order to best represent the function $f(t)$. In an attempt to answer this question, let us integrate the series (8.3.1) over one cycle and examine the result.

$$\int_{t_0}^{t_0+T} f(t)\, dt = \int_{t_0}^{t_0+T} \left[\frac{a_0}{2} + \sum_{n=1}^{\infty} a_n \cos n\omega_1 t + b_n \sin n\omega_1 t\right] dt \tag{8.3.2}$$

For commonly encountered engineering functions, we may interchange the order of integration and summation to give

$$\int_{t_0}^{t_0+T} f(t)\, dt = \frac{a_0 T}{2} + \sum_{n=1}^{\infty} a_n \int_{t_0}^{t_0+T} \cos n\omega_1 t\, dt + b_n \int_{t_0}^{t_0+T} \sin n\omega_1 t\, dt \tag{8.3.3}$$

Since the integral of the sinusoids of one or more periods is zero, the following expression can be derived for the average value of the waveform over a period:

$$\frac{a_0}{2} = \frac{1}{T}\int_{t_0}^{t_0+T} f(t)\, dt \tag{8.3.4}$$

This may be thought of as the "area under the curve $f(t)$" divided by the period of the cycle. Often the easiest way to evaluate $a_0/2$ for waveforms made solely of straight-line segments is to divide the net area by the period.

To evaluate the coefficients a_n and b_n, we proceed in the following manner. First multiply both sides of expression (8.3.1) by $\cos m\omega_1 t$ and integrate over a complete period:

$$\int_{t_0}^{t_0+T} f(t) \cos m\omega_1 t \, dt = \frac{a_0}{2} \int_{t_0}^{t_0+T} \cos m\omega_1 t \, dt$$
$$+ \sum_{n=1}^{\infty} a_n \int_{t_0}^{t_0+T} \cos m\omega_1 t \cos n\omega_1 t \, dt$$
$$+ b_n \int_{t_0}^{t_0+T} \cos m\omega_1 t \sin n\omega_1 t \, dt \qquad \textbf{(8.3.5)}$$

Upon applying the orthogonality relations (Eqs. (8.2.1) and (8.2.3)), and noting that the integral over an integer number of cycles of a sinusoid is zero, we find the following:

$$\int_{t_0}^{t_0+T} f(t) \cos m\omega_1 t \, dt = a_m \int_{t_0}^{t_0+T} \cos^2 m\omega_1 t \, dt \qquad \textbf{(8.3.6)}$$

Now applying relation (8.2.1) to the right side, the result is simply

$$a_m = \frac{2}{T} \int_{t_0}^{t_0+T} f(t) \cos m\omega_1 t \, dt \qquad m = 1, 2, \ldots \qquad \textbf{(8.3.7)}$$

Multiplying through by $\sin m\omega_1 t$ and integrating over one period while employing similar arguments to those used above results in

$$b_m = \frac{2}{T} \int_{t_0}^{t_0+T} f(t) \sin m\omega_1 t \, dt \qquad m = 1, 2, \ldots \qquad \textbf{(8.3.8)}$$

To apply the above analysis to a specific problem one need not repeat the above work but simply use the results (Eqs. (8.3.4), (8.3.7), and (8.3.8)). These results, for convenience, are repeated below.

Summary: A function $f(t)$ which is periodic on the interval $(t_0, t_0 + T)$ may be represented by a Fourier series of the form

$$f(t) = \frac{a_0}{2} + \sum_{n=1}^{\infty} a_n \cos n\omega_1 t + b_n \sin n\omega_1 t \qquad \textbf{(8.3.9)}$$

in which the Fourier coefficients are given by

$$\frac{a_0}{2} = \frac{1}{T} \int_{t_0}^{t_0+T} f(t) \, dt \qquad \textbf{(8.3.10)}$$

$$a_n = \frac{2}{T} \int_{t_0}^{t_0+T} f(t) \cos n\omega_1 t \, dt \qquad n = 0, 1, 2, \ldots \qquad \textbf{(8.3.11)}$$

and

$$b_n = \frac{2}{T} \int_{t_0}^{t_0+T} f(t) \sin n\omega_1 t \, dt \qquad n = 1, 2, \ldots \qquad \textbf{(8.3.12)}$$

Note that relation (8.3.11) will also yield the correct value for a_0 and hence this is the reason for calling the constant term of the series $a_0/2$ rather than a_0. In fact, relation (8.3.11) gives the correct $a_0(n=0)$ and will therefore be used to find all a_n, $n = 0, 1, 2, \ldots$. It is important to note that t_0 can often be chosen to simplify the evaluation of the integrals.

Example 8.1

Let us consider as an example the sawtooth function shown in Fig. 8.2. The period of the waveform is 10 milliseconds and hence the fundamental radian frequency is

$$\omega_1 = \frac{2\pi}{T} = \frac{2\pi}{10 \times 10^{-3}} = \frac{2\pi}{10^{-2}} = 628 \text{ r/s}$$

The description of the wave for one-half period either side of the temporal origin is as follows:

$$f(t) = \frac{8}{5 \times 10^{-3}} t \qquad -5 \text{ ms} \le t \le 5 \text{ ms}$$

Since we may integrate over any time interval one cycle in length, we may integrate from -5 ms to 5 ms, i.e., $t_0 = -5 \times 10^{-3}$ seconds will be used. Also, note that this waveform has no dc component; hence

$$\frac{a_0}{2} = 0$$

Now evaluate the cosine coefficients:

$$a_n = \frac{2}{10^{-2}} \int_{-5\times10^{-3}}^{5\times10^{-3}} \frac{8}{5 \times 10^{-3}} t \cos n628t \, dt$$

$$= \frac{16}{5 \times 10^{-5}} \int_{-5\times10^{-3}}^{5\times10^{-3}} t \cos n628t \, dt$$

$$= \frac{16}{5 \times 10^{-5}} \left[\frac{1}{(628n)^2} \cos n628t + \frac{t}{(628n)} \sin n628t \right]_{-5\times10^{-3}}^{5\times10^{-3}}$$

$$= \frac{16}{5 \times 10^{-5}} \left[\frac{5 \times 10^{-3}}{628n} (\sin n\pi + \sin -n\pi) \right] = 0 \text{ for all } n$$

This shows that the Fourier series for the waveform of Fig. 8.2 has no cosine terms. The sine coefficients are

$$b_n = \frac{2}{10^{-2}} \int_{-5\times10^{-3}}^{5\times10^{-3}} \frac{8}{5 \times 10^{-3}} t \sin n628t \, dt$$

$$= \frac{16}{5 \times 10^{-5}} \left[\frac{1}{(628n)^2} \sin n628t - \frac{1}{(628n)} t \cos n628t \right]_{-5\times10^{-3}}^{5\times10^{-3}}$$

$$= \frac{16}{5 \times 10^{-5}} \left\{ -\frac{5 \times 10^{-3}}{628n} [\cos n\pi + \cos(-n\pi)] \right\} = \frac{16}{n\pi} (-1)^{n+1}$$

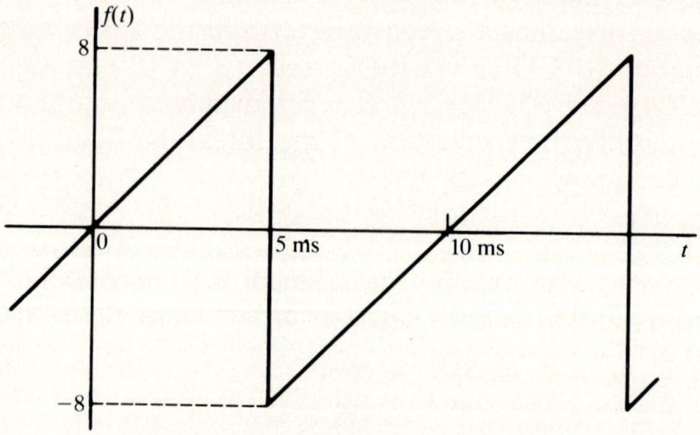

FIGURE 8.2 Sawtooth waveform.

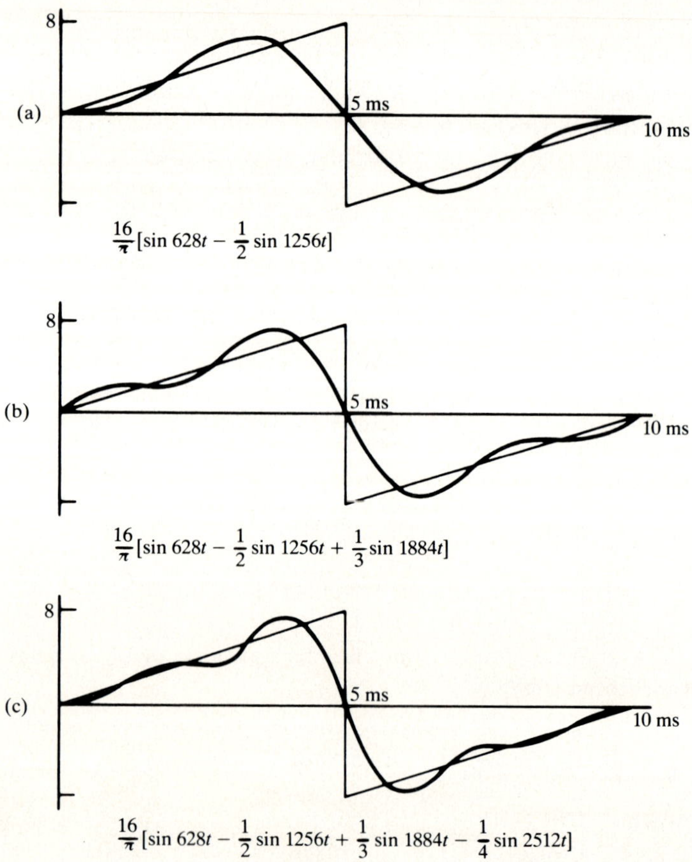

FIGURE 8.3 Synthesis of a sawtooth waveform with Fourier components.

The resulting Fourier series for the function is

$$f(t) = \sum_{n=1}^{\infty} \frac{16}{n\pi} (-1)^{n+1} \sin n628t$$

We may examine the effect of each term of the series as follows. Figure 8.3(a) illustrates the first two harmonics and their sum. Figure 8.3(b) shows the sum of the first three harmonics and Fig. 8.3(c) shows the sum of the first four harmonics and, as one can see, this rapidly approaches the original sawtooth function.

Illustrated in Appendix B are the Fourier coefficients for a wide range of commonly encountered periodic functions.

8.4 FOURIER SERIES AS A BEST LEAST-SQUARES FIT

It is often practical to approximate a periodic time function, $f(t)$, by a *finite* summation of terms rather than represent it exactly as an infinite summation. Practically speaking, one would often want to choose this approximation so that the error between $f(t)$ and its approximation would have the least average power. The problem is: How do we choose the coefficients of the approximation so as to minimize the average power in the error?

Specifically, we know that $f(t)$, for most engineering purposes, can be represented as

$$f(t) = \frac{a_0}{2} + \sum_{n=1}^{\infty} (a_n \cos n\omega_1 t + b_n \sin n\omega_1 t) \tag{8.4.1}$$

where $\omega_1 = 2\pi/T$ and T is the fundamental period.

Let us now approximate $f(t)$ as $\hat{f}_N(t)$, which is a finite trigonometric series or

$$\hat{f}_N(t) = \frac{c_0}{2} + \sum_{n=1}^{N} (c_n \cos n\omega_1 t + d_n \sin n\omega_1 t) \tag{8.4.2}$$

Our problem is to choose the c_n and d_n so that the average power in the error is a minimum.

For a general signal $f(t)$, the error signal is $f(t) - \hat{f}_N(t)$ or

$$e(t) = f(t) - \hat{f}_N(t) \tag{8.4.3}$$

Since the power varies as $f^2(t)$, the average power of the error or the mean square error (MSE) is given as

$$\text{MSE} = \frac{1}{T} \int_0^T e^2(t)\, dt \tag{8.4.4}$$

This MSE is a minimum when its partial derivatives with respect to the cs and ds are equal to zero, i.e.,

$$\frac{\partial \text{MSE}}{\partial c_i} = \frac{1}{T}\int_0^T 2e(t)\,\frac{\partial e(t)}{\partial c_i}\,dt = 0 \qquad i = 0, 1, \ldots N \tag{8.4.5}$$

$$\frac{\partial \text{MSE}}{\partial d_i} = \frac{1}{T}\int_0^T 2e(t)\,\frac{\partial e(t)}{\partial d_i}\,dt = 0 \qquad i = 1, \ldots N \tag{8.4.6}$$

Since we can easily see from above that $\partial e/\partial c_i = -\cos(i\omega_1 t)$, Eq. (8.4.5) becomes the following after substituting in Eq. (8.4.3):

$$-\frac{1}{T}\int_0^T 2[f(t) - \hat{f}_N(t)]\cos(i\omega_1 t)\,dt = 0 \tag{8.4.7}$$

We can then write

$$\frac{1}{T}\int_0^T f(t)\cos(i\omega_1 t)\,dt = \frac{1}{T}\int_0^T \hat{f}_N(t)\cos(i\omega_1 t)\,dt \tag{8.4.8}$$

The left side of Eq. (8.4.8) is equal to $a_i/2$. The right side of Eq. (8.4.8) is equal to $c_i/2$. Therefore:

$$c_i = a_i \qquad i = 1, 2, \ldots N \tag{8.4.9}$$

Likewise, it can be shown that

$$d_i = b_i \qquad i = 1, 2, \ldots N \tag{8.4.10}$$

and

$$c_0 = a_0 \tag{8.4.11}$$

Therefore, we reach the following important and practical conclusion: A finite Fourier approximation to $f(t)$, given in the form of

$$\hat{f}_N(t) = \frac{a_0}{2} + \sum_{n=1}^{N} (a_n \cos n\omega_1 t + b_n \sin n\omega_1 t) \tag{8.4.12}$$

will be the best approximation in the mean-square sense. Another way of saying this is that a finite Fourier series approximation to $f(t)$ is the "best" (in the least-mean-square sense) possible representation of $f(t)$ given that the approximation is to be made up of a finite sum of trigonometric functions up to and including N terms.

8.5 ESTIMATION OF ERROR DUE TO SERIES TRUNCATION

In engineering practice we cannot take an infinite number of terms in the series representing some periodic function. Therefore, it would be useful to say something quantitative about the errors introduced in the representation of the function by a Fourier series truncated to a finite number of terms. Let us define the mean square value of the function $f(t)$ to be

$$F^2 = \frac{1}{T}\int_0^T f^2(t)\,dt \tag{8.5.1}$$

Also, we may similarly define the mean square value of $\hat{f}_N(t)$ by

$$F_N^2 = \frac{1}{T}\int_0^T \hat{f}_N^2(t)\,dt \tag{8.5.2}$$

If we substitute relation (8.4.12) for $\hat{f}_N(t)$ and invoke the orthogonality property of the sines and cosines, it is easy to show that the mean square value of the truncated approximation is

$$F_N^2 = \frac{a_0^2}{4} + \frac{1}{2}\sum_{n=1}^{N}(a_n^2 + b_n^2) \tag{8.5.3}$$

This relation is known as the Parseval relation. The error in the approximation at any time t is

$$e(t) = f(t) - \left[\frac{a_0}{2} + \sum_{n=1}^{N}(a_n \cos n\omega_1 t + b_n \sin n\omega_1 t)\right] \tag{8.5.4}$$

The mean square error is defined as

$$\text{MSE}_N = \frac{1}{T}\int_0^T e^2(t)\,dt \tag{8.5.5}$$

Substituting Eq. (8.5.4) into Eq. (8.5.5) and expanding the integrand, noting the orthogonality of the sines and cosines, yields

$$\text{MSE}_N = F^2 - F_N^2 \tag{8.5.6}$$

where F^2 and F_N^2 are defined, respectively, in relations (8.5.1) and (8.5.3). A popular way of measuring the relative error is to normalize the MSE with respect to the mean square value of the original waveform, or the normalized mean square error is

$$\text{NMSE}_N = \frac{F^2 - F_N^2}{F^2}$$

One way to assess quantitatively the effect of series truncation is to evaluate NMSE_N and NMSE_{N+1} to give the relative contribution, in the mean square sense, of the $(N+1)$th harmonic to the total waveform.

Example 8.2

Consider the sawtooth waveform of Example 8.1, wherein the resultant Fourier series was found to be

$$f(t) = \sum_{n=1}^{\infty} \frac{16}{n\pi}(-1)^{n+1} \sin n628t$$

The mean square value of the waveform is given by relation (8.5.1) to be

$$F^2 = \frac{1}{10^{-2}} \int_{-5\times 10^{-3}}^{5\times 10^{-3}} \left(\frac{8}{5\times 10^{-3}}\, t\right)^2 dt$$

$$= \frac{64}{25\times 10^{-8}} \left.\frac{t^3}{3}\right|_{-5\times 10^{-3}}^{5\times 10^{-3}} = \frac{(128)(125\times 10^{-9})}{(25\times 10^{-8})(3)}$$

$$= 21.33 \text{ (units of } f^2)$$

If we truncate the series to four terms,

$$F_4^2 = \frac{1}{2}\left(\frac{16}{\pi}\right)^2 \left(1 + \frac{1}{4} + \frac{1}{9} + \frac{1}{16}\right)$$

$$F_4^2 = 18.462$$

And if the series is truncated at five terms the result is

$$F_5^2 = 18.98$$

and one can see the contribution of the fifth harmonic to the total waveform. The normalized mean square errors for truncation at four and five terms are

$$\text{NMSE}_4 = \frac{21.33 - 18.462}{21.33} = 0.134$$

and

$$\text{NMSE}_5 = \frac{21.33 - 18.98}{21.33} = 0.110$$

It is clear that, to decrease the error greatly, a considerable number of terms will be needed. This slow convergence is largely due to the discontinuities (jumps) in the sawtooth waveform.

8.6 SYMMETRY CONDITIONS ON FOURIER SERIES

In this section we shall discuss methods by which we may reduce the labor of integration when the waveform has certain symmetry properties.

8.6.1 Even Function Symmetry

An even function is one which possesses a symmetry about the temporal origin. The definition of an even function $f(t)$ is one which obeys

$$f(t) = f(-t) \tag{8.6.1}$$

Examples of even functions are given below.

$f(t) = \text{constant}$

$f(t) = t^2, t^4$, etc. (all even-ordered polynomials)

$f(t) = \cos \alpha t$ (all cosine functions)

and there are many others.

8.6.2 Odd Function Symmetry

An odd function, in contrast to an even function, is one which possesses a special asymmetry with respect to the temporal origin or more specifically

$$f(t) = -f(-t)$$

Examples of odd functions are

$f(t) = t, t^3$, etc. (all odd-ordered polynomials)

$f(t) = \sin \alpha t$ (all sine functions)

At this point, one might get the idea that all functions are either odd or even functions but that is not the case. There are many functions which are neither even nor odd. One should mention at this point that the product or quotient of even functions is an even function, the product or quotient of odd functions is also an even function, and the product or quotient of an odd and even function is an odd function, where it is to be noted that division by zero is obviously excluded.

8.6.3 Symmetry Shortcuts

If one examines the elementary form of the trigonometric Fourier series, it is clear that it is composed of even functions (the constant term and all the cosine terms) and odd functions (the sine terms). It would seem possible that, if we are expanding an even function in a Fourier series, it should not contain any sine terms since they are odd functions. We can prove this by evaluating the b_n coefficients or

$$b_n = \frac{2}{T} \int_{-T/2}^{T/2} f(t) \sin n\omega_1 t \, dt \tag{8.6.2}$$

where the interval of integration has been taken over the period symmetric with respect to the origin. If we examine the integrand of Eq. (8.6.2) we find that it is a product of an even and an odd function and therefore an odd function. The integral of all odd functions on the interval $(-T/2, T/2)$ will be zero and, thus, all the b_n will be zero for an even function.

A similar argument will indicate that all the cosine terms and the $a_0/2$ term will be zero for all expansions of odd functions $f(t)$. From the above arguments it is clear that all odd functions have no average value for that is what $a_0/2$ is. We should also examine those functions which result in odd functions after subtraction of the average value from them. By the arguments just made, these functions will contain only sine functions and an $a_0/2$ term. In the case of even or odd symmetry, one need only integrate over the first half period and double the result to get each nonzero coefficient.

Example 8.3

If we examine the waveform of Example 8.1 we see that this waveform is an odd function; hence, the Fourier series will contain only sine terms. In Example 8.1 we found this to be true by brute force.

8.6.4 Half-Wave Symmetry

We have previously defined several symmetry conditions which will, in practice, save us considerable labor in the evaluation of the Fourier coefficients. Another such condition is that of half-wave symmetry, which is defined by

$$f(t) = -f\left(t + \frac{T}{2}\right) \tag{8.6.3}$$

In this case, if we carefully explore the integrals for the Fourier coefficients by substitution of relation (8.6.3), we will find that all the coefficients for even values of n (even harmonics) will vanish. Also, it is easy to show that one need only integrate over the first half-cycle and double the result to evaluate the integrals.

If a function is either even or odd *and* possesses half-wave symmetry, it can be shown that the appropriate previous terms will vanish. Also, the remaining coefficients may be given by integration of the first quarter cycle ($0 \le t \le T/4$) and multiplication of the integral by $8/T$ in place of the usual $2/T$.

Example 8.4

Based on the discussion in Section 8.6, examine the terms present in the Fourier series expansions for the three waveforms given in Fig. 8.4.

Waveform (a) is an even function and hence the sine terms will not be present. The $a_0/2$ term will be present since the average value is 0.75. All harmonics will be present since the function does not possess half-wave symmetry about the average value.

Waveform (b) is not an even function and has an average value of unity and as such cannot be an odd function in the pure sense. However, the function is an odd function about the average value and as a result it will consist of an $a_0/2$ term and sine terms. The function also has the half-wave symmetry property about the average value and thus only the odd terms ($n = 1, 3, 5, \ldots$) will be present.

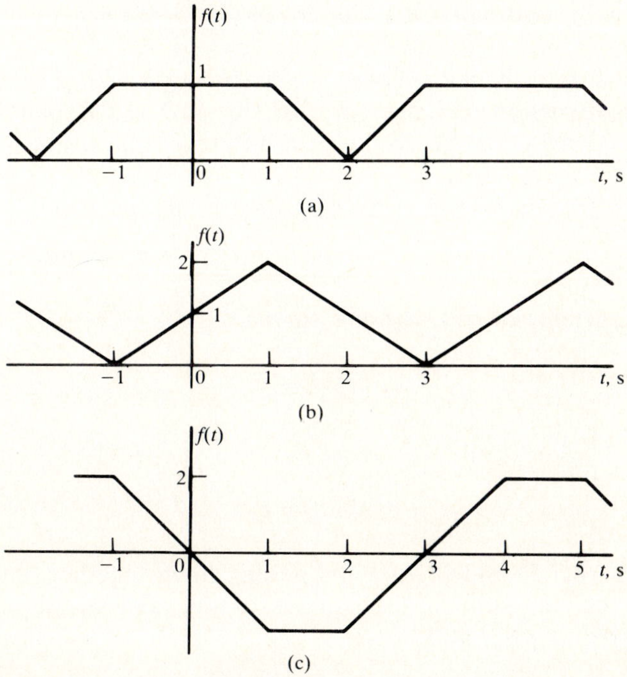

FIGURE 8.4 Three waveforms with a variety of symmetry conditions.

Finally, waveform (c) has no average value and thus $a_0/2 = 0$. The waveform is also an odd function and thus only sine terms will be present. The function also has half-wave symmetry which will then indicate that the Fourier series representation will consist of only the odd sine terms.

8.7 AN ALTERNATIVE FORM OF THE FOURIER SERIES

We have thus far considered a series of the form

$$f(t) = \frac{a_0}{2} + \sum_{n=1}^{\infty} a_n \cos n\omega_1 t + b_n \sin n\omega_1 t \tag{8.7.1}$$

in which there are, at each frequency $n\omega_1$, two sinusoids. The nth term of the series looks like

$$f_n(t) = a_n \cos n\omega_1 t + b_n \sin n\omega_1 t \tag{8.7.2}$$

We know that the sum of two sinusoids of a single frequency can be written as a single sinusoid at the same frequency of the form

$$c_n \cos(n\omega_1 t - \theta_n) = a_n \cos n\omega_1 t + b_n \sin n\omega_1 t \tag{8.7.3}$$

(We could just as well have chosen a sine function with a phase but will continue our discussion with the form of Eq. (8.7.3).) Expanding the left side with the identity for the cosine of a difference the result is

$$c_n[\cos n\omega_1 t \cos \theta_n + \sin n\omega_1 t \sin \theta_n] = a_n \cos n\omega_1 t + b_n \sin n\omega_1 t \tag{8.7.4}$$

Equating coefficients of $\cos n\omega_1 t$ and $\sin n\omega_1 t$,

$$c_n \cos \theta_n = a_n \tag{8.7.5}$$

and

$$c_n \sin \theta_n = b_n \tag{8.7.6}$$

Squaring Eqs. (8.7.5) and (8.7.6) and adding them together, and recognizing the fact that the sum of the squares of sine and cosine is unity, the relation between a_n, b_n, and c_n is

$$c_n = \sqrt{a_n^2 + b_n^2} \tag{8.7.7}$$

Division of expression (8.7.6) by Eq. (8.7.5) yields

$$\frac{b_n}{a_n} = \tan \theta_n \tag{8.7.8}$$

and solving for the phase angle yields

$$\theta_n = \tan^{-1}\left(\frac{b_n}{a_n}\right) + k\pi \qquad k = \begin{cases} 0 & \text{for } a_n > 0 \\ 1 & \text{for } a_n < 0 \end{cases} \tag{8.7.9}$$

This yields a form of the series with a single cosine at each frequency with amplitude c_n and phase θ_n.

$$f(t) = c_0 + \sum_{n=1}^{\infty} c_n \cos(n\omega_1 t - \theta_n) \tag{8.7.10}$$

where c_0 is the average value previously denoted as $a_0/2$.

Example 8.5

Characterize the sawtooth wave of Example 8.1 by a Fourier series of the form of Eq. (8.7.10). The original series was composed of sine terms or

$$f(t) = \sum_{n=1}^{\infty} \frac{16}{n\pi} (-1)^{n+1} \sin n628t$$

The coefficients c_n are given by Eq. (8.7.7) to be

$$c_n = \sqrt{a_n^2 + b_n^2} = \frac{16}{n\pi}$$

while the phase angles θ_n are found by Eq. (8.7.8) to be

$$\theta_n = \tan^{-1}(\pm\infty) \qquad \begin{matrix} +\text{if } n \text{ is odd} \\ -\text{if } n \text{ is even} \end{matrix}$$

Hence

$$\theta_n = \pm 90° = (-1)^{n+1}\frac{\pi}{2}$$

The series is then

$$f(t) = \sum_{n=1}^{\infty} \frac{16}{n\pi} \cos(n628t - (-1)^{n+1}\pi/2)$$

Here we have been careful to make the arguments of the cosine function to be dimensionally consistent and expressed in radians.

8.8 THE CONCEPT OF THE FREQUENCY SPECTRUM OF A SIGNAL

We would like some way to display graphically the frequency content of a periodic signal such that it would be easy to interpret. We may accomplish this simply by plotting, in line-graph fashion, the magnitude of the sinusoidal component c_n at each frequency as a function of frequency. The waveform is written as a Fourier series as

$$f(t) = c_0 + \sum_{n=1}^{\infty} c_n \cos(n\omega_1 t - \theta_n) \tag{8.8.1}$$

To plot the spectrum of a signal, we simply plot c_n as a function of ω, yielding a plot which appears as in Fig. 8.5.

The value of this representation is that one can learn to tell by inspection which frequency bands contain the energy of a signal. This spectral concept has become very popular in the fields of communication systems, vibration data analysis, acoustics, and geophysics, as well as others.

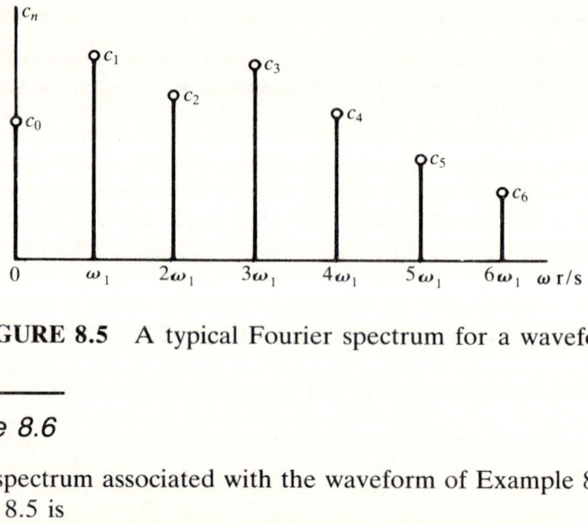

FIGURE 8.5 A typical Fourier spectrum for a waveform.

Example 8.6

Plot the spectrum associated with the waveform of Example 8.5. The series of Example 8.5 is

$$f(t) = \sum_{n=1}^{\infty} \frac{16}{n\pi} \cos(n628t - (-1)^{n+1}\pi/2)$$

The Fourier coefficients are

$$c_n = \frac{16}{n\pi} \qquad n = 1, 2, \ldots$$

These magnitudes are plotted in Fig. 8.6 as a function of their respective frequency and these data are referred to as the spectrum or, more correctly, the magnitude spectrum of the sawtooth function given.

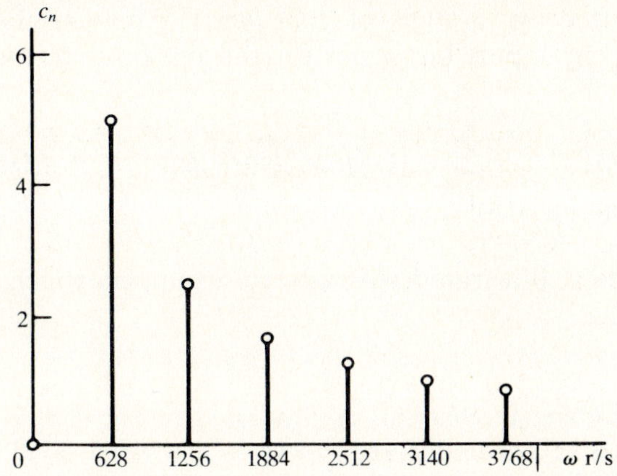

FIGURE 8.6 Spectrum of a sawtooth waveform.

Note that there is no spectral line at $\omega = 0$ because there is no dc component in the original sawtooth waveform. A spectral line is an indication that signal energy is present at that frequency.

8.9 TRANSMISSION OF PERIODIC SIGNALS IN LINEAR SYSTEMS

Let us consider a stable linear system with an input $y(t)$, response $x(t)$, and transfer function $G(s)$, as illustrated in Figure 8.7.

Since the input signal is periodic, it may be represented by a Fourier series of the form

$$y(t) = c_0 + \sum_{n=1}^{\infty} c_n \cos(n\omega_1 t - \theta_n) \tag{8.9.1}$$

Since the system is linear, we may apply the principle of superposition and treat the transmission of the input signal one term at a time and sum the results. We know that a single sinusoid of frequency ω (r/s) passed through a linear system will have its amplitude attenuated (or amplified) by the factor $|G(j\omega)|$ and its phase shifted by an angle $\angle G(j\omega)$. (It has been assumed that a steady-state output exists and all transients have died out.)

If we employ the principle of superposition to the input, we see that the nth input amplitude will be attenuated by a factor of $|G(jn\omega_1)|$ and the

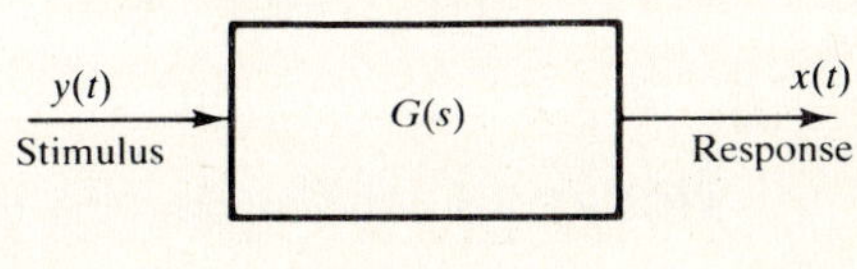

FIGURE 8.7 Linear system with periodic input.

phase will be shifted by an angle of $\angle G(jn\omega_1)$. Applying the above to the Fourier series of the input, the series for the response $x(t)$ is

$$x(t) = c_0|G(j0)| + \sum_{n=1}^{\infty} c_n|G(jn\omega_1)|\cos(n\omega_1 t - \theta_n + \angle G(jn\omega_1)) \tag{8.9.2}$$

These ideas are best illustrated with several examples which follow.

Example 8.7

A shock-absorbing mechanism, as illustrated in Fig. 8.8, is subjected to a periodic series of sawtooth force shocks as illustrated. Find the steady-state response $y(t)$. Using the data from Appendix B, it is easy to show that the Fourier series of the forcing function is

$$f(t) = \sum_{n=1}^{\infty} \frac{200}{n\pi} \cos(n2\pi t - (-1)^{n+1}\pi/2)$$

The force to displacement transfer function is given by summation of forces. Laplace transformation of the equation of motion gives a transfer function of

$$G(s) = \frac{Y(s)}{F(s)} = \frac{10^{-3}}{s+10}$$

The frequency response function is then

$$G(j\omega) = \frac{10^{-3}}{j\omega + 10}$$

and its magnitude and angle are, respectively,

$$|G(j\omega)| = \frac{10^{-3}}{\sqrt{\omega^2 + 100}}$$

and

$$\angle G(j\omega) = -\tan^{-1}\left(\frac{\omega}{10}\right) = -\tan^{-1}\left(\frac{n\pi}{5}\right)$$

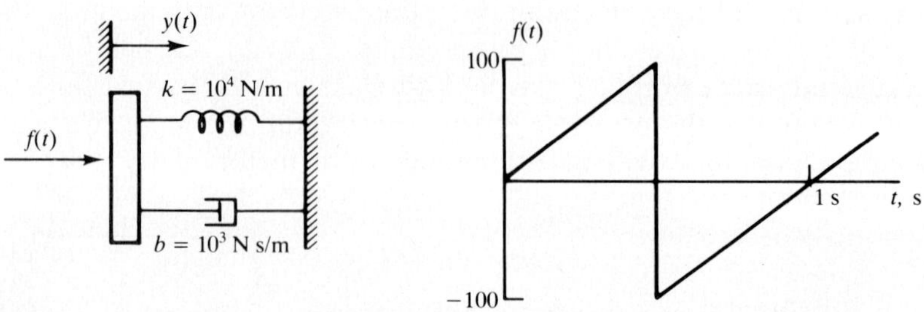

FIGURE 8.8 Shock-absorbing mechanism driven by a sawtooth force.

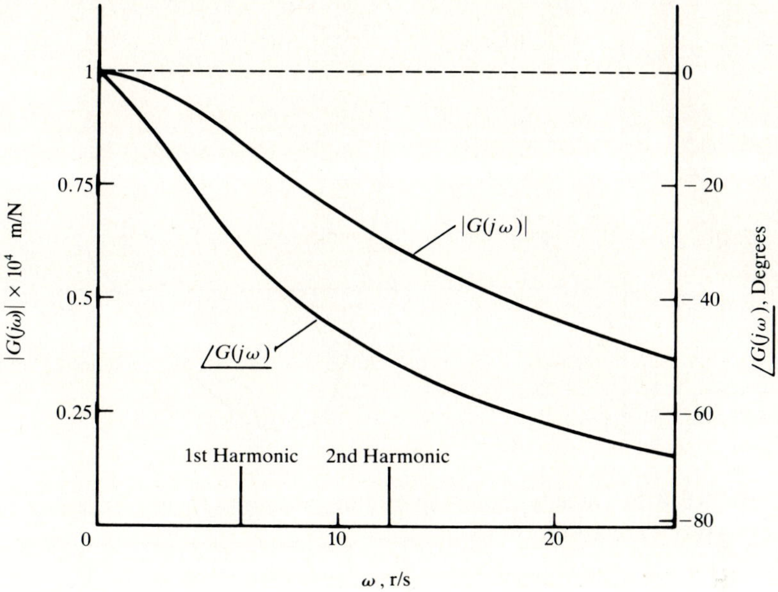

FIGURE 8.9 Frequency response data for shock-absorbing mechanism.

which are plotted in Fig. 8.9. If we evaluate these quantities at the various harmonics of the forcing function we get the data of Table 8-1. The closed form of the response $y(t)$ is written as

$$y(t) = \sum_{n=1}^{\infty} \frac{0.2}{n\pi\sqrt{(n2\pi)^2 + 100}} \cos\left[n2\pi t - \frac{\pi}{2}(-1)^{n+1} - \tan^{-1}\left(\frac{n\pi}{5}\right)\right]$$

From the response illustrated in Fig. 8.10, note that the response is nearly sinusoidal because of the high attenuation of the higher harmonics of the forcing function due to the lowpass nature of the transfer function.

TABLE 8-1 Fourier Response Data for Shock Absorber

n	$n\omega_1$	F_n	θ_n	$\lvert G(jn\omega_1)\rvert$	$\angle G(jn\omega_1)$ rad	Y_n
1	2π	63.66	$\pi/2$	8.47×10^{-5}	-0.561	5.98×10^{-3}
2	4π	31.83	$-\pi/2$	6.23×10^{-5}	-0.899	1.98×10^{-3}
3	6π	21.22	$\pi/2$	4.69×10^{-5}	-1.083	0.99×10^{-3}
4	8π	15.92	$-\pi/2$	3.70×10^{-5}	-1.192	0.59×10^{-3}
5	10π	12.73	$\pi/2$	3.03×10^{-5}	-1.263	0.38×10^{-3}

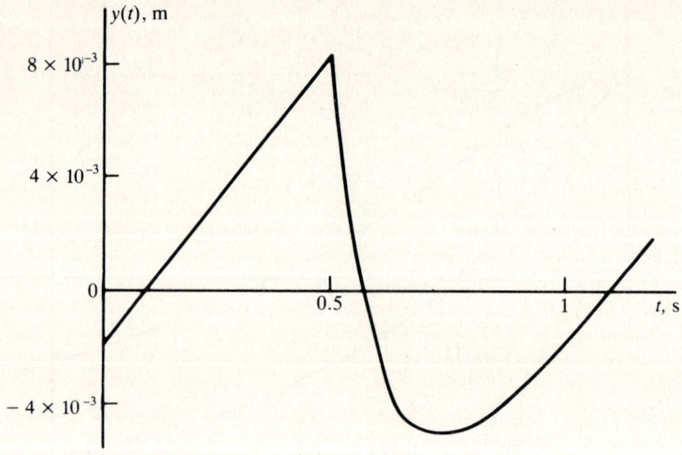

FIGURE 8.10 Steady-state response of shock-absorbing mechanism to sawtooth forcing function.

Example 8.8

Consider the bandpass filter circuit configuration shown in Fig. 8.11. Let the circuit have the following parameters:

$$L = 10^{-4}\text{ H},\ C = 10^{-6}\text{ F},\ R = 0.5\ \Omega$$

To investigate the signal transmission through this system, we must first find the transfer function of the system, which is easily done, employing Kirchhoff's voltage law around the left loop and the expression for the voltage drop across the resistor:

$$G(s) = \frac{V_2(s)}{V_1(s)} = \frac{R}{R + Ls + \dfrac{1}{sC}}$$

and the frequency response function is

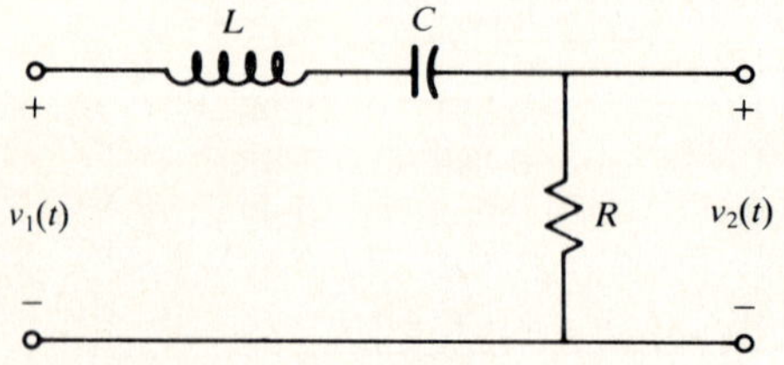

FIGURE 8.11 R-L-C bandpass filter.

$$G(j\omega) = \frac{R}{R + j\left(\omega L - \dfrac{1}{\omega C}\right)} = \frac{jR\omega C}{jR\omega C + (1 - \omega^2 LC)}$$

Substitution of the system parameters into this equation yields

$$G(j\omega) = \frac{j5 \times 10^{-7}\omega}{j5 \times 10^{-7}\omega + (1 - 10^{-10}\omega^2)}$$

The magnitude of this function is

$$|G(j\omega)| = \frac{5 \times 10^{-7}\omega}{\sqrt{25 \times 10^{-14}\omega^2 + (1 - 10^{-10}\omega^2)^2}} \quad \textbf{(a)}$$

and the angle of the complex function is

$$\underline{/G(j\omega)} = \frac{\pi}{2} - \tan^{-1}\left(\frac{5 \times 10^{-7}\omega}{1 - 10^{-10}\omega^2}\right) \quad \textbf{(b)}$$

Let us consider the input voltage to this circuit $v_1(t)$ to be a square wave, as illustrated in Fig. 8.12. The Fourier series for this input voltage is

$$v_1(t) = \sum_{n=1,3,5,\ldots}^{\infty} \frac{32}{n\pi} \cos\left(n3 \times 10^4 t - \frac{\pi}{2}\right)$$

The spectrum of this voltage is shown in Fig. 8.13, along with the magnitude of the frequency response function for the filter. It can be seen from this plot that the filter should readily pass the third harmonic of the input and should attenuate all others to a greater degree. Applying expression (8.9.2) to this particular problem yields, after using Eqs. **(a)** and **(b)**,

$$v_2(t) = \sum_{n=1,3,5}^{\infty} \frac{32 \times 10^{-7} n\omega_1}{n\pi\sqrt{25 \times 10^{-14} n^2\omega_1^2 + (1 - 10^{-10}\omega_1^2 n^2)^2}}$$
$$\times \cos\left(n\omega_1 t + \left(-\frac{\pi}{2} + \frac{\pi}{2} - \tan^{-1}\left[\frac{5 \times 10^{-7} n\omega_1}{1 - 10^{-10}\omega_1^2 n^2}\right]\right) \Big/ 57.3°/\text{rad}\right)$$

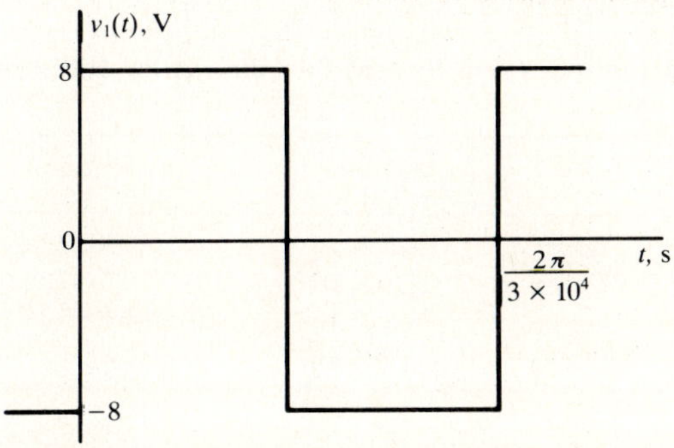

FIGURE 8.12 Square wave input voltage for an R-L-C filter.

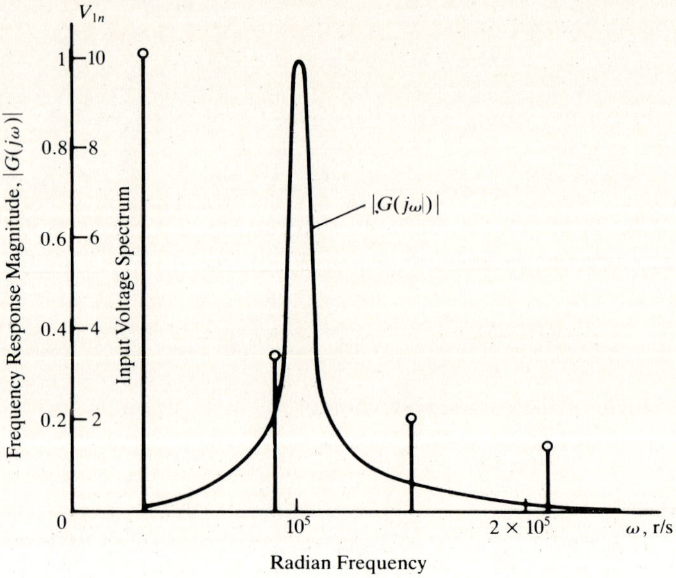

FIGURE 8.13 Frequency response function and input voltage spectrum as functions of frequency.

Table 8-2 gives values of the frequencies present in $v_1(t)$, along with the magnitude and angle of $G(j\omega)$ for the first seven harmonics along with input and output voltage amplitudes. Figure 8.14 illustrates the sum of the series through the seventh harmonic and, as one may easily observe, the third harmonic is quite prominent. This fact is further borne out by examination of the Fourier spectrum of the output voltage $v_2(t)$ in Fig. 8.15. This bears out the original hypothesis.

TABLE 8-2 Fourier Amplitudes and System Parameters for Different Harmonics of the Input Waveform

n	$n\omega_1$	V_{1n}	$\|G(jn\omega_1)\|$	$\angle G(jn\omega_1)$	V_{2n}
1	3×10^4	$32/\pi$	1.65×10^{-2}	90°	0.168
3	9×10^4	$32/3\pi$	23×10^{-2}	76.6°	0.783
5	15×10^4	$32/5\pi$	6×10^{-2}	−59°	0.1220
7	21×10^4	$32/7\pi$	3.17×10^{-2}	−88.2°	0.046

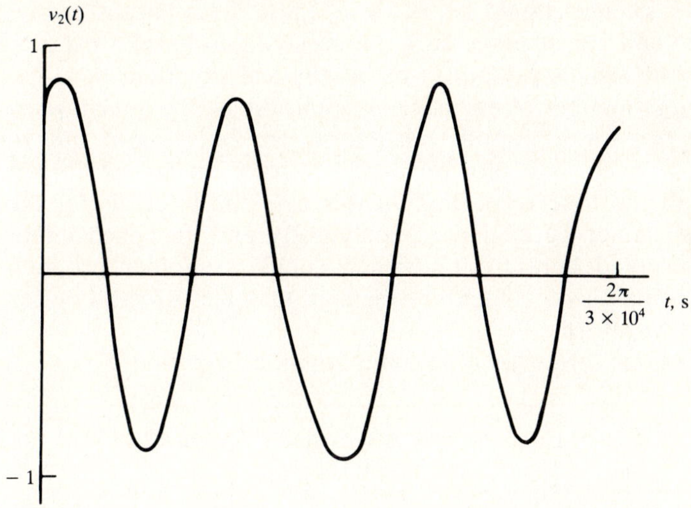

FIGURE 8.14 Sum of seven terms of output voltage series.

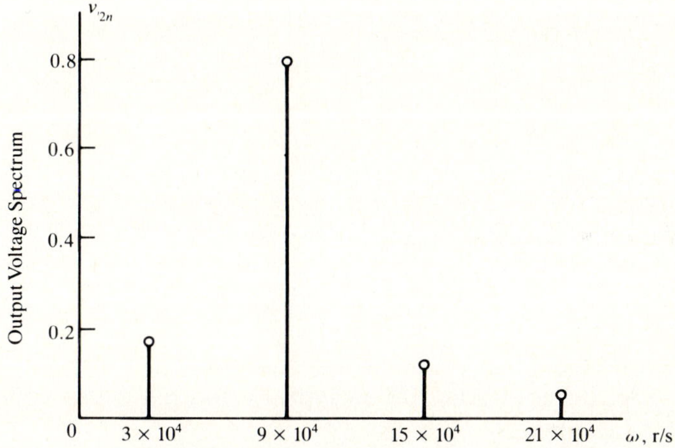

FIGURE 8.15 Output voltage spectrum.

8.10 NUMERICAL FOURIER ANALYSIS

Any modern coverage of Fourier analysis and applications would be remiss if it did not provide the student with an introduction to discrete Fourier analysis. It is from this introduction that comes the discrete Fourier transform (DFT) and the computationally more efficient fast Fourier transform

(FFT). The student should be aware of these ideas because many modern instruments and signal processors use these algorithms to extract information about the signals present in many physical situations.

In a large number of engineering applications, the analytical expression or expressions for a periodic waveform are not known but rather the waveform is given as a plot on paper, a trace on an oscilloscope, or an analog or digital tape recording. In this case the integrals for the Fourier coefficients cannot be evaluated analytically and an approximate method will be presented here from which a simple computer program can be written in a high-level computer language to solve for the Fourier coefficients of the waveform.

The Fourier series for a periodic function of period T is

$$f(t) = \frac{a_0}{2} + \sum_{n=1}^{\infty} [a_n \cos(n2\pi f_1 t) + b_n \sin(n2\pi f_1 t)] \qquad \textbf{(8.10.1)}$$

where the fundamental frequency f_1 in hertz is the reciprocal of the period T.

The integrals for the Fourier coefficients were shown to be

$$a_n = \frac{2}{T}\int_0^T f(t)\cos(n2\pi f_1 t)\,dt \qquad n = 0, 1, 2, \ldots \qquad \textbf{(8.10.2)}$$

and

$$b_n = \frac{2}{T}\int_0^T f(t)\sin(n2\pi f_1 t)\,dt \qquad n = 1, 2, 3, \ldots \qquad \textbf{(8.10.3)}$$

while the average value is given for $n = 0$ to be

$$\frac{a_0}{2} = \frac{1}{T}\int_0^T f(t)\,dt \qquad \textbf{(8.10.4)}$$

We, in this case, do not know $f(t)$ analytically and must now approximate the integrals numerically.

Consider the periodic function $f(t)$ which is periodic on the interval $(0, T)$ as shown in Fig. 8.16.

Subdivide the interval $(0, T)$ into N subintervals each of width Δ s. Denote the sample of $f(t)$ at time $i\Delta$ to be f_i, as shown. We can approximate the integral for the average value as

$$\frac{a_0}{2} \simeq \frac{1}{N\Delta}\sum_{i=1}^{N} f_i\Delta \qquad \textbf{(8.10.5)}$$

$$\simeq \frac{1}{N}\sum_{i=1}^{N} f_i \qquad \textbf{(8.10.6)}$$

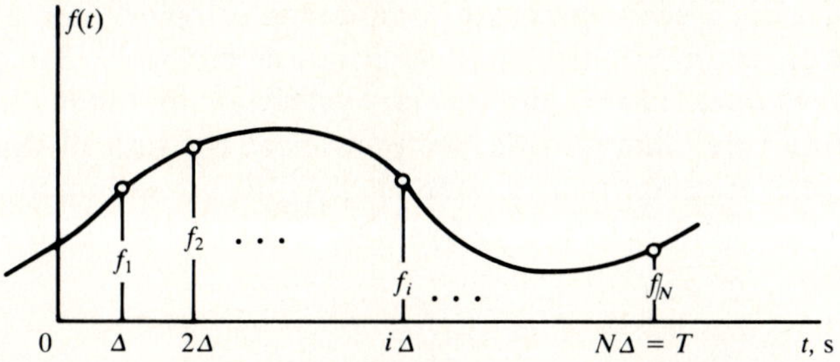

FIGURE 8.16 A periodic function.

which is simply the arithmetic mean of the samples taken from the waveform. Similarly, we can approximate the integral for a_n as

$$a_n \simeq \frac{2}{N\Delta} \sum_{i=1}^{N} f_i \cos\left(\frac{n2\pi i\Delta}{N\Delta}\right)\Delta \tag{8.10.7}$$

Cleaning up the algebra yields a simple sum for the estimate of a_n, which we shall term a_n [†]

$$\hat{a}_n = \frac{2}{N} \sum_{i=1}^{N} f_i \cos(2\pi in/N) \tag{8.10.8}$$

By a similar approximation, the estimated sine coefficients are

$$\hat{b}_n = \frac{2}{N} \sum_{i=1}^{N} f_i \sin(2\pi in/N) \tag{8.10.9}$$

Equations (8.10.6), (8.10.8), and (8.10.9) form simple algorithms for evaluation of Fourier coefficients.

We also know that the alternative series coefficients are given by

$$\hat{c}_n = \sqrt{\hat{a}_n^2 + \hat{b}_n^2} \tag{8.10.10}$$

and the estimated phase angle s by

$$\hat{\theta}_n = \tan^{-1}\left(\frac{\hat{b}_n}{\hat{a}_n}\right) \tag{8.10.11}$$

Because the trigonometric functions are periodic in frequency (as well as time) an attempt to evaluate more than $N/2$ harmonics from N samples per cycle of the waveform will yield repeating amplitudes and conjugate

[†]The circumflexes in these expressions denote the estimate of the actual quantity.

phase angles. This is a consequence of employing sampled data and cannot be remedied with more sophisticated integration methods. To get higher frequency spectral data, more samples per cycle must be taken. In fact, the sampling theorem (see Cheng in the References at the end of this chapter) states that the maximum frequency content, f_{max}, to be estimated in a signal is related to the sampling interval Δ by

$$\Delta < \frac{1}{2f_{max}} \tag{8.10.12}$$

Example 8.9

Consider the square wave shown in Fig. 8.17. The samples to be used in a 20-point ($N = 20$) harmonic analysis are given below as

i	1	2	3 . . . 9	10	11 . . . 19	20
f_i	1	1	1 . . . 1	0	−1 . . . −1	0

A simple BASIC program similar to that given in Appendix D will yield the results displayed in Table 8-3, along with those established by analytical integration.

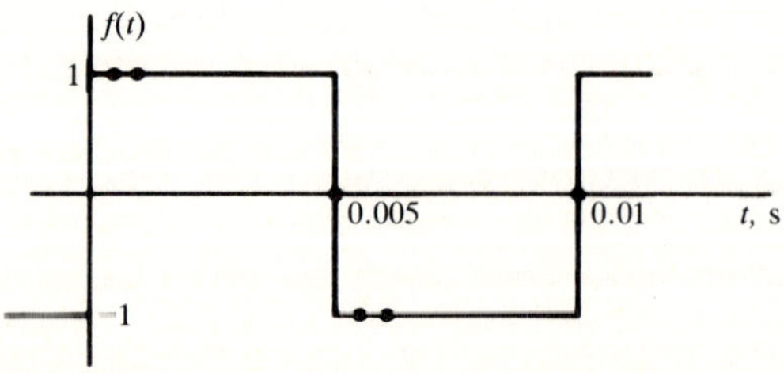

FIGURE 8.17 Sampled square wave.

TABLE 8-3 Numerical Fourier Data ($N = 20$)

n	nf_1	$\hat{a}_n$	a_n	$\hat{b}_n$	b_n
1	100	2.98×10^{-6}	0	1.263	1.273
2	200	3.58×10^{-7}	0	-1.431×10^{-7}	0
3	300	3.50×10^{-6}	0	0.393	0.424
4	400	-1.11×10^{-6}	0	-2.146×10^{-6}	0
5	500	1.22×10^{-6}	0	0.20	0.255
6	600	1.156×10^{-6}	0	6.91×10^{-7}	0

Note that the values which were supposed to be small are, in fact, small. However, there is some significant discrepancy between the b_n coefficients and their estimates $\hat{b}_n$. This discrepancy is due to the fact that the integrals for the coefficients were approximated with rectangular integration. The cure for this problem is to sample the waveform more frequently so that there are more samples per cycle of the waveform.

8.11 SUMMARY

In this chapter we have developed the Fourier series as a technique for the decomposition of periodic signals into a sum of sinusoids. This representation is valuable since functions which are often discontinuous are represented by a sum of continuous functions. The least-squares best-fit property is discussed and quantitative methods for assessing errors incurred due to truncation of the series are developed.

The steady-state response of a linear system driven by a periodic function is investigated and the Fourier series for the response function is given. A brief development of discrete-time Fourier analysis for periodic functions and the algorithm for approximate calculation of the Fourier coefficients are given.

REFERENCES

Cheng, D. K. *Analysis of Linear Systems*. Addison-Wesley, Reading, MA, 1959.

Nilsson, J. W. *Electric Circuits*. Addison-Wesley, Reading, MA, 1983.

Stanley, W. D., Dougherty, G. R. and Dougherty, R. *Digital Signal Processing*, 2d ed., Reston, Englewood Cliffs, NJ, 1984.

PROBLEMS

8-1. Evaluate the following integrals employing the orthogonality relations where they are applicable

(a) $$I_1 = \int_0^{\frac{2\pi}{10}} \cos 10t \cos 10t \, dt$$

(b) $$I_2 = \int_0^{\frac{2\pi}{10}} \sin 10t \cos 100t \, dt$$

(c) $$I_3 = \int_0^{\frac{2\pi}{10}} \sin 10t \cos 15t \, dt$$

8-2. For the sawtooth waveform shown below, prove that $a_0/2 = 1$, $a_1 = 0$, and $b_1 = -2/\pi$.

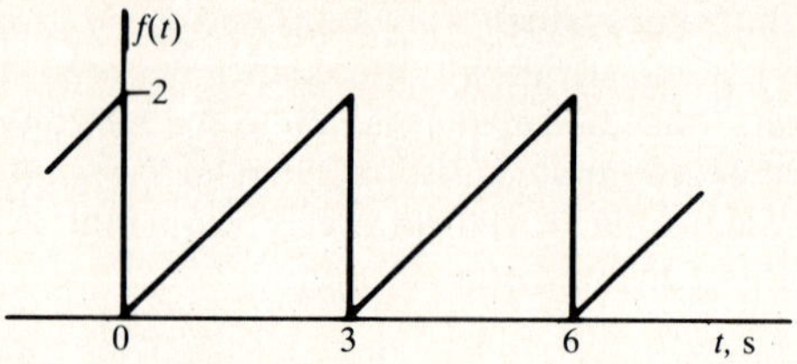

8-3. Find the Fourier coefficients for the triangular waveform illustrated below.

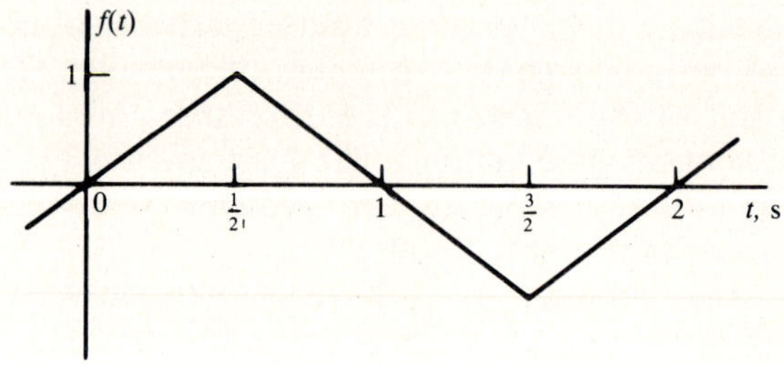

8-4. For the waveform below, find the Fourier coefficients $a_0/2$, a_1, a_2, a_3, b_1, and b_2. If you employ the symmetry conditions you must state which condition you apply.

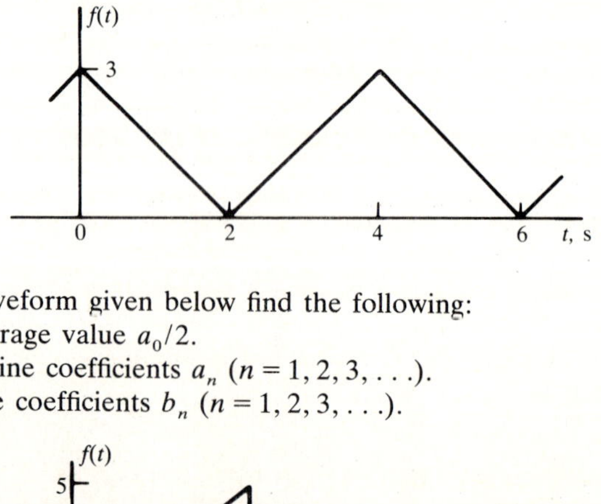

8-5. For the waveform given below find the following:
(a) The average value $a_0/2$.
(b) The cosine coefficients a_n $(n = 1, 2, 3, \ldots)$.
(c) The sine coefficients b_n $(n = 1, 2, 3, \ldots)$.

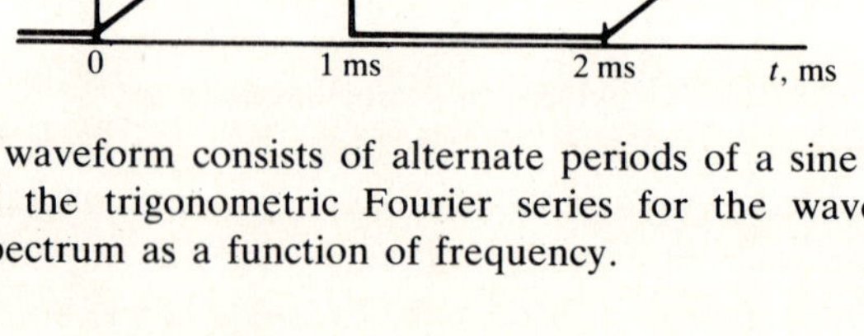

8-6. A recurrent waveform consists of alternate periods of a sine wave, as shown below. Find the trigonometric Fourier series for the waveform. Plot the amplitude spectrum as a function of frequency.

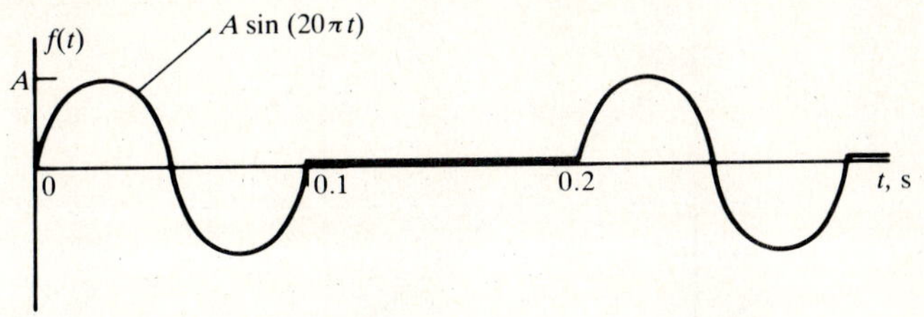

8-7. Find the trigonometric Fourier series for the half-wave rectified sine wave shown below. Sketch the spectrum by plotting c_n as a function of frequency.

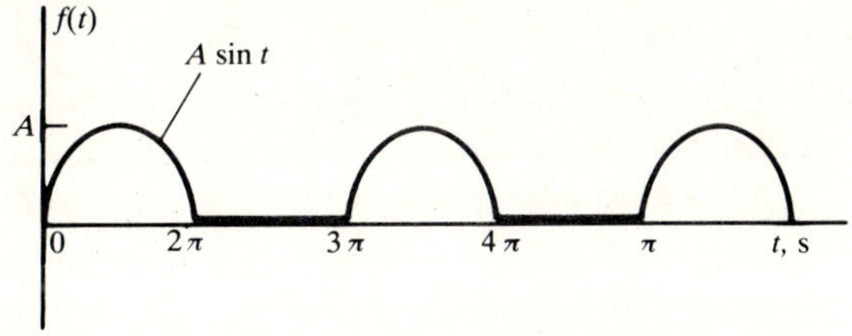

8-8. For the function made up of parabolic segments, find the Fourier coefficients $a_0/2$, a_n, and b_n. Also find c_0, c_n, and θ_n.

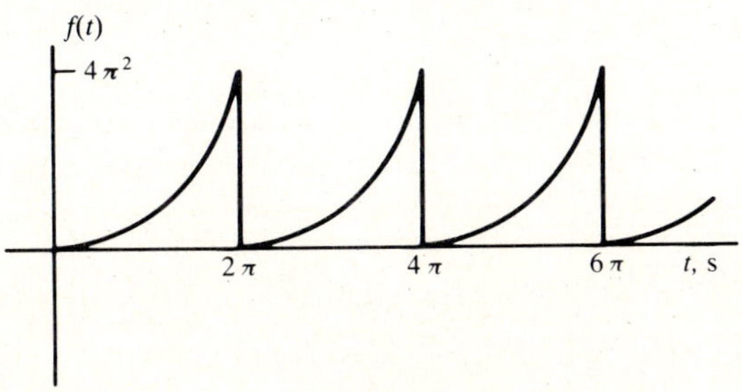

8-9. For the function shown, find the following:

(a) The mean square value F^2.

(b) The Fourier series for the waveform is

$$f(t) = \sum_{n=1}^{\infty} \frac{4}{n\pi} \sin n\pi t$$

Find the normalized mean square error if the series is truncated at one term. Repeat your calculations for a series truncated at two terms.

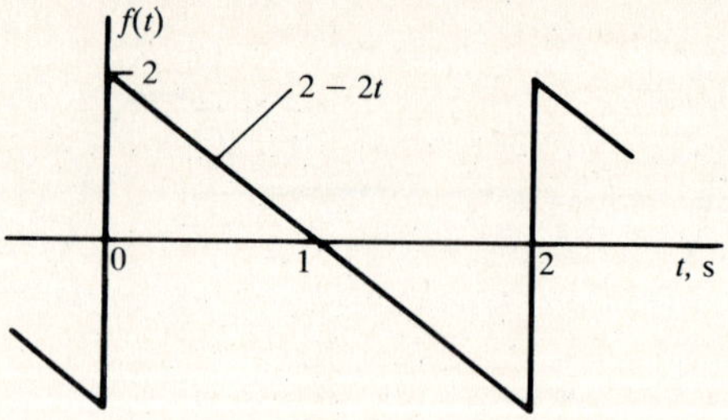

8-10. Comment on whether each of the following has even, odd, or half-wave symmetry.

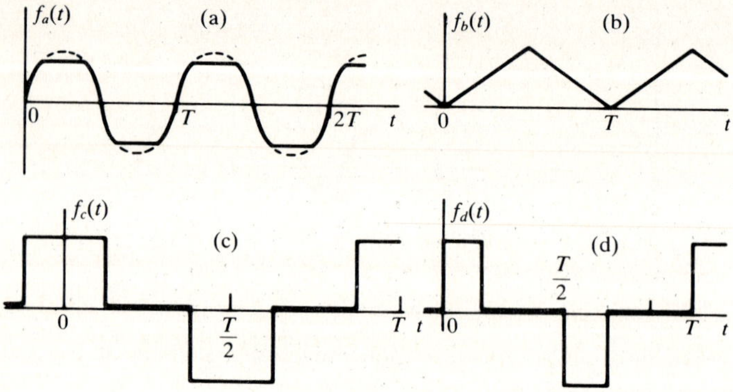

8-11. The full-wave rectified waveform shown below is described by the Fourier series

$$f(t) = \frac{2A}{\pi}\left[1 + \frac{2}{3}\cos(3141t) - \frac{2}{15}\cos(6283t) + \frac{2}{35}\cos(9423t)\ldots\right]$$

Find the alternate form of the Fourier series.

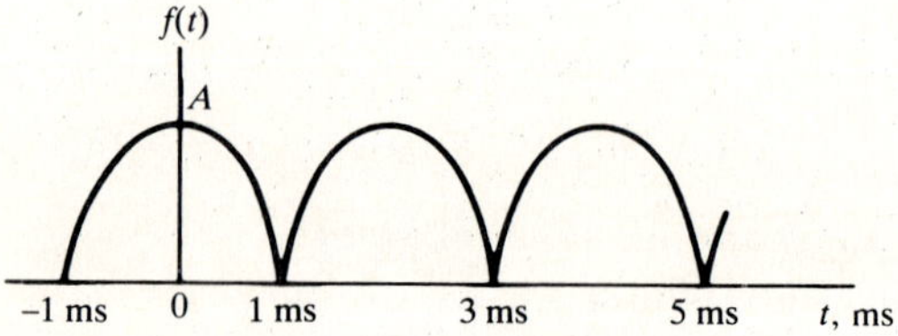

$$f(t) = c_0 + \sum_{n=1}^{\infty} c_n \cos(n\omega_1 t - \theta_n)$$

where the c_n are positive, given by

$$c_n = \sqrt{a_n^2 + b_n^2}$$

8-12. The square wave shown below is described by the Fourier series

$$y(t) = \frac{4}{\pi}\cos t + \frac{4}{3\pi}\cos(3t - \pi) + \frac{4}{5\pi}\cos(5t) + \cdots$$

(a) Find the Fourier series of the steady-state output of the integrator shown if the output $x(t)$ at $t = 0$ is zero.

(b) Sketch the output waveform $x(t)$ being sufficiently quantitative that values of $x(t)$ are known for particular values of t.

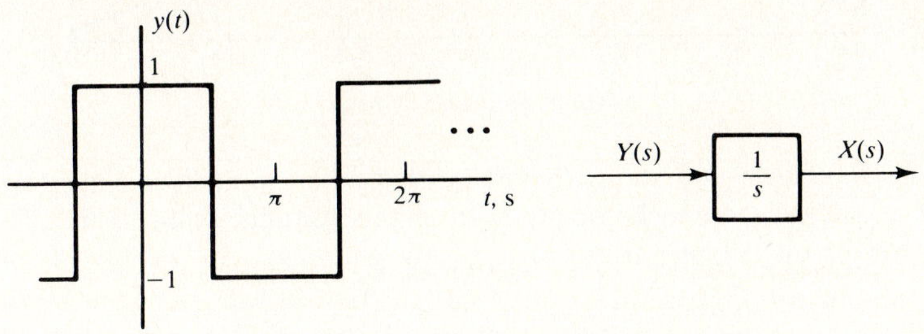

8-13. The square waveform shown below is represented by the Fourier series

$$v_1(t) = 4 + \sum_{n=1,3,5,\ldots}^{\infty} \frac{16}{n\pi}\cos\left(n50t - \frac{\pi}{2}\right)$$

This waveform is filtered by the lowpass filter shown. Find the Fourier series of the output waveform $v_2(t)$.

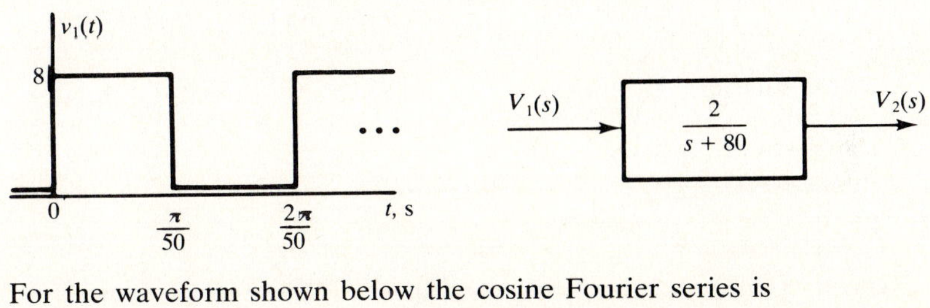

8-14. For the waveform shown below the cosine Fourier series is

$$y(t) = 2 + \sum_{n=1,3,5,\ldots}^{\infty} \frac{4}{n\pi}\cos\left(n800t + \frac{\pi}{2}\right)$$

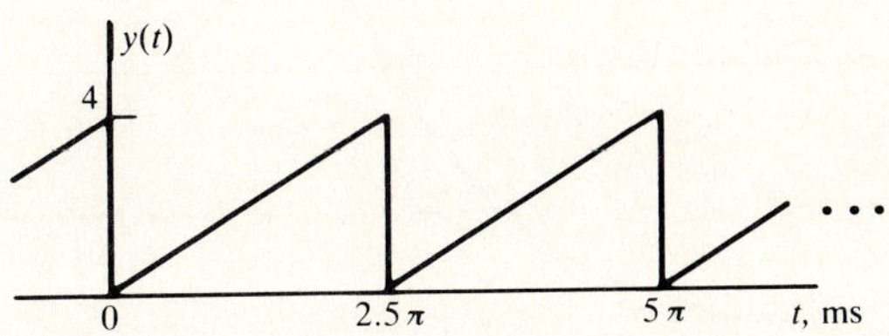

This waveform is to be filtered by a dynamic system with a frequency response function $G(j\omega)$ illustrated below. Write out the first three nonzero terms of the Fourier series of the steady-state filter output $x(t)$.

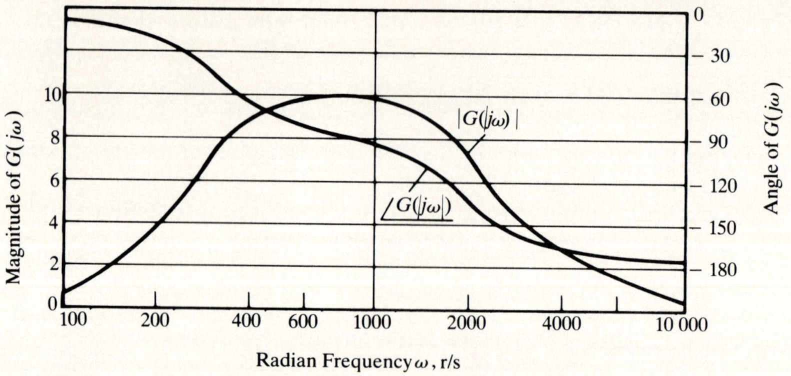

8-15. The waveform $y(t)$ shown below is used as an input to a linear system with frequency response character given in the accompanying figure. The Fourier series of the waveform is

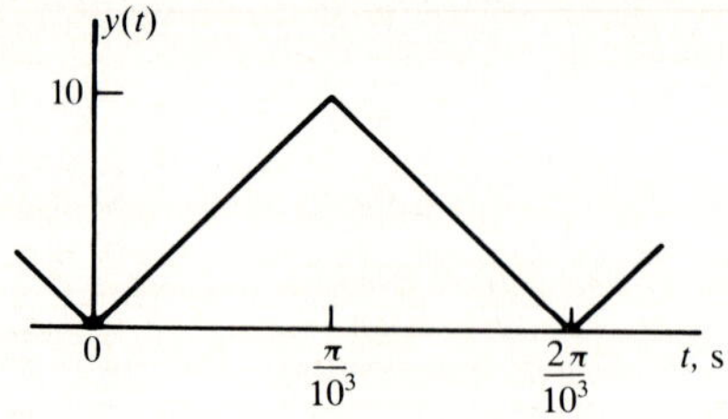

$$y(t) = 5 + \frac{40}{\pi^2}\left[\cos(1000t - \pi) + \frac{1}{9}\cos(3000t - \pi) + \frac{1}{25}\cos(5000t - \pi) + \cdots\right]$$

Find the Fourier series of the steady-state system response up through the fifth harmonic.

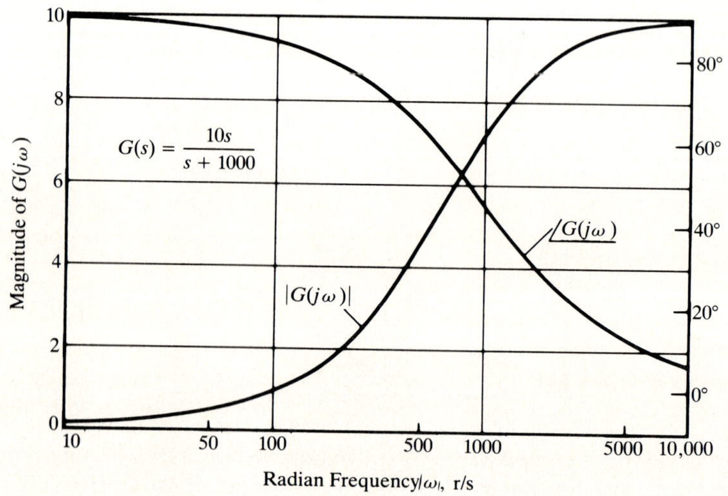

8-16. Using the numerical Fourier analysis program given in Appendix D you are to perform an approximate Fourier analysis on the waveform of Problem 8-5, employing 20 samples. Calculate the a_n, b_n, and c_n coefficients using the following samples as input data. Also plot the Fourier spectrum for the function.

i	1	2	3	4	5	6	7	8	9	10
f_i	0.5	1	1.5	2	2.5	3	3.5	4	4.5	2.5
i	11	12	13	14	15	16	17	18	19	20
f_i	0	0	0	0	0	0	0	0	0	0

8-17. When a person exercises strenuously, his or her pulse rate increases radically. Often a physiologist can assess the condition of the heart by examining the electrocardiogram (EKG). A more rational and scientific method of examination of the waveform involves the use of Fourier analysis. A typical exercise EKG signal is given below. **(a)** Perform a numerical Fourier analysis on the waveform samples given below employing the program given in Appendix D. **(b)** Plot the Fourier spectrum as a function of frequency in hertz. **(c)** Using this spectrum, indicate the bandwidth of an amplifier necessary to amplify this signal.

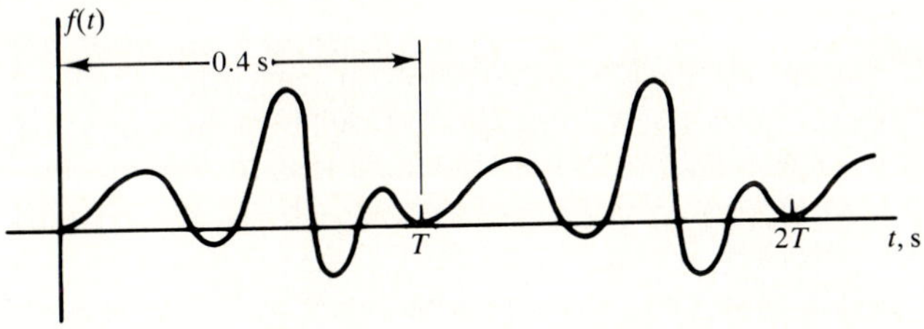

Data

Samples of the waveform taken every 0.01333 seconds (30 per cycle) are given below.

i	f_i	i	f_i	i	f_i
1	0.2	11	0.4	21	−0.1
2	0.35	12	0.2	22	0.2
3	0.5	13	−0.05	23	0.32
4	0.63	14	−0.15	24	0.4
5	0.75	15	0.0	25	0.5
6	0.85	16	0.6	26	0.5
7	0.9	17	2.0	27	0.4
8	0.85	18	0.65	28	0.18
9	0.72	19	0.05	29	0.0
10	0.6	20	−0.4	30	0.0

9

STATE VARIABLE REPRESENTATIONS IN LINEAR SYSTEMS

9.1 INTRODUCTION

In the late 1950s, new analytical tools were introduced to the field of dynamic systems, particularly in the study of automatic control systems. The state variable representation of a dynamic system introduced in this chapter is an alternative to the transfer function representation discussed in all previous chapters. This state representation provides a sound structure on which to base the system optimization or system simulation problem. This formulation is especially important because it is in a form readily applicable to the digital computer. Extensive software libraries exist to perform the computations needed in the solution of state variable problems and a number of special-purpose software packages are also available for certain types of these problems.

9.2 SOME STATE VARIABLE FORMULATIONS

We saw in earlier chapters that the transfer function representation was always equivalent to a differential equation of the same order as the denominator of the transfer function. We shall show here that the system may also be represented by a set of first-order differential equations which may then be written in matrix-vector form. A valuable mathematical tool for the study of such differential equations is linear algebra and matrix theory. The elements of matrix algebra are presented in Appendix C, which may be useful to those readers who have not previously used matrix notation.

A system with m input functions, $u_1(t), \ldots, u_m(t)$, can be described by the set of differential equations

$$\begin{aligned}
\dot{x}_1 &= a_{11}x_1 + a_{12}x_2 + \cdots + a_{1n}x_n + b_{11}u_1 + \cdots + b_{1m}u_m \\
\dot{x}_2 &= a_{21}x_1 + a_{22}x_2 + \cdots + a_{2n}x_n + b_{21}u_1 + \cdots + b_{2m}u_m \\
&\vdots \\
\dot{x}_n &= a_{n1}x_1 + a_{n2}x_2 + \cdots + a_{nn}x_n + b_{m1}u_1 + \cdots + b_{nm}u_m
\end{aligned} \tag{9.2.1}$$

The variables $x_i(t)$ are called the state variables. Generally the r output variables $y_i(t)$ will be linear combinations of the n state variables the m input functions or

$$\begin{aligned} y_1 &= c_{11}x_1 + \cdots + c_{1n}x_n + d_{11}u_1 + \cdots + d_{1m}u_m \\ y_2 &= c_{21}x_1 + \cdots + c_{2n}x_n + d_{21}u_1 + \cdots + d_{2m}u_m \\ &\vdots \\ y_r &= c_{r1}x_1 + \cdots + c_{rn}x_n + d_{r1}u_1 + \cdots + d_{rm}u_m \end{aligned} \quad (9.2.2)$$

In compact matrix-vector notation, equations (9.2.1) may be written as

$$\dot{\mathbf{x}}(t) = \mathbf{A}\mathbf{x}(t) + \mathbf{B}\mathbf{u}(t) \quad (9.2.3)$$

with initial condition vector $\mathbf{x}(0) = \mathbf{x}_0$. The output variables in vector form are

$$\mathbf{y}(t) = \mathbf{C}\mathbf{x}(t) + \mathbf{D}\mathbf{u}(t) \quad (9.2.4)$$

For a given problem, there is no unique way to assign the state variables but, once the state variables are chosen, the elements of the output matrix **C** are fixed. There are a number of standard techniques for assigning state variables (see Takahashi or Brogan in the References at the end of this chapter) for both transfer functions and for integro-differential equation descriptions.

To clear up some of the obscurity of the preceding discussion, several examples will serve us well.

Example 9.1

Let us derive a state variable model for the simple mechanical system shown in Fig. 9.1.

The equation of motion for this system is given by Newton's second law to be

$$m\ddot{y} + b\dot{y} + ky = u(t)$$

Let us solve for the highest order derivative to give

$$\ddot{y} = -\frac{k}{m}y - \frac{b}{m}\dot{y} + \frac{1}{m}u(t) \quad \textbf{(a)}$$

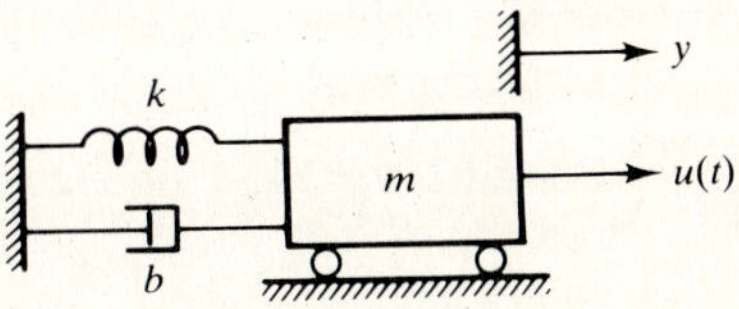

FIGURE 9.1 One degree of freedom mechanical system.

Since this is a second-order system, we will need two state variables. Let us define the first state variable to be the displacement or

$$x_1 = y \qquad \textbf{(b)}$$

and the second state variable to be the velocity or

$$x_2 = \dot{y} \qquad \textbf{(c)}$$

If we differentiate Eq. **(b)** and equate it to Eq. **(c)**, we get

$$\dot{x}_1 = x_2 \qquad \textbf{(d)}$$

We may now substitute the state variables into Eq. **(a)** to get

$$\dot{x}_2 = -\frac{k}{m}x_1 - \frac{b}{m}x_2 + \frac{1}{m}u(t) \qquad \textbf{(e)}$$

Relations **(d)** and **(e)** are the state representation of the system dynamics, which may be written in matrix-vector form as

$$\frac{d}{dt}\begin{bmatrix} x_1 \\ x_2 \end{bmatrix} = \begin{bmatrix} 0 & 1 \\ -k/m & -b/m \end{bmatrix}\begin{bmatrix} x_1 \\ x_2 \end{bmatrix} + \begin{bmatrix} 0 \\ 1/m \end{bmatrix} u(t)$$

It is now possible to identify the matrices **A** and **B** as, using Eq. (9.2.3),

$$\mathbf{A} = \begin{bmatrix} 0 & 1 \\ -k/m & -b/m \end{bmatrix} \qquad \mathbf{B} = \begin{bmatrix} 0 \\ 1/m \end{bmatrix}$$

If the output variable for this system is the position $y(t)$, then the output relation is

$$y(t) = [1 \quad 0]\begin{bmatrix} x_1(t) \\ x_2(t) \end{bmatrix}$$

This is the output relation similar to that of expression (9.2.4) so the **C** and **D** matrices are

$$\mathbf{C} = [1 \quad 0] \qquad \mathbf{D} = 0$$

In mechanical systems it is common to choose the state variables as the displacements and velocities, although in some problems other state variable assignments may prove more useful. State variable techniques are also useful in the modeling and simulation of electrical networks, although a general treatment of this is beyond the scope of this text. An example will illustrate a simple but typical application.

Example 9.2

Consider the two-loop electrical network shown in Fig. 9.2. If we employ the method of loop currents, using the loop currents assigned in Fig. 9.2 the loop equations are

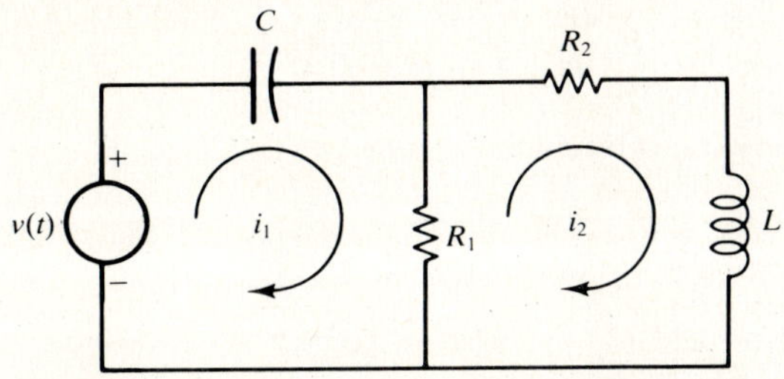

FIGURE 9.2 Two-loop electrical network.

$$\frac{1}{C}\int i_1\,dt + R_1(i_1 - i_2) = v(t)$$

$$R_2 i_2 + L\,\frac{di_2}{dt} + R_1(i_2 - i_1) = 0$$

These expressions are a perfectly adequate description of the network and would provide any information desired about the response of the network to a specified source voltage $v(t)$. Although it is seldom clear how to assign the state variables, one useful technique is that of letting the states be the charge on the capacitor and the current in the inductor. For the capacitor charge we have

$$x_1 = \int i_1\,dt$$

and for the inductor current we have

$$x_2 = i_2$$

Substitution of these relations into the original loop equations gives

$$\frac{1}{C}x_1 + R_1\left(\frac{dx_1}{dt} - x_2\right) = v(t) \qquad \textbf{(a)}$$

and

$$R_2 x_2 + L\,\frac{dx_2}{dt} + R_1\left(x_2 - \frac{dx_1}{dt}\right) = 0 \qquad \textbf{(b)}$$

The second of these equations is not in the correct form for a state equation since the derivatives of both state variables appear. This difficulty can be overcome by solving Eq. **(a)** for dx_1/dt or

$$\frac{dx_1}{dt} = -\frac{1}{R_1 C}x_1 + x_2 + \frac{1}{R_1}v(t) \qquad \textbf{(c)}$$

which is the correct form for a state equation. This expression can be substituted into Eq. **(b)** for dx_1/dt to give

$$R_2x_2 + L\frac{dx_2}{dt} + R_1\left[x_2 + \frac{1}{R_1C}x_1 - x_2 - \frac{1}{R_1}v(t)\right] = 0$$

which can now be solved for dx_2/dt.

$$\frac{dx_2}{dt} = \frac{-1}{LC}x_1 - \frac{R_2}{L}x_2 + \frac{1}{L}v(t) \quad \textbf{(d)}$$

Expressions **(c)** and **(d)** are now in correct state variable form which can be written in matrix form as

$$\frac{d}{dt}\begin{bmatrix} x_1 \\ x_2 \end{bmatrix} = \begin{bmatrix} \frac{-1}{R_1C} & 1 \\ \frac{-1}{LC} & \frac{R_2}{L} \end{bmatrix}\begin{bmatrix} x_1 \\ x_2 \end{bmatrix} + \begin{bmatrix} \frac{1}{R_1} \\ \frac{1}{L} \end{bmatrix} v(t)$$

The output could be any of the branch currents or voltages in the network. In some cases, to get an output equation of the form of Eq. (9.2.4), some substitution of the state equations may be necessary to eliminate the rate (d/dt) terms present.

9.3 LAPLACE TRANSFORM SOLUTION OF THE STATE EQUATION

9.3.1 Solution for Unforced Systems

Let us first consider the solution to the homogeneous (unforced) state equation which has response due only to initial stored energy reflected in a nonzero initial state vector. The system is

$$\dot{\mathbf{x}} = \mathbf{A}\mathbf{x} \qquad \mathbf{x}(0) = \mathbf{x}_0 \quad \textbf{(9.3.1)}$$

If we take the Laplace transform of this equation we get

$$s\mathbf{X}(s) - \mathbf{x}_0 = \mathbf{A}\mathbf{X}(s) \quad \textbf{(9.3.2)}$$

where $\mathbf{X}(s)$ is the Laplace transform of $\mathbf{x}(t)$. Upon solving for $\mathbf{X}(s)$ the result is

$$\mathbf{X}(s) = (s\mathbf{I} - \mathbf{A})^{-1}\mathbf{x}_0 \quad \textbf{(9.3.3)}$$

where $\mathbf{I}$ is the $n \times n$ identity matrix. Since the initial condition vector $\mathbf{x}_0$ is independent of s, it is sufficient that we find the inverse transform of the elements of the $(s\mathbf{I} - \mathbf{A})^{-1}$ matrix. We shall define the inverse Laplace transform of this matrix as the *state transition matrix* $\mathbf{\Phi}(t)$ or

$$\mathbf{\Phi}(t) = \mathcal{L}^{-1}[(s\mathbf{I} - \mathbf{A})^{-1}] \quad \textbf{(9.3.4)}$$

The solution to the differential equation (9.3.1) is then

$$\mathbf{x}(t) = \mathbf{\Phi}(t)\mathbf{x}_0 \tag{9.3.5}$$

We see from this expression that the matrix $\mathbf{\Phi}(t)$ dictates how the initial state $\mathbf{x}_0$ evolves into the current state $\mathbf{x}(t)$ and hence the name state transition matrix. An example should aid in understanding the exact mechanics of obtaining and using the state transition matrix.

Example 9.3

Consider the system of Example 9.1 with no forcing function $[u(t) = 0]$ and the following numerical values for the system parameters:

$$m = 1 \text{ kg}$$

$$b = 9 \text{ N} \cdot \text{s/m}$$

$$k = 20 \text{ N/m}$$

The $\mathbf{A}$ matrix from Example 9.1 is

$$\mathbf{A} = \begin{bmatrix} 0 & 1 \\ -20 & -9 \end{bmatrix}$$

Then

$$(s\mathbf{I} - \mathbf{A}) = \begin{bmatrix} s & -1 \\ 20 & s+9 \end{bmatrix}$$

The determinant of the matrix is

$$\Delta(s) = \begin{vmatrix} s & -1 \\ 20 & s+9 \end{vmatrix} = s^2 + 9s + 20$$

This polynomial may be written in factored form to give

$$\Delta(s) = (s+4)(s+5)$$

Then the matrix inverse from Appendix C is

$$(s\mathbf{I} - \mathbf{A})^{-1} = \frac{1}{\Delta(s)} \begin{bmatrix} s+9 & 1 \\ -20 & s \end{bmatrix}$$

and with substitution of $\Delta(s)$ the result is

$$(s\mathbf{I} - \mathbf{A})^{-1} = \begin{bmatrix} \dfrac{s+9}{(s+4)(s+5)} & \dfrac{1}{(s+4)(s+5)} \\ \\ \dfrac{-20}{(s+4)(s+5)} & \dfrac{s}{(s+4)(s+5)} \end{bmatrix}$$

If we invert each of the indicated Lapace transforms by making four separate partial fraction expansions, the result is

$$\mathcal{L}^{-1}[(s\mathbf{I}-\mathbf{A})^{-1}] = \mathbf{\Phi}(t) = \begin{bmatrix} 5e^{-4t} - 4e^{-5t} & e^{-4t} - e^{-5t} \\ -20e^{-4t} + 20^{-5t} & -4e^{-4t} + 5e^{-5t} \end{bmatrix}$$

Thus, if the initial condition vector were known, the solution to the problem would be simply a matter of multiplication of $\mathbf{\Phi}(t)$ times $\mathbf{x}_0$. If we take the following initial conditions $\mathbf{x}_0 = \begin{bmatrix} 1 \\ 2 \end{bmatrix}$, the solution vector is

$$\begin{bmatrix} x_1(t) \\ x_2(t) \end{bmatrix} = \begin{bmatrix} 7e^{-4t} - 6e^{-5t} \\ -28e^{-4t} + 30e^{-5t} \end{bmatrix}$$

These solutions are shown in Fig. 9.3.

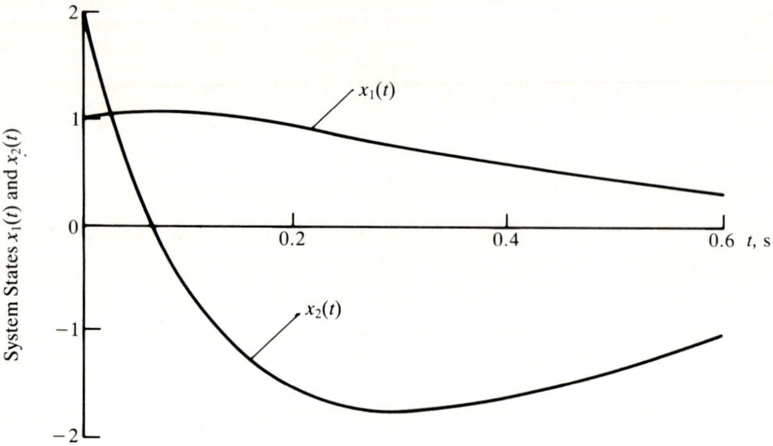

FIGURE 9.3 Responses of mass-spring-damper system to initial conditions $x_1(0) = 1$, $x_2(0) = 2$.

9.3.2 Solution for Forced Systems

Consider a system described by the forced state equation

$$\dot{\mathbf{x}} = \mathbf{A}\mathbf{x} + \mathbf{B}\mathbf{u} \tag{9.3.6}$$

As in the unforced problem, take the Laplace transform of Eq. (9.3.6) to give

$$s\mathbf{X}(s) - \mathbf{x}_0 = \mathbf{A}\mathbf{X}(s) + \mathbf{B}\mathbf{U}(s) \tag{9.3.7}$$

Now solve for $\mathbf{X}(s)$ to give

$$\mathbf{X}(s) = (s\mathbf{I} - \mathbf{A})^{-1}\mathbf{x}_0 + (s\mathbf{I} - \mathbf{A})^{-1}\mathbf{B}\mathbf{U}(s) \tag{9.3.8}$$

Note that the first term of this expression is the same as that given by the unforced system in relation (9.3.3). Now we must invert the indicated transforms using the definition of Eq. (9.3.4). We note that the second term of Eq. (9.3.8) is the product of two transform functions. The convolution theorem examined in Section 4.6 stated that the inverse transform of the product of two s-domain functions is the convolution of the inverse transform of one function into the inverse transform of the other or, upon using relation (9.3.4),

$$\mathbf{x}(t) = \mathbf{\Phi}(t)\mathbf{x}_0 + \int_0^t \mathbf{\Phi}(\tau)\mathbf{B}\mathbf{u}(t-\tau)\,d\tau \tag{9.3.9}$$

Now if the forcing function vector $\mathbf{u}(t)$ is known, the response of the state variables can be calculated by performing the integrations indicated by expression (9.3.9). An example will serve to clarify this matter.

Example 9.4

Consider the system examined in Examples 9.1 and 9.3, except now let the forcing function be a step force with a magnitude of 0.1 newtons.

$$u(t) = \begin{cases} 0.1\text{ N} & t \geq 0 \\ 0 & t < 0 \end{cases} \tag{a}$$

and initial conditions $x_1(0) = x_2(0) = 0$.

Let us first compute the matrix product in the integrand of relation (9.3.9):

$$\begin{aligned}\mathbf{\Phi}(\tau)\mathbf{B} &= \begin{bmatrix} 5e^{-4\tau} - 4e^{-5\tau} & e^{-4\tau} - e^{-5\tau} \\ -20e^{-4\tau} + 20e^{-5\tau} & -4e^{-4\tau} + 5e^{-5\tau} \end{bmatrix}\begin{bmatrix} 0 \\ 1 \end{bmatrix} \\ &= \begin{bmatrix} e^{-4\tau} - e^{-5\tau} \\ -4e^{-4\tau} + 5e^{-5\tau} \end{bmatrix}\end{aligned} \tag{b}$$

The forcing function which appears in the convolution theorem is $u(t-\tau)$ which, from Eq. **(a)**, is

$$u(t-\tau) = \begin{cases} 0.1\text{ N} & \tau \leq t \\ 0 & \tau > t \end{cases}$$

With the domain of integration being $0 \leq \tau \leq t$, the forcing function is 0.1 N everywhere within the interval of integration, so we may integrate Eq. **(b)** to yield

$$\begin{bmatrix} x_1(t) \\ x_2(t) \end{bmatrix} = 0.1 \int_0^t \begin{bmatrix} e^{-4\tau} - e^{-5\tau} \\ -4e^{-4\tau} + 5e^{-5\tau} \end{bmatrix} d\tau$$

$$= 0.1 \begin{bmatrix} \frac{1}{4}(1 - e^{-4t}) - \frac{1}{5}(1 - e^{-5t}) \\ -(1 - e^{-4t}) + (1 - e^{-5t}) \end{bmatrix}$$

These responses are shown in Fig. 9.4, where it must be noted that the final value of the displacement $x_1(\infty) = 5 \times 10^{-3}$ m.

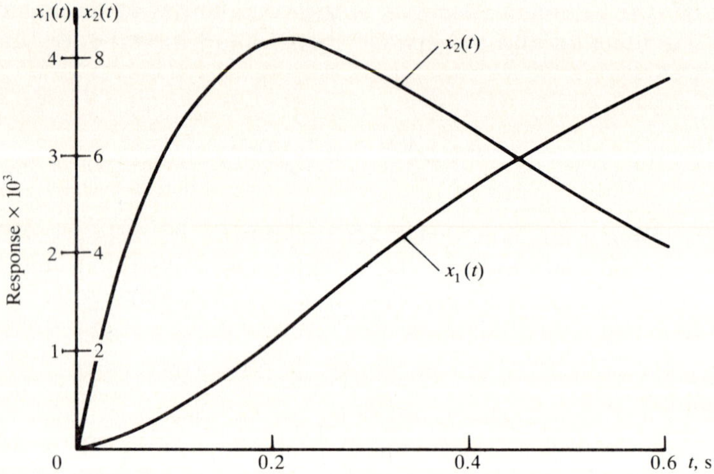

FIGURE 9.4 State responses of mass-spring-damper system with a stop input force.

In this example we have seen that it does not take a very complicated problem to lead to analytical difficulty employing transform methods. As a result, methods of solution of the state equation employing the digital computer must be sought. These techniques are the topics of several subsequent sections of this chapter.

In the example just examined the characteristic roots were real, thus making the evaluation of the state transition matrix rather straightforward. If, on the other hand, the roots were complex, the work required to evaluate the transition matrix would be roughly doubled.

9.4 A SIMPLE STATE VARIABLE DIGITAL SIMULATION SCHEME

In the previous section we saw that when the system was of high order the matrix inversion and transform inversions required may not be possible when doing hand calculations. If the state variable method is to be truly useful it must be such that it can be implemented for automatic computation by the digital computer.

We are still interested in the solution of

$$\dot{\mathbf{x}} = \mathbf{A}\mathbf{x} + \mathbf{B}\mathbf{u}(t) \tag{9.4.1}$$

subject to the initial condition $\mathbf{x}(0) = \mathbf{x}_0$. We shall now divide the time axis into small equal increments of T seconds in duration. We wish to evaluate the state variables at times T, $2T$, $3T$, ... We shall use sampled values of the forcing function, namely $\mathbf{u}(0)$, $\mathbf{u}(T)$, $\mathbf{u}(2T)$, . . . assuming that T is small enough that any fluctuations in $\mathbf{u}(t)$ are adequately represented in the samples.

If the time increment T is small, the derivative at time $t = kT$ can be approximated by the forward difference (Euler's approximation) so expression (9.4.1) can be approximated by

$$\frac{\mathbf{x}((k+1)T) - \mathbf{x}(kT)}{T} = \mathbf{A}\mathbf{x}(kT) + \mathbf{B}\mathbf{u}(kT) \tag{9.4.2}$$

If we multiply by T the result is

$$\mathbf{x}((k+1)T) - \mathbf{x}(kT) = \mathbf{A}T\mathbf{x}(kT) + \mathbf{B}T\mathbf{u}(kT) \tag{9.4.3}$$

Solving for $\mathbf{x}[(k+1)T]$ gives

$$\mathbf{x}((k+1)T) = (\mathbf{I} + \mathbf{A}T)\mathbf{x}(kT) + \mathbf{B}T\mathbf{u}(kT) \tag{9.4.4}$$

where $\mathbf{I}$ is the $n \times n$ identity matrix. If $\mathbf{x}(0)$ is specified and $\mathbf{u}(0)$, $\mathbf{u}(T)$, $\mathbf{u}(2T)$, . . . are known, then relation (9.4.4) can be solved sequentially forward in time to produce the sequence of states $\mathbf{x}(T)$, $\mathbf{x}(2T)$, . . . , which will be the approximate solution of the state equation (Eq. (9.4.1)). For the approximation of the derivative in Eq. (9.4.2) to be valid, the time increment T must be chosen to be a fraction of the fastest characteristic (time constant or natural period). However, if T is chosen too small, excessive computation time will be needed to get the solution over a useful span of time. We shall now demonstrate the use of expression (9.4.4) to evaluate the state variable responses for a simple problem.

Example 9.5

Consider the system examined in Examples 9.1 and 9.3. Let us use the same parameters for the system as in Example 9.3, so the **A** and **B** matrices are, respectively,

$$\mathbf{A} = \begin{bmatrix} 0 & 1 \\ -20 & -9 \end{bmatrix} \qquad \mathbf{B} = \begin{bmatrix} 0 \\ 1 \end{bmatrix}$$

In Example 9.3 we found the characteristic roots of this system to be at $s = -4, -5$, so the two associated time constants are $0.25s$ and $0.2s$. We thus pick a good representative sample time to be $T = 0.02s$. The first matrix in Eq. (9.4.4) is

$$(\mathbf{I} + \mathbf{A}T) = \begin{bmatrix} 1 & 0.02 \\ -0.4 & 0.82 \end{bmatrix}$$

In Example 9.3, the forcing function $u(t)$ was taken to be constant $u(t) = 0.1\,\text{N}$, so the last term of Eq. (9.4.4) is

$$\mathbf{B}Tu(t) = \begin{bmatrix} 0 \\ 0.002 \end{bmatrix}$$

Employing these matrices to solve for the states for zero initial conditions, the responses of Fig. 9.5 were generated by a simple computer program. Note that they are reasonably close to the analytical solutions generated in Example 9.4. If one wishes a closer approximation, all that must be done is to reduce the time increment. It should be noted that, if this system were lightly damped, a much smaller time increment T would be needed to obtain results which compare favorably with analytical solutions.

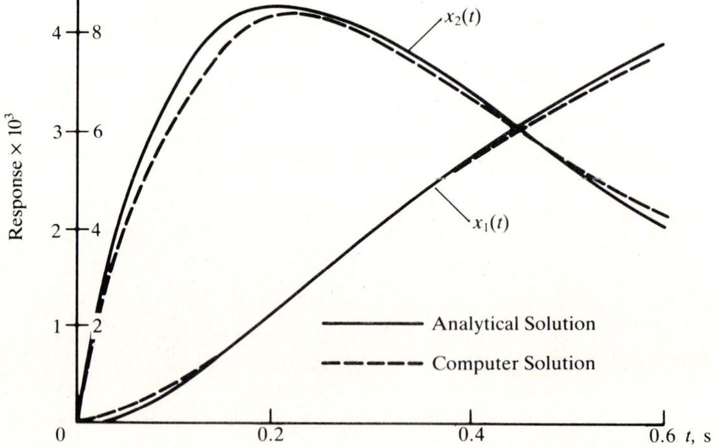

FIGURE 9.5 State variable responses of the forced overdamped vibratory system. Dashed curves represent actual analytical solutions.

9.5 AN INFINITE SERIES APPROACH TO THE SOLUTION OF THE STATE EQUATION

Let us first consider the solution of the unforced state equation where the response occurs because of stored energy reflected in an initial state vector which reflects the system being away from the equilibrium position ($\mathbf{x} = \mathbf{0}$). The problem is

$$\dot{\mathbf{x}} = \mathbf{A}\mathbf{x} \tag{9.5.1}$$

subject to a known initial condition $\mathbf{x}(0) = \mathbf{x}_0$. Most readers will have been exposed to the use of infinite series in the solution of scalar ordinary differential equations and a parallel approach will be taken here. Let us assume a power series solution to Eq. (9.5.1) of the form

$$\mathbf{x}(t) = \mathbf{a}_0 + \mathbf{a}_1 t + \mathbf{a}_2 t^2 + \cdots \tag{9.5.2}$$

where each of the vectors $\mathbf{a}_i$ is an unknown n-dimensional vector. If we differentiate Eq. (9.5.2), we get

$$\dot{\mathbf{x}}(t) = \mathbf{a}_1 + 2\mathbf{a}_2 t + 3\mathbf{a}_3 t^2 + \cdots \tag{9.5.3}$$

Substitution of Eq. (9.5.2) and Eq. (9.5.3) into the differential equation (Eq. (9.5.1)) yields

$$\mathbf{a}_1 + 2\mathbf{a}_2 t + 3\mathbf{a}_3 t^2 + \cdots = \mathbf{A}[\mathbf{a}_0 + \mathbf{a}_1 t + \mathbf{a}_2 t^2 + \cdots] \tag{9.5.4}$$

If this solution is to be valid for all t, then the like powers of the series must be equal so

$$\begin{aligned} t^0&: \mathbf{a}_1 = \mathbf{A}\mathbf{a}_0 \\ t^1&: 2\mathbf{a}_2 = \mathbf{A}\mathbf{a}_1 \\ t^2&: 3\mathbf{a}_3 = \mathbf{A}\mathbf{a}_2 \end{aligned} \tag{9.5.5}$$

and in general

$$t^{n-1}: n\mathbf{a}_n = \mathbf{A}\mathbf{a}_{n-1} \tag{9.5.6}$$

We now know $\mathbf{a}_1$ in terms of $\mathbf{a}_0$:

$$\mathbf{a}_1 = \mathbf{A}\mathbf{a}_0 \tag{9.5.7}$$

and now we can get $\mathbf{a}_2$ in terms of $\mathbf{a}_0$ or

$$\mathbf{a}_2 = \frac{1}{2}\mathbf{A}^2\mathbf{a}_0 \tag{9.5.8}$$

and in general

$$\mathbf{a}_n = \frac{1}{n!}\mathbf{A}^n\mathbf{a}_0 \tag{9.5.9}$$

and thus the solution is

$$\mathbf{x}(t) = \left[\mathbf{I} + \mathbf{A}t + \frac{\mathbf{A}^2t^2}{2!} + \frac{\mathbf{A}^3t^3}{3!} + \cdots\right]\mathbf{a}_0 \tag{9.5.10}$$

We may now evaluate the unknown vector $\mathbf{a}_0$ in terms of the initial condition vector $\mathbf{x}_0$ so at $t = 0$

$$\mathbf{a}_0 = \mathbf{x}_0 \tag{9.5.11}$$

The solution is then

$$\mathbf{x}(t) = \left[\mathbf{I} + \mathbf{A}t + \frac{1}{2!}\mathbf{A}^2t^2 + \frac{1}{3!}\mathbf{A}^3t^3 + \cdots\right]\mathbf{x}_0 \tag{9.5.12}$$

If we examine the form of this series and compare it to the scalar McLaurin series for an exponential function which is

$$e^{ax} = 1 + ax + \frac{1}{2!}a^2x^2 + \frac{1}{3!}a^3x^3 + \cdots \tag{9.5.13}$$

we see that they are of the same form and for that reason the series in the brackets of Eq. (9.5.12) is commonly called the *matrix exponential series* and given the symbol $\exp[\mathbf{A}t]$. We may then write the differential equation solution as

$$\mathbf{x}(t) = \exp[\mathbf{A}t]\mathbf{x}_0 \tag{9.5.14}$$

If we compare Eq. (9.5.14) with Eq. (9.3.5), we see that the matrix $\exp[\mathbf{A}t]$ and the state transition matrix $\mathbf{\Phi}(t)$ are one and the same or

$$\mathbf{\Phi}(t) = \exp[\mathbf{A}t] = \sum_{i=0}^{\infty}\frac{\mathbf{A}^it^i}{i!} \tag{9.5.15}$$

It can be shown that for any fixed t the series is convergent and thus, beyond some upper limit, it may be truncated for any desired degree of accuracy. Unfortunately, if the solution $\mathbf{x}(t)$ is desired at a large value of t, more terms will be necessary in the series, making the series method, so far, unattractive. Also it is not generally possible to recognize the closed forms of individual matrix elements which are themselves infinite series. Let us write expression (9.5.14) between t and some arbitrary starting point t_0 or

$$\mathbf{x}(t) = \exp[\mathbf{A}(t - t_0)]\mathbf{x}(t_0) \tag{9.5.16}$$

Now let us divide the temporal axis into intervals T seconds long as we did in Section 9.4. Now let us apply Eq. (9.5.16) between point t_k and point t_{k+1}, separated by an interval of T seconds (i.e., $t_{k+1} - t_k = T$) so

$$\mathbf{x}((k+1)T) = \exp[\mathbf{A}T]\mathbf{x}(kT) = \mathbf{\Phi}(T)\mathbf{x}(kT) \tag{9.5.17}$$

Now the value of T in the matrix exponential series is fixed, so we can evaluate $\exp[\mathbf{A}T]$ by digital computer from the definition of Eq. (9.5.15) or

$$\exp[\mathbf{A}T] = \mathbf{I} + \mathbf{A}T + \frac{1}{2!}\mathbf{A}^2T^2 + \frac{1}{3!}\mathbf{A}^3T^3 + \cdots \tag{9.5.18}$$

and the series can now be safely truncated at some fixed number of terms. The exact number of terms is often arrived at experimentally by comparing the elements of successively larger numbers of terms until the changes in terms are less than some specific value.

Now let us consider the forced problem where $\mathbf{u}(t)$ is no longer zero. Let us assume that the input functions are sampled and held at the sampled value throughout the interval of T seconds. If T is small enough, the sampled and held version is an accurate representation of $\mathbf{u}(t)$, as illustrated in Fig. 9.6. With this simplification, the input may be treated as constant on each interval. Expression (9.3.9) was derived for calculation of the forced response of the state equation by convolution.

Let us apply this expression across the interval from $t = kT$ to $t = (k+1)T$ so

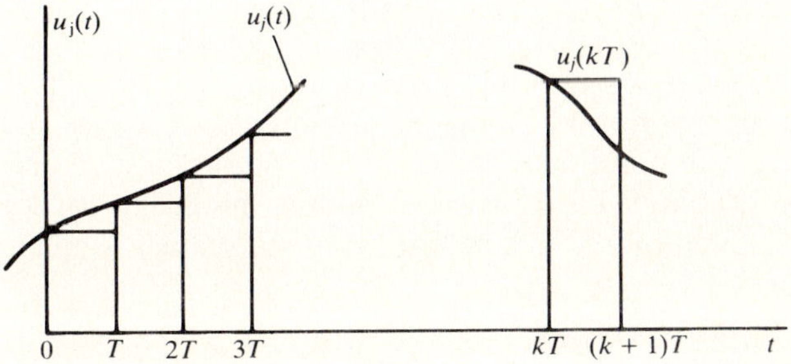

FIGURE 9.6 Rectangular pulse approximation of a forcing function $u_j(t)$.

$$\mathbf{x}((k+1)T) = \exp(\mathbf{A}T)\mathbf{x}(kT) + \int_{kT}^{(k+1)T} \exp(\mathbf{A}\tau)\mathbf{B}\mathbf{u}((k+1)T - \tau)\, d\tau \tag{9.5.19}$$

Let us assume that each element $u_j(t)$ of the forcing function vector can be approximated by a series of rectangular pulses, as illustrated in Fig. 9.6. This will be a good approximation if the time interval T is smaller than the time scale of variation of $u_j(t)$. If $\mathbf{u}(t)$ is assumed fixed over the interval $kT < t < (k+1)T$, then the term $\mathbf{Bu}(kT)$ may be factored from the integral to the right to give

$$\mathbf{x}((k+1)T) = \exp(\mathbf{A}T)\mathbf{x}(kT) + \int_{kT}^{(k+1)T} \exp(\mathbf{A}\tau)\, d\tau\, \mathbf{B}\mathbf{u}(kT) \tag{9.5.20}$$

Let us define a matrix $\mathbf{\Gamma}(T)$ to be

$$\mathbf{\Gamma}(T) = \int_{kT}^{(k+1)T} \exp(\mathbf{A}\tau)\, d\tau\, \mathbf{B} \tag{9.5.21}$$

Now let us make a change of variable by letting

$$\tau = \lambda + kT \tag{9.5.22}$$

and make the appropriate change in the limits to get

$$\mathbf{\Gamma}(T) = \int_{0}^{T} \exp(\mathbf{A}\lambda)\, d\lambda\, \mathbf{B} \tag{9.5.23}$$

If we substitute the series of Eq. (9.5.18) and integrate term-by-term, the result is[†]

$$\mathbf{\Gamma}(T) = \sum_{i=0}^{\infty} \frac{1}{(i+1)!} \mathbf{A}^i T^{i+1} \mathbf{B} \tag{9.5.24}$$

Now expression (9.5.20) may be rewritten as the following matrix-vector difference equation:

$$\mathbf{x}(k+1) = \mathbf{\Phi}(T)\mathbf{x}(k) + \mathbf{\Gamma}(T)\mathbf{u}(k) \tag{9.5.25}$$

where T has been omitted from the arguments of $\mathbf{x}$ and $\mathbf{u}$ for a concise notation. The computational steps in simulating a dynamic system are illustrated in Table 9-1. A microcomputer program to solve the state equation for a single forcing function ($u(t)$ is a scalar) is given in Appendix D.

[†]To integrate term by term we must interchange the order of summation and integration, which is valid in this case because we can show that the series is convergent.

TABLE 9-1 Steps in Simulation of a Dynamic System Employing Relation (9.5.25)

1. Input sampling interval T and the number of series terms N.
2. Input **A** and **B** matrices.
3. Calculate $\mathbf{\Phi}(T)$ and $\mathbf{\Gamma}(T)$ using Eqs. (9.5.18) and (9.5.24).
4. Input initial conditions $\mathbf{x}(0)$ and number of steps $k_{\max}$.
5. Input samples $\mathbf{u}(0)$, $\mathbf{u}(1)$, $\mathbf{u}(2), \ldots, \mathbf{u}(k_{\max}-1)$.
6. Let $k = 0$.
7. Let $k = k + 1$.
8. Evaluate $\mathbf{x}(k) = \mathbf{\Phi}(T)\mathbf{x}(k-1) + \mathbf{\Gamma}(T)\mathbf{u}(k-1)$.
9. Output kT, $\mathbf{x}_1(k), \ldots, \mathbf{x}_n(k)$.
10. If $k \le k_{\max}$ go to step 7.
11. End.

Example 9.6

Consider the problem solved in Example 9.5 for which the system matrices were

$$\mathbf{A} = \begin{bmatrix} 0 & 1 \\ -20 & -9 \end{bmatrix} \qquad \mathbf{B} = \begin{bmatrix} 0 \\ 1 \end{bmatrix}$$

Find the state transition matrix $\mathbf{\Phi}(T)$ and the matrix $\mathbf{\Gamma}(T)$ for a sampling interval of $T = 0.02$ s. Using the computer program of Appendix D.7 and a seven-term matrix exponential series the results are

$$\mathbf{\Phi}(0.02) = \begin{bmatrix} 0.9962 & 0.0183 \\ -0.3656 & 0.8317 \end{bmatrix}$$

and

$$\mathbf{\Gamma}(0.02) = \begin{bmatrix} 1.884 \times 10^{-4} \\ 0.01827 \end{bmatrix}$$

These matrices are better approximations than the two crude approximations given in Example 9.5.

If expression (9.5.25) is employed to solve the problem of Example 9.5 for which the forcing function is a step of 0.1 N, the numerical results are indistinguishable from the data presented for the exact solution in Fig. 9.4.

9.6 STATE VARIABLE REPRESENTATION OF TRANSFER FUNCTIONS

In Section 5.6 we considered the analog computer simulation of transfer functions of the form

$$G(s) = \frac{Y(s)}{U(s)} = \frac{b_{n-1}s^{n-1} + \cdots + b_1 s + b_0}{s^n + a_{n-1}s^{n-1} + \cdots + a_1 s + a_0} \tag{9.6.1}$$

We demonstrated in Section 5.6 that this problem could be decomposed into two problems, namely a differential equation to be associated with the denominator

$$w^n + a_{n-1}w^{n-1} - 1 + \cdots + a_1\dot{w} + a_0 w = u(t) \tag{9.6.2}$$

and an output relation associated with the numerator or

$$y(t) = b_0 w + b_1 \dot{w} + \cdots + b_{n-1}w^{n-1} \tag{9.6.3}$$

Let us now define the state variables to be $w(t)$ and its successive derivatives up to order $n-1$ or

$$x_1 = w \tag{9.6.4}$$

$$x_2 = \dot{w} = \dot{x}_1 \tag{9.6.5}$$

$$x_3 = \ddot{w} = \dot{x}_2 \tag{9.6.6}$$

and finally

$$x_n = w^{n-1} = \dot{x}_{n-1} \tag{9.6.7}$$

With these definitions we may write the family of state equations as

$$\dot{x}_1 = x_2 \tag{9.6.8}$$

$$\dot{x}_2 = x_3 \tag{9.6.9}$$

$$\vdots$$

$$\dot{x}_{n-1} = x_n \tag{9.6.10}$$

and substitution of relations (9.6.4) through (9.6.7) into relation (9.6.2) gives the final state equation

$$\dot{x}_n = -a_0 x_1 - a_1 x_2 - \cdots - a_{n-1}x_n + u(t) \tag{9.6.11}$$

From relation (9.6.3) the output relation can be written as

$$y(t) = b_0 x_1 + b_1 x_2 + \cdots + b_{n-1}x_n \tag{9.6.12}$$

Relations (9.6.8) through (9.6.11) may be written in matrix form as

$$\frac{d}{dt}\begin{bmatrix} x_1 \\ x_2 \\ \vdots \\ x_n \end{bmatrix} = \begin{bmatrix} 0 & 1 & 0 & 0 & \cdots & 0 \\ 0 & 0 & 1 & 0 & \cdots & 0 \\ \vdots & \vdots & & & & \vdots \\ 0 & & & & 0 & 1 \\ -a_0 & -a_1 & & \cdots & & -a_{n-1} \end{bmatrix} \begin{bmatrix} x_1 \\ x_2 \\ \vdots \\ x_n \end{bmatrix} + \begin{bmatrix} 0 \\ 0 \\ \vdots \\ 0 \\ 1 \end{bmatrix} u(t) \tag{9.6.13}$$

and output relation (9.6.12) may be written as

$$y(t) = [b_0 b_1 \ldots b_{n-1}] \begin{bmatrix} x_1 \\ x_2 \\ \vdots \\ x_n \end{bmatrix} \tag{9.6.14}$$

Note that relations (9.6.13) and (9.6.14) are the same as relations (9.2.3) and (9.2.4), respectively, with $\mathbf{D} = \mathbf{0}$. Thus for any transfer function of order n, the **A**, **B**, and **C** matrices are of the form

$$\mathbf{A} = \begin{bmatrix} 0 & 1 & 0 & 0 & \cdots & 0 \\ 0 & 0 & 1 & 0 & \cdots & 0 \\ \vdots & \vdots & & & & \vdots \\ 0 & & & & 0 & 1 \\ -a_0 & -a_1 & -a_2 & \cdots & & -a_{n-1} \end{bmatrix} \tag{9.6.15}$$

$$\mathbf{B} = \begin{bmatrix} 0 \\ 0 \\ \vdots \\ 0 \\ 1 \end{bmatrix} \tag{9.6.16}$$

$$\mathbf{C} = [b_0 b_1 \ldots b_{n-1}] \tag{9.6.17}$$

A state variable model of any transfer function may now be written down by inspection, thus enabling us to simulate an arbitrary transfer function using the program given in Appendix D.

This method is just one of several commonly employed methods for assigning the state variables for a transfer function. In fact, there are an infinite number of ways to assign the states but this method is easier to apply than most others.

Example 9.7

Consider the transfer function example examined in Example 5.3 for which the transfer function $G(s)$ was

$$G(s) = \frac{Y(s)}{U(s)} = \frac{4s + 3}{s^2 + 2s + 5}$$

We may by inspection find the state variable matrices to be

$$\mathbf{A} = \begin{bmatrix} 0 & 1 \\ -5 & -2 \end{bmatrix}$$

and

$$\mathbf{B} = \begin{bmatrix} 0 \\ 1 \end{bmatrix}$$

with an output matrix

$$\mathbf{C} = [3 \quad 4]$$

Let us now simulate the step response of this transfer function, starting from zero initial condition and using the program given in Appendix D. Employing the **A**, **B**, and **C** matrices the $\mathbf{\Phi}(T)$ and $\mathbf{\Gamma}(T)$ matrices for $T = 0.5$ s for seven terms of the exponential series are

$$\mathbf{\Phi}(0.5) = \begin{bmatrix} 0.583 & 0.255 \\ -1.276 & 0.0725 \end{bmatrix}$$

$$\mathbf{\Gamma}(0.5) = \begin{bmatrix} 0.0834 \\ 0.2550 \end{bmatrix}$$

9.7 SUMMARY

In this chapter we have introduced the concept of the state space as a useful alternative representation of dynamic systems. This representation has been shown to be especially powerful because it lends itself to implementation on the digital computer for digital solutions of the governing differential equations. The role of the state transition matrix is explored and it is shown that for digital computer implementation the truncated matrix exponential series is useful for simulation purposes. The state space representation of transfer functions is also given and shown to be useful in the simulation of transfer functions. This exposure to the state space concept has been deliberately introductory to give the beginning student an idea of the power of these techniques.

REFERENCES

Dorf, R. C. *Time-Domain Analysis and Design of Control Systems*. Addison-Wesley, Reading, MA, 1965.

Takahashi, Y., Rabins, M. J. and Auslander, D. M. *Control and Dynamic Systems*. Addison-Wesley, Reading, MA, 1970.

DeRusso, P. M., Roy, R. F., and Close, C. M. *State Variables for Engineers*. John Wiley, New York, 1965.

Brogan, W. L. *Modern Control Theory*, 2nd ed. Prentice-Hall, Englewood Cliffs, N.J., 1985.

PROBLEMS

9-1. Given the differential equation

$$\ddot{y} + 4\dot{y} + 13y = u(t)$$

Find the state variable description by letting $x_1 = y$, and $x_2 = \dot{y}$. Find the state transition matrix $\mathbf{\Phi}(t)$ for the system.

9-2. Given the spring-mass system shown, write the two differential equations of motion. Make the following state variable assignments: $x_1 = y_1$, $x_2 = y_2$, $x_3 = \dot{y}_1$, and $x_4 = \dot{y}_2$; find the state variable description. Assume that the system outputs are the displacements $y_1(t)$ and $y_2(t)$.

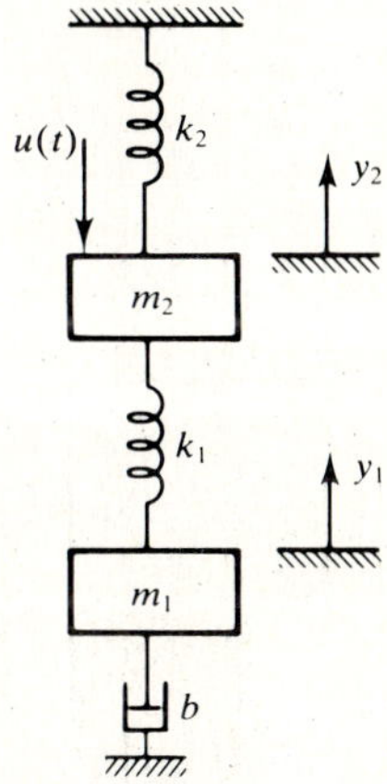

9-3. For the network shown, write the node voltage equation for the nodes labeled $v_1(t)$ and $v_2(t)$. By letting the node voltages be the state variables, find the state variable description for the network by assuming that $v_2(t)$ is the system output $y(t)$.

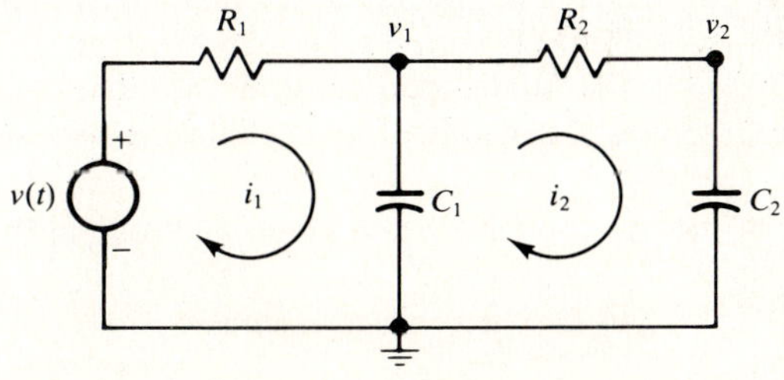

9-4. For the network given in Problem 9-3 write the two loop equations employing the mesh currents shown. Define the state variables to be the charges on the respective capacitors such that

$$x_1 = q_1 = \int i_1 - i_2 \, dt$$

and

$$x_2 = q_2 = \int i_2 \, dt$$

Eliminate the currents from the loop equations employing these definitions. Find a proper state description of the system by substitution of one equation into the other to get a single rate (d/dt) term in each equation. Assume that the system output is node voltage $v_2(t)$.

9-5. In Chapter 7 we explored electromechanical concepts and derived the governing equations for the dc motor shown below which are

$$J \frac{d\omega}{dt} + b\omega = K_t i - T_L(t)$$

$$L \frac{di}{dt} + Ri + K_b \omega = v(t)$$

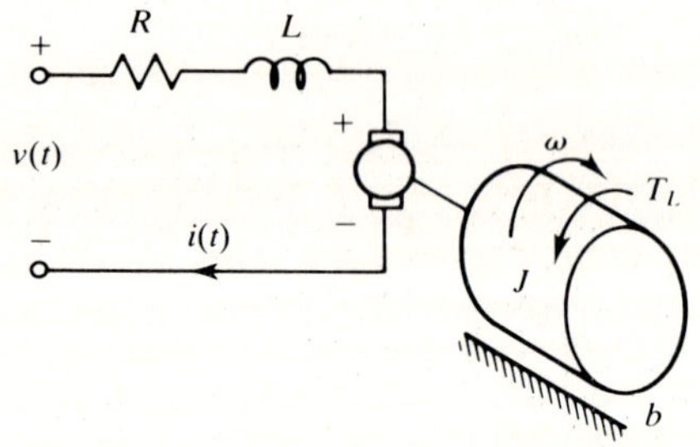

Assign the state variables to be $x_1 = \omega$ and $x_2 = i$. This version of the problem has two forcing functions, namely the applied armature voltage $v(t)$ and load torque $T_L(t)$. Find a state variable model for the motor.

9-6. Develop a computer program to solve the state equations derived in Problem 9-1 employing the Euler or forward-difference method examined in Section 9.4. Assume that the input function is $u(t) = 2e^{-t}$ and that the initial conditions are $y(0) = \dot{y}(0) = 0$. Since the undamped natural frequency is $\sqrt{13}$ r/s and the associated period is about 2 seconds, it makes sense to choose a time interval of $T = 0.1$ s.

9-7. The equations of motion for the system given in Problem 9-2 are

$$m_1\ddot{y}_1 + k_1(y_1 - y_2) + b\dot{y}_1 = 0$$

$$m_2\ddot{y}_2 + k_1(y_2 - y_1) + k_2 y_2 = u(t)$$

Put these equations in state variable form by assigning the states as $x_1 = y_1$, $x_2 = y_2$, $x_3 = \dot{y}_1$, and $x_4 = \dot{y}_2$. For the following parameter values formulate the numerical state variable model:

$$m_1 = 1\ \text{kg} \qquad k_1 = 1000\ \text{N/m} \qquad b = 400\ \text{N}\cdot\text{s/m}$$

$$m_2 = 2\ \text{kg} \qquad k_2 = 2000\ \text{N/m}$$

Simulate the system on the digital computer for a constant forcing function of $u(t) = 20$ N and zero initial conditions in all the state variables. Use a time interval of $T = 0.02$ s.

9-8. For the dc motor considered in Problem 9-5, the following are typical parameters:

$$J = 0.05\ \text{kg}\cdot\text{m}^2$$

$$b = 0.03\ \text{N}\cdot\text{m}\cdot\text{s/r}$$

$$K_t = 0.4\ \text{N}\cdot\text{m/A}$$

$$K_b = 0.4\ \text{V}\cdot\text{s/r}$$

$$L = 20\ \text{mH}$$

$$R = 5\ \Omega$$

Write a digital computer program to simulate this system for zero-load torque and a suddenly applied armature voltage of 20 V. Use a time interval of $T = 0.001$ s.

APPENDIX A

TABLES OF LAPLACE TRANSFORMS AND PROPERTIES

TABLE A-1 Laplace Transforms of Operations

No.	Operation	$f(t)$	$F(s) = \mathscr{L}[f(t)]$
1.	Definition	$f(t)$	$\int_0^\infty f(t)e^{-st}\,dt$
2.	First Derivative	$\frac{df}{dt}$	$sF(s) - f(0)$
3.	Second Derivative	$\frac{d^2f}{dt^2}$	$s^2F(s) - sf(0) - f'(0)$
4.	nth Derivative	$\frac{d^nf}{dt^n}$	$s^nF(s) - \sum_{i=0}^{n-1} s^i \frac{d^{n-i-1}f(0)}{dt^{n-i-1}}$
5.	Indefinite Integral	$\int f(t)\,dt$	$\frac{1}{s}[F(s) + f^{-1}(0)]$†
6.	Definite Integral	$\int_0^t f(t)\,dt$	$\frac{1}{s}F(s)$
7.	Multiplication by t	$tf(t)$	$-\frac{d}{ds}F(s)$
8.	Multiplication by e^{-at}	$e^{-at}f(t)$	$F(s+a)$
9.	Change of Scale	$f(at)$	$\frac{1}{a}F\left(\frac{s}{a}\right)$
10.	Delay	$f(t-a)u(t-a)$	$e^{-as}F(s)$
11.	Initial Value	$\lim_{t\to 0} f(t)$	$\lim_{s\to\infty} sF(s)$
12.	Final Value	$\lim_{t\to\infty} f(t)$	$\lim_{s\to 0} sF(s)$
13.	Convolution	$\int_0^t f_1(\tau)f_2(t-\tau)\,d\tau$	$F_1(s)F_2(s)$
14.	Periodic Function	$f(t) = \sum_{n=0}^{\infty} f_1(t-nT)$	$\frac{F_1(s)}{1-e^{-Ts}}$

†The term $f^{-1}(0)$ is defined by the relation

$$f^{-1}(0) = \int_{-\infty}^{0} f(t)\,dt$$

TABLE A-2 Laplace Transforms of Functions

No.	$f(t)$, $t>0$	$F(s)=\mathscr{L}[f(t)]$
1.	$\delta(t)$	1
2.	$u(t)$	$\dfrac{1}{s}$
3.	t	$\dfrac{1}{s^2}$
4.	t^n	$\dfrac{n!}{s^{n+1}}$
5.	e^{-at}	$\dfrac{1}{s+a}$
6.	$\dfrac{1}{a}(1-e^{-at})$	$\dfrac{1}{s(s+a)}$
7.	$\sin \omega t$	$\dfrac{\omega}{s^2+\omega^2}$
8.	$\cos \omega t$	$\dfrac{s}{s^2+\omega^2}$
9.	te^{-at}	$\dfrac{1}{(s+a)^2}$
10.	$\dfrac{1}{b-a}[e^{-at}-e^{-bt}]$	$\dfrac{1}{(s+a)(s+b)}$
11.	$\dfrac{1}{a-b}[ae^{-at}-be^{-bt}]$	$\dfrac{s}{(s+a)(s+b)}$
12.	$\dfrac{t^{n-1}}{(n-1)!}e^{-at}$	$\dfrac{1}{(s+a)^n}$
13.	$e^{-at}\sin \omega t$	$\dfrac{\omega}{(s+a)^2+\omega^2}$
14.	$e^{-at}\cos \omega t$	$\dfrac{s+a}{(s+a)^2+\omega^2}$
15.	$t\sin \omega t$	$\dfrac{2\omega s}{(s^2+\omega^2)^2}$
16.	$t\cos \omega t$	$\dfrac{s^2-\omega^2}{(s^2+\omega^2)^2}$

APPENDIX B

FOURIER COEFFICIENTS FOR COMMONLY ENCOUNTERED PERIODIC FUNCTIONS

1. Square Wave

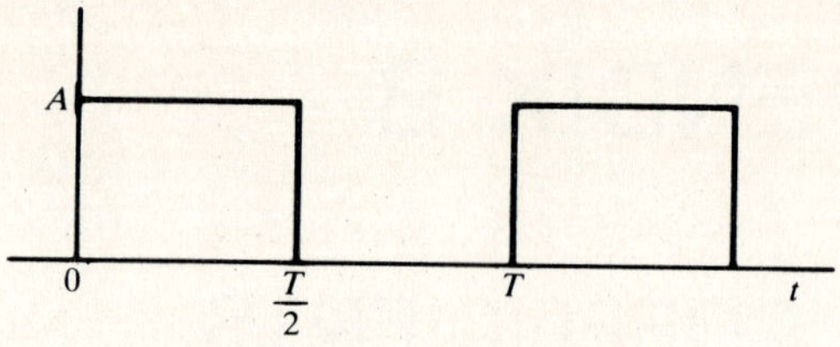

$$f(t) = \begin{cases} 0 & -\dfrac{T}{2} \le t \le 0 \\ A & 0 \le t \le \dfrac{T}{2} \end{cases}$$

$$\frac{a_0}{2} = \frac{A}{2}$$

$$a_n = 0$$

$$b_n = \begin{cases} \dfrac{2A}{n\pi} & n = 1, 3, 5, \ldots \\ 0 & n = 2, 4, 6, \ldots \end{cases}$$

2. Triangular Wave

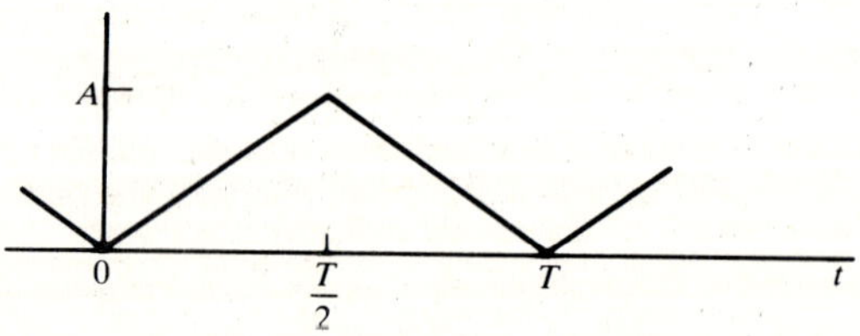

$$f(t) = \begin{cases} -\dfrac{2A}{T}\,t & -\dfrac{T}{2} \le t \le 0 \\ \dfrac{2A}{T}\,t & 0 \le t \le \dfrac{T}{2} \end{cases}$$

$$\frac{a_0}{2} = \frac{A}{2}$$

$$a_n = \begin{cases} \dfrac{-4A}{n^2\pi^2} & n = 1, 3, 5, \ldots \\ 0 & n = 2, 4, 6, \ldots \end{cases}$$

$$b_n = 0 \qquad \text{for all } n$$

3. Sawtooth Wave

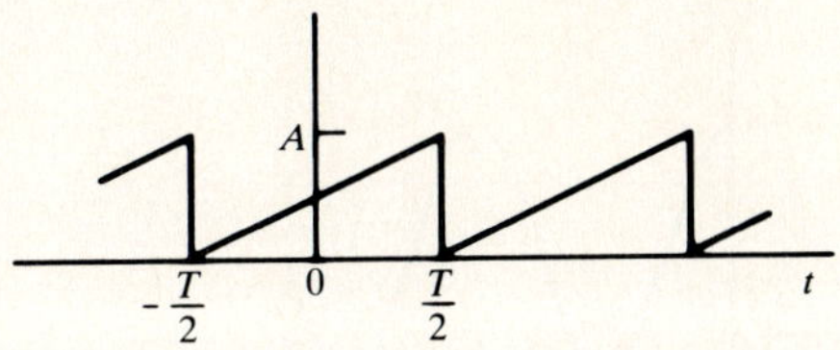

$$f(t) = \frac{At}{T} + \frac{A}{2} \qquad -\frac{T}{2} \le t \le \frac{T}{2}$$

$$\frac{a_0}{2} = \frac{A}{2}$$

$$a_n = 0 \qquad \text{for all } n$$

$$b_n = \frac{A}{n\pi}(-1)^{n+1} \qquad n = 1, 2, 3, \ldots$$

4. Sine Wave

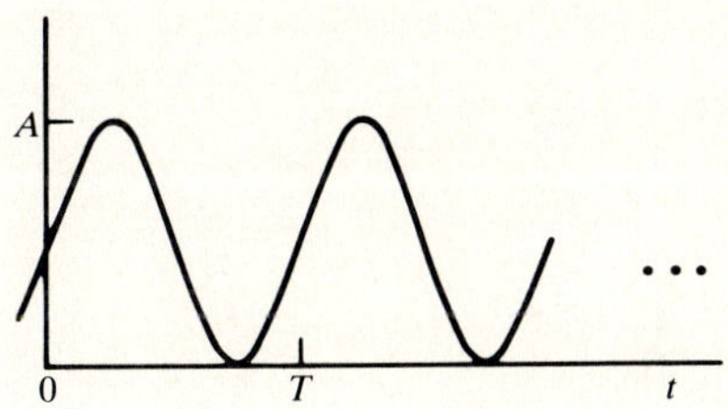

$$f(t) = \frac{A}{2} + \frac{A}{2} \sin \frac{2\pi}{T} t$$

$$\frac{a_0}{2} = \frac{A}{2}$$

$$a_n = 0$$

$$b_n = \begin{cases} \frac{A}{2} & n = 1 \\ 0 & n > 1 \end{cases}$$

5. Cosine Wave

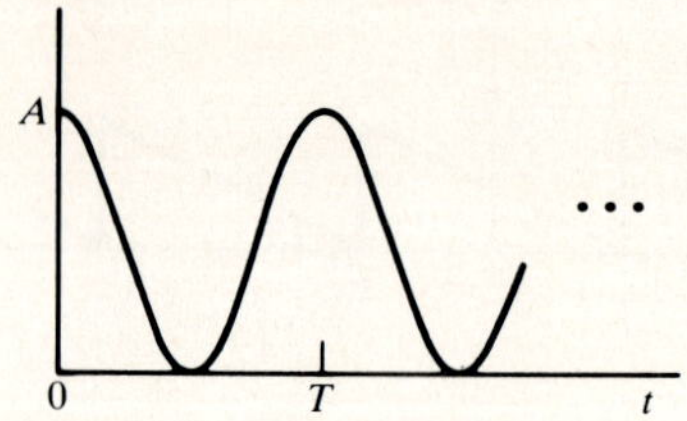

$$f(t) = \frac{A}{2} + \frac{A}{2}\cos\frac{2\pi}{T}t$$

$$\frac{a_0}{2} = \frac{A}{2}$$

$$a_n = \begin{cases} \dfrac{A}{2} & n = 1 \\ 0 & n > 1 \end{cases}$$

$$b_n = 0 \qquad \text{for all } n$$

6. Half-Wave Rectified Cosine Wave

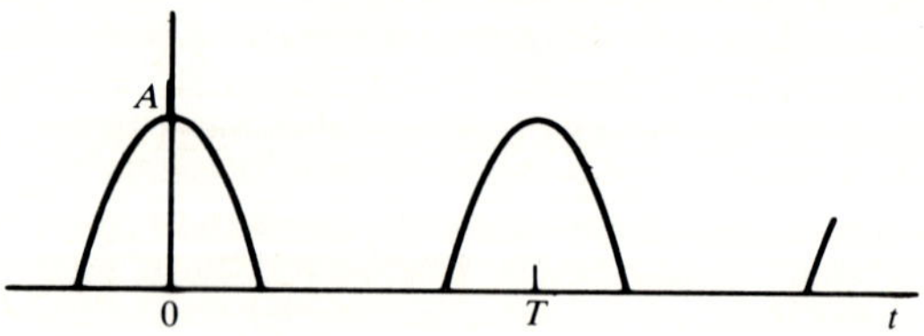

$$f(t) = \begin{cases} A\cos\dfrac{2\pi}{T}t & -\dfrac{T}{4} \le t \le \dfrac{T}{4} \\ 0 \text{ Otherwise} & \end{cases}$$

$$\frac{a_0}{2} = \frac{A}{\pi}$$

$$a_1 = \frac{A}{2}$$

$$a_n = \begin{cases} \dfrac{2A(-1)^{(n/2-1)}}{\pi(n^2-1)} & n = 2, 4, 6, \ldots \\ 0 & n = 3, 5, 7, \ldots \end{cases}$$

$$b_n = 0 \qquad \text{for all } n$$

7. Full-Wave Rectified Cosine Wave

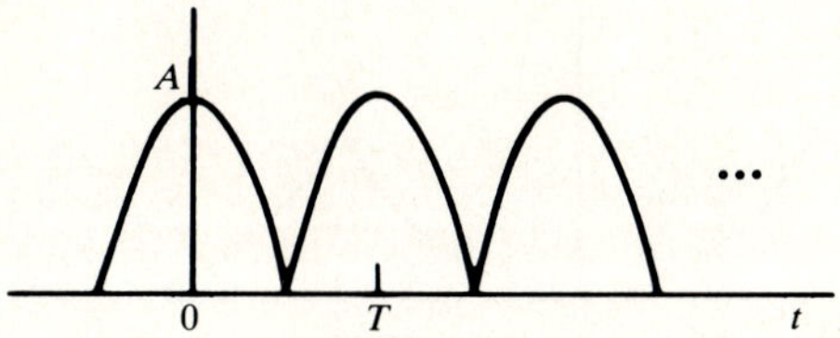

$$f(t) = A \cos \frac{\pi t}{T} \qquad -\frac{T}{2} \leq t \leq \frac{T}{2}$$

$$\frac{a_0}{2} = \frac{2A}{\pi}$$

$$a_n = \frac{4A(-1)^{n-1}}{\pi(4n^2 - 1)} \qquad n = 1, 2, 3, \ldots$$

$$b_n = 0 \qquad \text{for all } n$$

8. Pulse Train

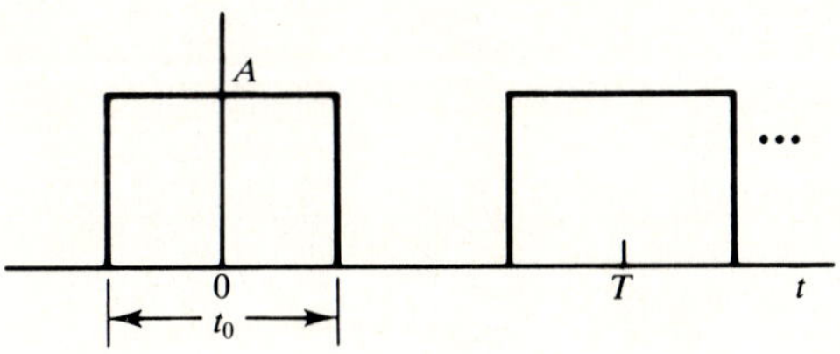

$$f(t) = \begin{cases} A & -\frac{t_0}{2} \leq t \leq \frac{t_0}{2} \\ 0 & \text{Otherwise} \end{cases}$$

$$\frac{a_0}{2} = \frac{At_0}{T}$$

$$a_n = \frac{2At_0}{T} \left[\frac{\sin(n\pi t_0/T)}{n\pi t_0/T} \right] \qquad n = 1, 2, 3, \ldots$$

$$b_n = 0 \qquad \text{for all } n$$

9. Parabolic Pulse Train

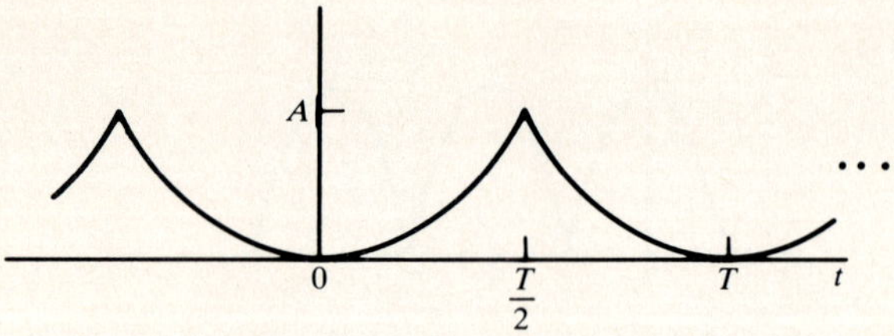

$$f(t) = \frac{4At^2}{T^2} \qquad -\frac{T}{2} \le t \le \frac{T}{2}$$

$$\frac{a_0}{2} = \frac{A}{3}$$

$$a_n = \frac{4A(-1)^n}{\pi^2 n^2} \qquad n = 1, 2, 3, \ldots$$

$$b_n = 0 \qquad \text{for all } n$$

APPENDIX C

ELEMENTS OF MATRIX ALGEBRA

C.1 INTRODUCTION

The purpose of this appendix is to provide an introduction to the reader who is unfamiliar with matrix operations. Matrix analysis has proven to be a powerful, concise representation with applications in linear algebraic equations, systems of linear differential equations, and many other applied engineering areas. Matrix representations are particularly adaptable to solution and manipulation on modern digital computers.

C.2 MATRICES

We shall introduce the ideas of matrix representation in the context of the linear equation problem. Consider the two simultaneous equations

$$\begin{aligned} a_{11}x_1 + a_{12}x_2 &= y_1 \\ a_{21}x_1 + a_{22}x_2 &= y_2 \end{aligned} \tag{C.1}$$

where the a_{ij} are commonly referred to as the coefficients. An example of a matrix is to arrange the coefficients in a rectangular array, which we shall call **A**, in the same manner as they occur above:

$$\mathbf{A} = \begin{bmatrix} a_{11} & a_{12} \\ a_{21} & a_{22} \end{bmatrix} \tag{C.2}$$

A matrix is defined as an array of elements with particular attention to the order of the elements. An $m \times n$ (read "m by n") matrix is an array of m rows and n columns:

$$\mathbf{A} = [a_{ij}] = \begin{bmatrix} a_{11} & a_{12} \ldots a_{1n} \\ a_{21} & a_{22} \ldots a_{2n} \\ \vdots & \\ a_{m1} & a_{m2} \quad a_{mn} \end{bmatrix} \tag{C.3}$$

The first subscript of an element denotes its row location, while the second subscript denotes the column location. A matrix of one dimension is commonly called a vector. A $1 \times n$ array is called a row vector from its appearance, while an $n \times 1$ matrix is referred to as a column vector.

C.3 MATRIX ADDITION AND SUBTRACTION

The sum of two matrices **A** and **B** is defined as the matrix resulting from the addition of corresponding elements of **A** and **B** or if **C** represents the sum of

A and **B** or if **A** and **B** are $m \times n$, then the $m \times n$ sum **C** has elements c_{ij} given by

$$c_{ij} = a_{ij} + b_{ij} \qquad i = 1, \ldots, m \qquad j = 1, \ldots, n \tag{C.4}$$

Example C.1

Find the sum of **A** and **B** given below:

$$\mathbf{A} = \begin{bmatrix} 1 & 3 \\ 2 & 4 \end{bmatrix} \qquad \mathbf{B} = \begin{bmatrix} 3 & 4 \\ 6 & 2 \end{bmatrix}$$

$$\mathbf{C} = \mathbf{A} + \mathbf{B} = \begin{bmatrix} 1+3 & 3+4 \\ 2+6 & 4+2 \end{bmatrix} = \begin{bmatrix} 4 & 7 \\ 8 & 6 \end{bmatrix}$$

Similarly, we can define the difference of two matrices $\mathbf{D} = \mathbf{A} - \mathbf{B}$ where the elements of **D** are given by

$$d_{ij} = a_{ij} - b_{ij} \tag{C.5}$$

Example C.2

Calculate the difference $\mathbf{D} = \mathbf{A} - \mathbf{B}$ for the two matrices given in Example C.1.

$$\mathbf{D} = \begin{bmatrix} 1-3 & 3-4 \\ 2-6 & 4-2 \end{bmatrix} = \begin{bmatrix} -2 & -1 \\ -4 & 2 \end{bmatrix}$$

From the preceding definitions it follows that

$$\mathbf{A} + \mathbf{B} = \mathbf{B} + \mathbf{A} \tag{C.6}$$

and for any scalar α

$$\alpha(\mathbf{A} + \mathbf{B}) = \alpha\mathbf{A} + \alpha\mathbf{B} \tag{C.7}$$

C.4 MATRIX MULTIPLICATION

The product $\mathbf{C} = \mathbf{AB}$ of two matrices **A** and **B** is defined as a matrix with elements c_{ij} given by

$$c_{ij} = a_{i1}b_{1j} + a_{i2}b_{2j} + \cdots + a_{in}b_{nj} \qquad i = 1, \ldots m \qquad j = 1, \ldots l \tag{C.8}$$

Note that the column dimension of **A** and the row dimension of **B** must both be n. The product matrix will have row dimension the same as that of **A** and column dimension the same as that of **B** or **C** is an $m \times l$ matrix.

Example C.3

Find the product $\mathbf{C} = \mathbf{AB}$ of the matrices $\mathbf{A}$ and $\mathbf{B}$ given below:

$$\mathbf{A} = \begin{bmatrix} 1 & 2 & 3 \\ 4 & 2 & 3 \end{bmatrix} \qquad \mathbf{B} = \begin{bmatrix} 3 & 1 \\ 2 & 2 \\ 4 & 6 \end{bmatrix}$$

$$\mathbf{C} = \mathbf{AB} = \begin{bmatrix} 1 & 2 & 3 \\ 4 & 2 & 3 \end{bmatrix} \begin{bmatrix} 3 & 1 \\ 2 & 2 \\ 4 & 6 \end{bmatrix} = \begin{bmatrix} 19 & 23 \\ 28 & 26 \end{bmatrix}$$

where we have circled the elements used to calculate c_{12}. Note that because of the dimensions of the two matrices the product $\mathbf{BA}$ is undefined. In general a matrix product, even if it is defined in either order, will not be the same or, in mathematical terms, a matrix product will not commute, i.e., $\mathbf{AB} \neq \mathbf{BA}$.

C.5 DETERMINANTS

Determinants play an important role in the solution of linear equations, especially where the elements of the coefficient matrix are functions of one or more symbolic variables, as is the case when solving simultaneous differential equations with Laplace transforms. The solution of linear equations by employing Cramer's rule was discussed in Chapter 1.

A determinant is defined as a number associated with a square matrix by a rule. This number is a function of all the elements in the matrix. The rule for calculation of a determinant for a general $n \times n$ matrix $\mathbf{A}$ is an algebraic sum of $n!$ products formed from the elements a_{ij} in such a way that one and only one element from each row i and each column j appears in each term. Each term shall be assigned an algebraic sign according to the number of permutations present in the first subscripts of the elements forming the product. This number of permutations is equal to the number of changes of numbers, two at a time, to put the first subscripts in numerical order $1\,2\,3 \ldots n$.

The definition is rather mathematical and is not much use in the solution of practical engineering problems where an organized procedure is required.

For 2×2 and 3×3 determinants, the rule is quite simple and will be presented separately for each case below.

For a 2×2 matrix the determinant is

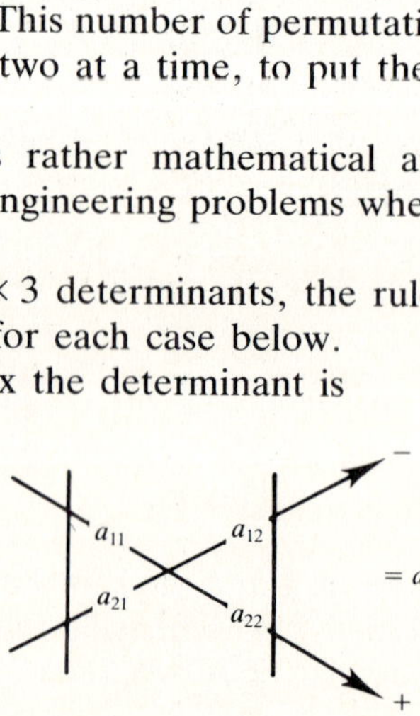

$$= a_{11}a_{22} - a_{21}a_{12} \qquad \textbf{(C.9)}$$

Note that the down and right product is positive, while the up and right product is negative. This rule for a 3×3 matrix will be given and then a general conclusion will be reached. For the 3×3 determinant the result is

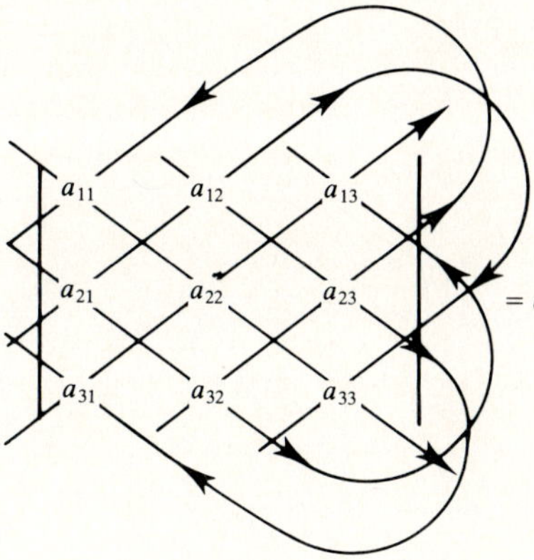

$$= a_{11}a_{22}a_{33} + a_{12}a_{23}a_{31} + a_{21}a_{32}a_{13} - (a_{31}a_{22}a_{13} + a_{32}a_{23}a_{11} + a_{21}a_{12}a_{33}) \quad \textbf{(C.10)}$$

Note that the down and right products indicated by the appropriate lines are all positive, while the up and right products are all negative. The arrow diagram below summarizes the sign convention for 2×2 and 3×3 determinants.

The results just given are not valid for determinants of orders greater than three and it is suggested that methods which involve determinants of high order are not as computationally efficient as other methods. This is particularly the case when solving linear equations with numerical coefficients where Gaussian reduction is a computationally efficient method for digital computation. An example of a program for solving a set of linear equations is given in Appendix D.1.

APPENDIX D

USEFUL COMPUTER PROGRAMS FOR SYSTEM ANALYSIS

The programs on the following pages are written in BASIC for the IBM personal computer or compatibles. They may require minor modification for other computers. Several of the programs employ standard IBM-BASIC graphics calls which require an IBM Standard or Paradise graphics card, commonly referred to as a CGA card.

D.1 LINEAR EQUATION SOLVER

This program solves a set of n linear equations for n unknowns by the method of Gaussian reduction with partial pivoting. The inputs are the number of equations (also the number of unknowns), the coefficients of the equations, and the constant terms. The outputs are the solutions for the unknowns.

```
10 CLS:PRINT:PRINT:PRINT"   THIS PROGRAM SOLVES THE SET OF LINEAR EQUATIONS"
20 PRINT"   IN MATRIX FORM"
30 PRINT:PRINT"                    A*X=Y"
40 PRINT:PRINT"   WHERE A IS A KNOWN NXN MATRIX,"
50 PRINT"   Y IS A KNOWN N-VECTOR AND"
60 PRINT"   X IS AN UNKNOWN N-VECTOR"
70 PRINT:PRINT"   INPUT THE SIZE OF THE PROBLEM N":INPUT N
80 DIM A(N,N+1),X(N)
90 CLS:PRINT:PRINT"   INPUT THE ELEMENTS OF THE A MATRIX"
100 FOR I=1 TO N:FOR J=1 TO N
110 PRINT  "   A(";I;",";J;")=":INPUT"       ";A(I,J)
120 NEXT J:NEXT I
130 PRINT:PRINT"   CHECK ELEMENTS OF THE A MATRIX"
140 FOR I=1 TO N:FOR J=1 TO N
150 PRINT"   A(";I;",";J;")=";A(I,J)
```

```
160 NEXT J:NEXT I
170 PRINT"   WANT TO CORRECT THEM?(Y OR N)"
180 INPUT X$
190 IF X$="Y" THEN 90
200 IF X$="N" THEN 220
210 GOTO 170
220 PRINT:PRINT"   INPUT THE ELEMENTS OF THE Y VECTOR"
230 FOR I=1 TO N
240 PRINT"   Y(";I;")=":INPUT"       ";A(I,N+1):NEXT I
250 PRINT:PRINT"   CHECK ELEMENTS OF THE Y VECTOR"
260 FOR I=1 TO N
270 PRINT"   Y(";I;")=";A(I,N+1):NEXT I
280 PRINT:PRINT"  WANT TO CORRECT THEM?(Y OR N)":INPUT X$
290 IF X$="Y" THEN 220
300 IF X$="N" THEN 320
310 GOTO 280
320 CLS:PRINT"   I'M SOLVING THE PROBLEM NOW"
330 M=N+1:L=N-1
340 FOR K=1 TO L:JJ=K:BI=ABS(A(K,K))
350 K1=K+1
360 FOR I=K1 TO N
370 AB=ABS(A(I,K)):IF BI>AB THEN 390
380 BI=AB:JJ=I
390 NEXT I
400 IF JJ=K THEN 440
410 FOR J=K TO M
420 TE=A(JJ,J):A(JJ,J)=A(K,J):A(K,J)=TE
430 NEXT J
440 FOR I=K1 TO N
450 QU=A(I,K)/A(K,K)
460 FOR J=K1 TO M
```

```
470 A(I,J)=A(I,J)-QU*A(K,J):NEXT J:NEXT I
480 FOR I=K1 TO N:A(I,K)=0:NEXT I
490 NEXT K
500 X(N)=A(N,M)/A(N,N)
510 FOR NN=1 TO L
520 S=0:I=N-NN:I1=I+1
530 FOR J=I1 TO N:S=S+A(I,J)*X(J):NEXT J
540 X(I)=(A(I,M)-S)/A(I,I)
550 NEXT NN
560 FOR I=1 TO N
570 PRINT "    X(";I;")=";X(I)
580 NEXT I
590 END
```

D.2 PARTIAL FRACTION EXPANSION PROGRAM

This program performs a partial fraction expansion on an *s*-domain function *F*(*s*) to give the inverse transform. The method used is outlined in Section 2.10 of the text. Inputs to the program are the number of origin poles, the number of real poles, the number of quadratic factors with complex poles, and the numerator polynomial coefficients. Other inputs are the real pole locations and the damping factor α_i and the undamped natural frequency ω_i for each set of complex poles. Outputs are the partial fractions coefficients, an equation for the inverse function, and a plot of the inverse function *f*(*t*).

```
10 SCREEN 0:KEY OFF
12 CLS:PRINT:PRINT"    THIS PROGRAM INVERTS LAPLACE TRANSFORMS BY"
14 PRINT:PRINT"    PARTIAL FRACTION EXPANSION"
16 PRINT:PRINT:PRINT"    WHEN PROMPTED PRESS ANY KEY TO CONTINUE"
18 GOSUB 422
20 PRINT:PRINT"    THE DENOMINATOR MUST BE IN FACTORED FORM WITH:"
22 PRINT:PRINT"     (A)-ORIGIN ROOTS OF THE FORM"
24 PRINT"          (S^NO)"
26 PRINT:PRINT"     (B)-REAL ROOTS OF THE FORM"
28 PRINT"          (S + BETA(I))"
30 PRINT:PRINT"     (C)-COMPLEX ROOTS LEFT IN THE QUADRATIC FORM"
32 PRINT"          (S^2 + 2*ALPHA(I)*S + OMEGA(I)^2)"
34 PRINT:INPUT"    NO. OF ORIGIN ROOTS NO=";NO
36 PRINT:INPUT"    NO. OF REAL ROOTS N1=";N1
38 PRINT:INPUT"    NO. OF QUADRATIC FACTORS N2=";N2
40 GOSUB 422
42 IF N1=0 THEN 50
44 DIM B(N1)
46 FOR I=1 TO N1
48 PRINT:PRINT"    BETA(";I;"))=":INPUT"                ";B(I):NEXT I
50 IF N2=0 THEN 62
```

```
52 DIM A(N2),W(N2)
54 FOR I=1 TO N2
56 PRINT:PRINT"   ALPHA(";I;")=":INPUT"            ";A(I)
58 PRINT:PRINT"   OMEGA(";I;")=":INPUT"            ";W(I)
60 NEXT I
62 GOSUB 422
64 R=NO+N1+2*N2:L=R-1:DIM H(L)
66 PRINT:PRINT:PRINT"   THE NUMERATOR IS OF THE FORM "
68 PRINT:PRINT"     H(";L;")*S^";L;"+ . . . +H(1)*S+H(0)"
70 PRINT:PRINT:PRINT"   INPUT THE NUMERATOR COEFFICIENTS"
72 FOR I=0 TO L:PRINT:PRINT"   H(";I;")=":INPUT"            ";H(I)
74 NEXT I
76 GOSUB 422
78 DIM P(R)
80 IF N1<>0 THEN 88
82 IF N2<>0 THEN 92
84 FOR I=1 TO R:P(I)=I:NEXT I
86 GOTO 94
88 FOR I=1 TO R:P(I)=B(1)*I:NEXT I
90 GOTO 94
92 FOR I=1 TO R:P(I)=W(1)*I:NEXT I:GOTO 94
94 DIM F(R)
96 FOR I=1 TO R:F(I)=H(0)
98 FOR K=1 TO R-1:F(I)=F(I)+H(K)*(P(I)^K):NEXT K
100 IF NO=0 THEN 104
102 F(I)=F(I)/(P(I)^NO)
104 IF N1=0 THEN 108
106 FOR K=1 TO N1:F(I)=F(I)/(P(I)+B(K)):NEXT K
108 IF N2=0 THEN 114
110 FOR K=1 TO N2:F(I)=F(I)/(P(I)^2+2*A(K)*P(I)+W(K)^2)
112 NEXT K
114 NEXT I
116 DIM Q(R,R),X(R)
```

```
118 FOR I=1 TO R
120 IF NO=0 THEN 124
122 FOR J=1 TO NO:Q(I,J)=1/(P(I)^J):NEXT J
124 IF N1=0 THEN 130
126 FOR J=NO+1 TO NO+N1:K=J-NO
128 Q(I,J)=1/(P(I)+B(K)):NEXT J
130 IF N2=0 THEN 148
132 FOR J=NO+N1+1 TO NO+N1+N2:K=J-NO-N1
134 D1=P(I)^2+2*A(K)*P(I)+W(K)^2
136 Q(I,J)=SQR(W(K)^2-A(K)^2)/D1
138 NEXT J
140 FOR J=R-N2+1 TO R
142 K=J-(NO+N1+N2):D1=P(I)^2+2*A(K)*P(I)+W(K)^2
144 Q(I,J)=(P(I)+A(K))/D1
146 NEXT J
148 NEXT I
150 REM DO ELIMINATION
152 L1=R-1
154 FOR K=1 TO L1
156 JJ=K:BI=ABS(Q(K,K)):K1=K+1
158 FOR I=K1 TO R
160 AB=ABS(Q(I,K))
162 IF BI>=AB THEN 166
164 BI=AB:JJ=I
166 NEXT I
168 IF JJ=K THEN 178
170 FOR J=K TO R
172 TE=Q(JJ,J):Q(JJ,J)=Q(K,J):Q(K,J)=TE
174 NEXT J
176 Z1=F(JJ):F(JJ)=F(K):F(K)=Z1
178 FOR I=K1 TO R
180 QU=Q(I,K)/Q(K,K)
```

```
182 FOR J=K1 TO R
184 Q(I,J)=Q(I,J)-QU*Q(K,J):NEXT J
186 F(I)=F(I)-QU*F(K):NEXT I
188 FOR I=K1 TO R:Q(I,K)=0:NEXT I
190 NEXT K
192 X(R)=F(R)/Q(R,R)
194 FOR NN=1 TO R-1
196 S=0:I=R-NN:I1=I+1
198 FOR J=I1 TO R:S=S+Q(I,J)*X(J):NEXT J
200 X(I)=(F(I)-S)/Q(I,I):NEXT NN
202 GOSUB 422
204 PRINT:PRINT:PRINT"   DO YOU WANT THE EQUATION OF THE INVERSE?(Y OR N)"
206 INPUT X$
208 IF X$="N" THEN 246
210 IF X$<>"Y" THEN 204
212 CLS:PRINT:PRINT"   THE INVERSE TRANSFORM IS"
214 PRINT:PRINT"                f(t)="
216 IF NO=0 THEN 230
218 S=1
220 FOR I=1 TO NO
222 IF I=1 THEN 226
224 S=S*(I-1)
226 PRINT:PRINT"   ";X(I)/S;"*(t^";I-1;") +"
228 NEXT I
230 IF N1=0 THEN 236
232 FOR I=1 TO N1
234 PRINT:PRINT"   ";X(I+NO);"*EXP(-";B(I);"*t) +":NEXT I
236 IF N2=0 THEN 246
238 FOR I=1 TO N2:G=SQR(W(I)^2-A(I)^2)
240 PRINT:PRINT"   ";X(I+N1+NO);"*EXP(-";A(I);"*t)*SIN(";G;"*t) +"
242 PRINT:PRINT"   ";X(I+NO+N1+N2);"*EXP(-";A(I);"*t)*COS(";G;"*t) +"
244 NEXT I
```

```
246 GOSUB 422
248 PRINT:PRINT:PRINT"   DO YOU WANT A PLOT OF THE INVERSE FUNCTION?(Y OR N)"
250 INPUT X$
252 IF X$="N" THEN 421
254 IF X$="Y" THEN 258
256 GOTO 248
258 DIM V(100)
260 SCREEN 0:WIDTH 80:INPUT"   INPUT THE LENGTH OF TIME DESIRED IN SEC.";T7
262 D=T7/100
264 PRINT:PRINT"   DON'T BE IMPATIENT I'M CALCULATING!!"
266 FOR K=0 TO 100:T=K*D:V(K)=0:S=1
268 IF NO=0 THEN 280
270 FOR J=1 TO NO
272 IF J=1 THEN 276
274 S=S*(J-1)
276 V(K)=V(K)+(X(J)*(T^(J-1))/S)
278 NEXT J
280 IF N1=0 THEN 288
282 FOR I=NO+1 TO NO+N1
284 L=I-NO
286 V(K)=V(K)+X(I)*EXP(-B(L)*T):NEXT I
288 IF N2=0 THEN 298
290 FOR J=1 TO N2:G=SQR(W(J)^2-A(J)^2)
292 V(K)=V(K)+X(NO+N1+N2+J)*EXP(-A(J)*T)*COS(G*T)
294 V(K)=V(K)+X(NO+N1+J)*EXP(-A(J)*T)*SIN(G*T)
296 NEXT J
298 NEXT K
300 GOSUB 422
302 REM PLOT ROUTINE
304 LO=1E+08:HI=-1E+08
306 FOR K=1 TO 100
308 IF V(K)<=LO THEN LO=V(K):KL=K
```

```
310 IF V(K)>=HI THEN HI=V(K):KH=K
312 NEXT K
314 HI1=ABS(HI):LO1=ABS(LO)
316 SCREEN 2:CLS
318 LOCATE 2,10
320 PRINT"MAX VALUE=";HI;"  Tmx=";KH*D
322 LOCATE 3,10
324 PRINT"MIN VALUE=";LO;"  Tmn=";KL*D
326 IF (HI>0) AND (LO>=0) THEN 348
328 IF (HI>0) AND (LO<0) THEN 366
330 D5=LO
332 Z=40:ZP=7
334 X7=(140*V(0))/D5+40
336 FOR K=0 TO 99
338 T5=6*K:T6=T5+6
340 X8=(140*V(K+1))/D5+40
342 LINE (T5,X7)-(T6,X8)
344 X7=X8:NEXT K
346 GOTO 386
348 D5=HI:Z=140:ZP=20
350 X7=140-(100*V(0))/D5
352 FOR K=0 TO 99
354 T5=6*K+40
356 T6=T5+6
358 X8=140-(100*V(K+1))/D5
360 LINE (T5,X7)-(T6,X8):X7=X8
362 NEXT K
364 GOTO 386
366 Z=100:ZP=15
368 IF HI1>LO1 THEN D5=HI
370 IF LO1>HI1 THEN D5=LO
372 X7=100-(60*V(0))/D5
```

```
374 FOR K=0 TO 99
376 T5=6*K+40
378 T6=T5+6
380 X8=100-(60*V(K+1))/D5
382 LINE (T5,X7)-(T6,X8)
384 X7=X8:NEXT K
386 LINE (40,Z)-(639,Z)
388 LINE (340,Z+10)-(340,Z)
390 LINE (40,40)-(40,140)
392 LOCATE ZP,42
394 PRINT T7/2
396 LOCATE ZP,70
398 PRINT"TIME,t"
400 LOCATE 6,1
402 PRINT"f(t)"
404 LOCATE 21,1
406 LOCATE ZP,2
408 PRINT 0
410 LOCATE 22,2:PRINT"   WANT TO SOLVE THE PROBLEM FOR ANOTHER LENGTH OF TIME?(Y
 OR N)"
412 INPUT X$
414 IF X$="Y" THEN 260
416 IF X$="N" THEN 420
418 GOTO 410
420 SCREEN 0:WIDTH 80
421 KEY ON:END
422 PRINT:PRINT"   PRESS RETURN TO CONTINUE":INPUT Q$:CLS:RETURN
```

D.3 POLYNOMIAL FACTORING PROGRAM

This program finds the roots of a polynomial of arbitrary degree in *s*. This is accomplished by employing an optimized Newton-Raphson technique. Inputs to the program are the polynomial degree and the polynomial coefficients. Outputs are the roots of the polynomial and the factors of the polynomial in which complex roots are represented by quadratic factors with real coefficients.

```
10 CLS:PRINT:PRINT:PRINT"   THIS PROGRAM SOLVES FOR ROOTS AND FACTORS OF A POLYN'
OMIAL"
20 PRINT"   OF DEGREE N"
30 PRINT:PRINT:PRINT"   INPUT THE DEGREE N":INPUT N
40 DIM A(N),R(N),I(N)
50 PRINT:PRINT"   THE POLYNOMIAL WILL BE OF THE FORM"
60 PRINT:PRINT"      A(";N;")*S^";N;"+. . . +A(1)*S+A(0)"
70 PRINT:PRINT"   INPUT THE COEFFICIENTS"
80 FOR I=0 TO N
90 PRINT"   A(";I;")=":INPUT A(I):NEXT I
100 AN=A(N)
110 GOSUB 1000
120 PRINT" THE ROOTS ARE"
130 FOR I=1 TO N
140 PRINT"   R(";I;")=";R(I);"   I(";I;")=";I(I)
150 NEXT I
160 PRINT:PRINT:PRINT"   DO YOU WANT THE POLYNOMIAL FACTORS?(Y OR N)":INPUT X$
170 IF X$="Y" THEN 200
180 IF X$="N" THEN 320
190 GOTO 160
200 I=1
```

```
210 IF I(I)<>0 THEN 260
220 PRINT:PRINT"    (S+";-R(I);")"
230 IF I=N THEN 310
240 I=I+1
250 GOTO 210
260 OM=SQR(I(I)^2+R(I)^2)
270 PRINT:PRINT"    (S^2+2*";-R(I);"*S+";OM;"^2)"
280 I=I+2
290 IF I>N THEN 310
300 GOTO 210
310 PRINT:PRINT"    WITH A CONSTANT FACTOR OF";AN
320 END
1000 REM SOLVES POLYNOMIAL
1010 FOR I=1 TO N
1020 R(I)=0
1030 I(I)=0
1040 NEXT I
1050 IF N=2 THEN 1760
1060 N1=N
1070 S1=-.707*(ABS(A(0)/A(N1)))^(1/N1)
1080 O1=-S1
1090 AL=-2*S1
1100 BE=S1^2+O1^2
1110 GOSUB 1260
1120 GOSUB 1360
1130 S5=S1
1140 O5=O1
1150 RA=(R4^2+I4^2)
1160 S1=S1-((R3*R4+I3*I4)/RA)
1170 O1=O1-((I3*R4-I4*R3)/RA)
1180 D=SQR((O1-O5)^2+(S1-S5)^2)
```

```
1190 IF D<.0001*(ABS(A(0)/A(N1))^(1/N1)) THEN 1210
1200 GOTO 1090
1210 GOSUB 1460
1220 IF N1>2 THEN 1070
1230 IF N1=2 THEN 1760
1240 IF N1=1 THEN 1730
1250 RETURN
1260 C(N1)=A(N1)
1270 C(N1-1)=A(N1-1)-AL*C(N1)
1280 FOR I=N1-2 TO 0 STEP -1
1290 C(I)=A(I)-AL*C(I+1)-BE*C(I+2)
1300 NEXT I
1310 R0=C(0)+AL*C(1)
1320 R1=C(1)
1330 R3=R0+S1*R1
1340 I3=O1*R1
1350 RETURN
1360 D(N1-1)=A(N1)*N1
1370 D(N1-2)=A(N1-1)*(N1-1)-AL*D(N1-1)
1380 FOR I=N1-3 TO 0 STEP -1
1390 D(I)=(I+1)*A(I+1)-AL*D(I+1)-BE*D(I+2)
1400 NEXT I
1410 R0=D(0)+AL*D(1)
1420 R1=D(1)
1430 R4=R0+S1*R1
1440 I4=R1*O1
1450 RETURN
1460 IF (ABS(O1)<.00001*(ABS(A(0)/A(N1))^(1/N1))) THEN 1620
1470 B(N1-2)=A(N1)
1480 B(N1-3)=A(N1-1)-AL*A(N1)
```

```
1490 IF N1=3 THEN 1530
1500 FOR K=4 TO N1
1510 B(N1-K)=A(N1-K+2)-BE*B(N1-K+2)-AL*B(N1-K+1)
1520 NEXT K
1530 FOR K=0 TO N1-2
1540 A(K)=B(K)
1550 NEXT K
1560 R(N1)=S1
1570 I(N1)=O1
1580 R(N1-1)=S1
1590 I(N1-1)=-O1
1600 N1=N1-2
1610 GOTO 1720
1620 B(N1-1)=A(N1)
1630 FOR K=2 TO N1
1640 B(N1-K)=A(N1-K+1)+B(N1-K+1)*S1
1650 NEXT K
1660 FOR K=0 TO N1-1
1670 A(K)=B(K)
1680 NEXT K
1690 R(N1)=S1
1700 I(N1)=0
1710 N1=N1-1
1720 RETURN
1730 R(1)=-A(0)/A(1)
1740 I(1)=0
1750 RETURN
1760 D7=A(1)*A(1)-4*A(2)*A(0)
1770 IF D7<0 THEN 1840
1780 S7=SQR(D7)
```

```
1790 R(2)=(-A(1)+S7)/(2*A(2))
1800 S(2)=0
1810 R(1)=(-A(1)-S7)/(2*A(2))
1820 S(1)=0
1830 GOTO 1880
1840 R(2)=-A(1)/(2*A(2))
1850 R(1)=R(2)
1860 I(2)=SQR(-D7)/(2*A(2))
1870 I(1)=-I(2)
1880 RETURN
```

D.4 TRANSFER FUNCTION SIMULATION PROGRAM

This program simulates a transfer function for possible step, ramp, and arbitrary sampled inputs. The associated differential equations are solved by a state variable matrix exponential series scheme. Inputs to the program are the transfer function coefficients, the type of input, and the integration time interval. The outputs are numerical values of the time response of the transfer function to the selected input function and a temporal plot of the response function.

```
2 CLS:PRINT:PRINT:PRINT"   THIS PROGRAM FINDS THE TIME RESPONSE OF A TRANSFER FU
NCTION"
4 PRINT"   WITH ZERO INITIAL CONDITIONS"
6 GOSUB 324
8 PRINT:PRINT:PRINT"   THE TRANSFER FUNCTION IS OF THE FORM"
10 PRINT:PRINT:PRINT"       G(S) = DO + [N(S)/D(S)]"
12 PRINT"   WHERE"
14 PRINT"      N(S)=B(N-1)*S^(N-1)+...+B(1)*S+B(O)"
16 PRINT"   AND":PRINT"       D(S)=S^N+A(N-1)*S^(N-1)+...+A(1)*S+A(O)"
18 PRINT"   INPUT THE SYSTEM ORDER N";:INPUT " ",N
20 DIM A1(N-1),B1(N-1),F(N,N),A(N,N),G(N),B(N),C(N),M(N,N),Q(N,N)
22 DIM Z(N),X(N),U(60),Y(60)
24 PRINT"   INPUT THE FEEDFORWARD GAIN DO";:INPUT " ",DO
26 PRINT"   INPUT THE DENOMINATOR COEFFICIENTS"
28 FOR I=O TO N-1
30 PRINT"   A(";I;")=";:INPUT " ",A1(I):NEXT I
32 PRINT"   INPUT THE NUMERATOR COEFFICIENTS"
34 FOR I=O TO N-1
36 PRINT"   B(";I;")=";:INPUT " ",B1(I):NEXT I
38 PRINT"    INPUT THE TIME INTERVAL STEP SIZE";:INPUT " ",T
```

```
40 FOR I=1 TO N-1:F(I,I+1)=1:NEXT I
42 FOR I=0 TO N-1:C(I+1)=B1(I):F(N,I+1)=-A1(I):NEXT I
44 G(N)=1:N1=7:I6=1
46 FOR I=1 TO N:Q(I,I)=1:P(I,I)=T:M(I,I)=1:NEXT I
48 FOR I1=1 TO N1
50 FOR I=1 TO N
52 FOR J=1 TO N
54 N(I,J)=0
56 FOR L=1 TO N
58 N(I,J)=N(I,J)+F(I,L)*M(L,J):NEXT L
60 NEXT J
62 NEXT I
64 FOR I=1 TO N:FOR J=1 TO N:M(I,J)=N(I,J):NEXT J:NEXT I
66 I6=I6*I1:I7=I6*(I1+1)
68 FOR I=1 TO N
70 FOR J=1 TO N
72 Q(I,J)=Q(I,J)+(M(I,J)*(T^I1)/I6)
74 P(I,J)=P(I,J)+(M(I,J)*(T^(I1+1))/I7)
76 NEXT J
78 NEXT I
80 NEXT I1
82 FOR I=1 TO N
84 B(I)=0
86 FOR I3=1 TO N
88 B(I)=B(I)+P(I,I3)*G(I3)
90 NEXT I3
92 NEXT I
94 GOSUB 324
96 CLS:PRINT:PRINT
98 CLS:PRINT"   THIS PROGRAM SEGMENT SELECTS THE FORCING FUNCTION"
100 PRINT:PRINT:PRINT"     PRESS 1 FOR A STEP INPUT"
```

```
102 PRINT:PRINT:PRINT"     PRESS 2 FOR A RAMP INPUT"
104 PRINT:PRINT:PRINT"     PRESS 3 FOR AN ARBITRARY SAMPLED INPUT UP TO 60 POINTS
106 PRINT:INPUT "          ",B$
108 IF B$="1" THEN 116
110 IF B$="2" THEN 122
112 IF B$="3" THEN 128
114 GOTO 98
116 PRINT:PRINT:PRINT"     INPUT  HEIGHT OF STEP";:INPUT " ",U(O)
118 FOR I=1 TO 60:U(I)=U(O):NEXT I
120 GOTO 138
122 PRINT:PRINT:PRINT"  INPUT SLOPE OF RAMP INPUT";:INPUT " ",S
124 FOR I=O TO 60:U(I)=I*S*T:NEXT I
126 GOTO 138
128 PRINT:PRINT:PRINT"   HOW MANY NONZERO INPUT SAMPLES ARE THERE";
130 INPUT " ",N3
132 FOR I=O TO N3-1:PRINT"  U(";I;")=";:INPUT " ",U(I):NEXT I
134 FOR I=N3 TO 59:U(I)=O:NEXT I
138 FOR I=1 TO N:X(I)=O:NEXT I
140 PRINT:PRINT:PRINT"   I'M CALCULATING THE RESPONSE NOW,  PLEASE WAIT!"
142 FOR I3=1 TO 60
144 FOR I=1 TO N
146 Z(I)=O
148 FOR J=1 TO N
150 Z(I)=Z(I)+Q(I,J)*X(J)
152 NEXT J
154 Z(I)=Z(I)+B(I)*U(I3-1)
156 NEXT I
158 FOR I=1 TO N:X(I)=Z(I):NEXT I
160 Y(I3)=DO*U(I3)
162 FOR I=1 TO N
164 Y(I3)=Y(I3)+C(I)*X(I)
```

```
166 NEXT  I
168 NEXT I3
170 Y(0)=D0*U(0)
172 N7=0
174 J=1:CLS
176 PRINT:PRINT"     TIME    RESPONSE"
178 N7=N7+1
180 T3=(N7+(J-1)*20)*T
182 PRINT T3,Y(N7+(J-1)*20)
184 IF(N7+(J-1)*20)=60 THEN 196
186 IF N7=20 THEN 190
188 GOTO 178
190 J=J+1:N7=0
192 GOSUB 324
194 CLS:GOTO 176
196 GOSUB 324
198 PRINT:PRINT"    WANT A SCREEN PLOT OF THE RESPONSE?(Y OR N)";
200 INPUT " ",X$
202 IF X$="H" THEN 210
204 IF X$="Y" THEN 208
206 GOTO 198
208 GOSUB 212
210 END
212 REM PLOT ROUTINE
214 LO=1E+08:HI=-1E+08
216 FOR K=1 TO 60
218 IF Y(K)<=LO THEN LO=Y(K):KL=K
220 IF Y(K)>=HI THEN HI=Y(K):KH=K
222 NEXT K
224 HI1=ABS(HI):LO1=ABS(LO)
226 SCREEN 2:CLS
```

```
228 LOCATE 2,10
230 PRINT "MAX VALUE=";HI;" Tmx=";KH*T
232 LOCATE 3,10
234 PRINT "MIN VALUE=";LO;" Tmn=";KL*T
236 IF (HI>0) AND (LO>=0) THEN 258
238 IF (HI>0) AND (LO<0)  THEN 276
240 D5=LO
242 Z=40:ZP=7
244 X7=(100*Y(0))/D5+40
246 FOR K=0 TO 59
248 T5=10*K+40:T6=T5+10
250 X8=(100*Y(K+1))/D5+40
252 LINE (T5,X7)-(T6,X8)
254 X7=X8:NEXT K
256 GOTO 296
258 D5=HI:Z=140:ZP=20
260 X7=140-(100*Y(0))/D5
262 FOR K=0 TO 59
264 T5=10*K+40
266 T6=T5+10
268 X8=140-(100*Y(K+1))/D5
270 LINE (T5,X7)-(T6,X8):X7=X8
272 NEXT K
274 GOTO 296
276 Z=100:ZP=15
278 IF HI1>LO1 THEN D5=HI
280 IF LO1>HI1 THEN D5=LO
282 X7=100-(60*Y(0))/D5
284 FOR K=0 TO 59
286 T5=10*K+40
288 T6=T5+10
```

```
290 X8=100-(60*Y(K+1))/D5
292 LINE (T5,X7)-(T6,X8)
294 X7=X8:NEXT K
296 LINE (40,Z)-(639,Z)
298 LINE (340,Z+10)-(340,Z)
300 LINE (40,40)-(40,140)
302 LOCATE ZP,42
304 PRINT 30*T
306 LOCATE ZP,70
308 PRINT"TIME,t"
310 LOCATE 6,1
312 PRINT"y(t)"
314 LOCATE 21,1
316 LOCATE ZP,2
318 PRINT 0
320 GOSUB 324
322 SCREEN 0:RETURN
324 PRINT:PRINT"---PRESS RETURN TO CONTINUE";:INPUT " ",S$:CLS:RETURN
```

D.5 FREQUENCY RESPONSE EVALUATION PROGRAM

This program evaluates the frequency response magnitude, phase, and log magnitude (dB) for a transfer function, over a decade in frequency. Inputs to the program are the coefficients of the transfer function and the starting point of frequency for the decade of interest. Outputs are a listing of magnitude, phase, and dB as functions of frequency. As many decades as needed may be evaluated.

```
5 CLS:PRINT:PRINT:PRINT"   THIS PROGRAM EVALUATES FREQUENCY RESPONSE OVER"
10 PRINT"   ONE DECADE IN FREQUENCY"
15 GOSUB 325
20 PRINT:PRINT"   THE TRANSFER FUNCTION IS OF THE FORM"
25 PRINT:PRINT"              B(N)*S^N + ... +B(1)*S + B(0)"
30 PRINT"   G(S) = ______________________________________________"
35 PRINT"              S^N + A(N-1)*S^(N-1) + ... +A(1)*S + A(0)"
40 PRINT:PRINT"   INPUT THE SYSTEM ORDER N":INPUT N
45 DIM W(29),A(10),B(10),S(4)
50 W(1)=1:W(2)=1.2:W(3)=1.4:W(4)=1.6:W(5)=1.8
55 W(6)=2:W(7)=2.5
60 FOR I=8 TO 15:W(I)=I-5:NEXT I
65 PRINT:PRINT"   INPUT THE DENOMINATOR COEFFICIENTS"
70 FOR I=0 TO N-1
75 PRINT"   A(";I;")=":INPUT"          ";A(I):NEXT I
80 PRINT:PRINT"   INPUT THE NUMERATOR COEFFICIENTS"
85 FOR I=0 TO N
90 PRINT"   B(";I;")=":INPUT"          ";B(I):NEXT I
95 A(N)=1
100 PRINT:PRINT:PRINT"   INPUT THE LOWER END OF THE DECADE IN RAD/SEC":INPUT W(0
)
```

```
105 CLS:PRINT:PRINT"   RAD/SEC       MAG              DB                 PHASE"
110 FOR I=1 TO 15:W=W(I)*W(0)
115 S(2)=0:S(1)=A(0)
120 S(4)=0:S(3)=B(0)
125 FOR J=2 TO (2*INT(N/2)) STEP 2
130 S(1)=S(1)+A(J)*(W^J*(-1)^(J/2))
135 S(3)=S(3)+B(J)*(W^J*(-1)^(J/2))
140 NEXT J
145 FOR J=1 TO ((2*INT((N-1)/2))+1) STEP 2
150 S(2)=S(2)+A(J)*(W^J*(-1)^((J+3)/2))
155 S(4)=S(4)+B(J)*(W^J*(-1)^((J+3)/2))
160 NEXT J
165 A=S(1):B=S(2)
170 GOSUB 270
175 P1=C
180 A=S(3):B=S(4)
185 GOSUB 270
190 P3=(C-P1)*57.296
195 P3=INT(1000*P3)/1000
200 MA=SQR(S(3)^2+S(4)^2)
205 MA=MA/SQR(S(1)^2+S(2)^2)
210 MA=INT(1000*MA)/1000
215 DB=8.6859*LOG(MA)
220 DB=INT(100*DB)/100
225 W=INT(1000*W)/1000
230 PRINT W,MA,DB,P3
235 NEXT I
240 PRINT:PRINT"   WANT ANOTHER DECADE?(Y OR N)"
245 INPUT X$
250 IF X$="Y" THEN 100
255 IF X$="N" THEN 265
```

```
260 GOTO 240
265 END
270 REM ARC TAN SUB
275 IF A=0 THEN 310
280 C=ATN(B/A)
285 IF A<0 AND B>0 THEN 300
290 IF A<0 AND B<0 THEN 305
295 GOTO 320
300 C=C+3.14159:GOTO 320
305 C=C-3.14159:GOTO 320
310 IF B<0 THEN C=-1.57079:GOTO 320
315 C=1.57079
320 RETURN
325 PRINT:PRINT:PRINT"---PRESS RETURN TO CONTINUE":INPUT X$:CLS:RETURN
```

D.6 RUNGE-KUTTA DIFFERENTIAL EQUATION SOLUTION

This program solves the second-order differential equation associated with the problem posed in Example 5.7. It employs the fourth-order Runge-Kutta method to generate 40 points on the solution trajectories for $x(t)$ and $v(t)$ for a computational interval of $T = 0.01$ s. The program could be easily modified to solve any pair of coupled first-order differential equations by modification of statements 30 through 65 and adjusting statement 10 for the appropriate time interval.

```
5 REM RUNGE-KUTTA 2ND ORDER EXAMPLE
10 T=.01
15 X=0:V=0
20 FOR I=1 TO 50
25 T1=I*T
30 K1=V*T
35 L1=(-20.13*V-2415*X+402.57)*T
40 K2=(V+L1/2)*T
45 L2=(-20.13*(V+L1/2)-2415*(X+K1/2)+402.57)*T
50 K3=(V+L2/2)*T
55 L3=(-20.13*(V+L2/2)-2415*(X+K2/2)+402.57)*T
60 K4=(V+L3)*T
65 L4=(-20.13*(V+L3)-2415*(X+K3)+402.57)*T
70 X1=X+((K1+2*K2+2*K3+K4)/6)
75 V1=V+((L1+2*L2+2*L3+L4)/6)
80 X=X1
85 V=V1
90 PRINT T1;X;V
95 NEXT I
100 END
```

D.7 FOURIER ANALYSIS PROGRAM

This program numerically calculates the Fourier coefficients from supplied samples of a periodic waveform. Inputs to the program are the number of samples needed, the sample, and the number of frequency points to be evaluated. The outputs are the sine and cosine coefficients of the waveform and the coefficients and phase angles for the alternative form of the series, as given in Section 8.7.

```
5 CLS:PRINT:PRINT:PRINT"                **** FOURIER ANALYSIS PROGRAM ****"
10 DIM A(20),B(20),C(20),P(20),F(100)
15 PRINT:PRINT:INPUT"   PERIOD IN SECONDS =";T
20 PRINT:PRINT:INPUT "   NUMBER OF SAMPLES PER CYCLE=";K
25 PRINT:PRINT:PRINT"   NUMBER OF HARMONICS MUST BE LESS THAN HALF THE NUMBER OF
 SAMPLES"
30 PRINT:PRINT:INPUT "   NUMBER OF HARMONICS=";M
35 PRINT:PRINT:PRINT"   TYPE IN SAMPLED DATA POINTS STARTING ONE POINT RIGHT OF
THE TIME ORIGIN"
40 S=0
45 FOR I=1 TO K
50 PRINT"   F(";I;")=";:INPUT "        ";F(I)
55 S=S+F(I)
60 NEXT I
65 CLS:PRINT:PRINT:PRINT"      SUMMARY OF INPUT DATA":PRINT:PRINT
70 PRINT"   THE PERIOD T=";T:PRINT:PRINT
75 PRINT"   THE SAMPLES ARE:"
80 FOR I=1 TO K
85 PRINT"   F(";I;")=";F(I)
90 NEXT I
95 PRINT:PRINT:PRINT"   ARE THESE DATA CORRECT?(Y OR N)"
100 INPUT X$
```

```
105 IF X$="Y" THEN 120
110 IF X$="N" THEN 40
115 GOTO 95
120 S=S/K:PRINT:PRINT"   THE AVERAGE VALUE A(0)/2=";S
125 PRINT:PRINT:PRINT"                    N          FREQ. IN HZ.       A(N)            B(N
)"
130 FOR J=1 TO M
135 S1=0:S2=0
140 FOR I=1 TO K
145 X=J*I*6.2832/K
150 S1=S1+F(I)*COS(X)
155 S2=S2+F(I)*SIN(X)
160 NEXT I
165 A(J)=S1*2/K:B(J)=S2*2/K
170 F1=J/T
175 A(J)=INT(A(J)*10000)/10000:B(J)=INT(B(J)*10000)/10000
180 PRINT"    ",J,F1,A(J),B(J)
185 NEXT J
190 PRINT:PRINT:PRINT"   DO YOU WANT THE C(N)'S AND THETA(N)'S?(Y OR N)":INPUT X
$
195 IF X$="Y" THEN 210
200 IF X$="N" THEN 260
205 GOTO 190
210 PRINT:PRINT:PRINT"                    N          FREQ. IN HZ.       C(N)            THETA
(N)"
215 PRINT"      ",0,0,S,"-----------"
220 FOR J=1 TO M
225 A=A(J):B=B(J):F1=J/T
230 GOSUB 265
235 P(J)=57.3*P
240 C(J)=SQR(A(J)^2+B(J)^2)
```

```
245 C(J)=INT(10000*C(J))/10000
250 PRINT"      ",J,F1,C(J),P(J)
255 NEXT J
260 END
265 IF A=0 THEN P=1.57079:GOTO 285
270 P=ATN(B/A)
275 IF A>0 THEN 295
280 IF B>0 THEN P=P+3.14159:GOTO 295
285 IF B<0 THEN P=P-3.14159:GOTO 295
290 IF B=0 THEN P=-3.14159:GOTO 295
295 RETURN
```

D.8 STATE EQUATION SOLVER

This program solves the state equation of order n for the single-input–single-output case by use of the matrix exponential series. Inputs to the program are the elements of the **A**, **b**, and $\mathbf{c}^T$ matrices, the sampling interval T, the type of input, and the number of terms desired in the matrix exponential series. Output is a tabulation of the output variable as a function of time and a plot of the output $y(t)$.

```
4 KEY OFF:CLS:PRINT:PRINT:PRINT" THIS PROGRAM SOLVES THE STATE EQUATION"
6 PRINT:PRINT"   DX/DT=A*X+B*U   X(O) KNOWN"
8 PRINT:PRINT"   WITH OUTPUT EQUATION
10 PRINT:PRINT"   Y=C*X"
12 PRINT:PRINT:PRINT"  MATRIX A IS NXN--INPUT  THE SYSTEM ORDER N"
14 INPUT N
16 DIM A(N,N),F(N,N),B(N),G(N),C(N),Q(N,N),N(N,N)
18 PRINT:PRINT:PRINT"     INPUT THE A MATRIX"
20 FOR I=1 TO N
22 FOR J=1 TO N
24 PRINT"   A(";I;",";J;")=";:INPUT F(I,J)
26 NEXT J
28 NEXT I
30 PRINT:PRINT:PRINT"     INPUT THE B MATRIX"
32 FOR I=1 TO N
34 PRINT:PRINT"   B(";I;")=";:INPUT G(I)
36 NEXT I
38 PRINT:PRINT:PRINT"INPUT ELEMENTS OF THE C MATRIX"
40 FOR I=1 TO N:PRINT"      C(";I;")=";:INPUT C(I):NEXT I
42 PRINT:PRINT:PRINT"INPUT THE TIME INTERVAL T=":INPUT T
44 PRINT:PRINT"  INPUT NUMBER OF TERMS IN THE SERIES":INPUT N1
```

```
46 I6=1
48 FOR I=1 TO N:Q(I,I)=1:NEXT I
50 FOR I1=N1 TO 2 STEP -1
52 F1=T/I1
54 FOR I=1 TO N
56 FOR J=1 TO N
58 N(I,J)=0
60 FOR K=1 TO N
62 N(I,J)=N(I,J)+F(I,K)*Q(K,J)*F1
64 NEXT K
66 IF I=J THEN N(I,I)=N(I,I)+1
68 NEXT J
70 NEXT I
72 FOR I=1 TO N:FOR J=1 TO N:Q(I,J)=N(I,J):NEXT J:NEXT I
74 NEXT I1
76 FOR I=1 TO N:FOR J=1 TO N:Q(I,J)=Q(I,J)*T:NEXT J:NEXT I
78 FOR I=1 TO N:FOR J=1 TO N
80 A(I,J)=0
82 FOR K=1 TO N
84 A(I,J)=A(I,J)+F(I,K)*Q(K,J)
86 NEXT K
88 IF I=J THEN A(I,I)=A(I,I)+1
90 NEXT J
92 NEXT I
94 FOR I=1 TO  N
96 B(I)=0
98 FOR K=1 TO N
100 B(I)=B(I)+Q(I,K)*G(K)
102 NEXT K
104 NEXT I
106 CLS:PRINT:PRINT
```

```
108 FOR I=1  TO N
110 FOR J=1  TO N
112 PRINT"    PHI(";I;",";J;")=";A(I,J)
114 NEXT J
116 NEXT I
118 FOR I=1  TO N
120 PRINT"    GAMMA(";I;")=";B(I):NEXT I
122 GOSUB 354
124 DIM U(60),Y(60),Z(N),X(N)
126 CLS:PRINT:PRINT
128 CLS:PRINT"    THIS PROGRAM SEGMENT SELECTS THE FORCING FUNCTION"
130 PRINT:PRINT:PRINT"     PRESS 1 FOR A STEP INPUT"
132 PRINT:PRINT:PRINT"     PRESS 2 FOR A RAMP INPUT"
134 PRINT:PRINT:PRINT"     PRESS 3 FOR AN ARBITRARY SAMPLED INPUT UP TO 60 POINTS
"
136 INPUT B$
138 IF B$="1" THEN 146
140 IF B$="2" THEN 152
142 IF B$="3" THEN 158
144 GOTO 128
146 PRINT:PRINT:PRINT"     INPUT  HEIGHT OF STEP":INPUT U(0)
148 FOR I=1  TO 59:U(I)=U(0):NEXT I
150 GOTO 166
152 PRINT:PRINT:PRINT"  INPUT SLOPE OF RAMP":INPUT":INPUTS
154 FOR I=0 TO 60:U(I)=I*S*T:NEXT I
156 GOTO 166
158 PRINT:PRINT:PRINT"    HOW MANY NONZERO INPUT SAMPLES ARE"
160 INPUT N3
162 FOR I=0 TO N3-1:PRINT"  U(";I;")=";:INPUT U(I):NEXT I
164 FOR I=N3 TO 59:U(I)=0:NEXT I
166 PRINT:PRINT:PRINT"    INPUT THE INITIAL STATE VECTOR"
```

```
168 FOR I=1 TO N:PRINT"  X(";I;")=";:INPUT X(I):NEXT I
170 PRINT:PRINT:PRINT"   I'M CALCULATING THE RESPONSE NOW,  PLEASE WAIT!"
172 Y(0)=0:FOR I=1 TO N:Y(0)=C(I)*X(I):NEXT I
174 FOR I3=1 TO 60
176 FOR I=1 TO N
178 Z(I)=0
180 FOR J=1 TO N
182 Z(I)=Z(I)+A(I,J)*X(J)
184 NEXT J
186 Z(I)=Z(I)+B(I)*U(I3-1)
188 NEXT I
190 FOR I=1 TO N:X(I)=Z(I):NEXT I
192 Y(I3)=0
194 FOR I=1 TO N
196 Y(I3)=Y(I3)+C(I)*X(I)
198 NEXT  I
200 NEXT I3
202 N7=0
204 J=1:CLS
206 PRINT:PRINT"    TIME   RESPONSE"
208 N7=N7+1
210 T3=(N7+(J-1)*20)*T
212 PRINT T3,Y(N7+(J-1)*20)
214 IF(N7+(J-1)*20)=60 THEN 226
216 IF N7=20 THEN 220
218 GOTO 208
220 J=J+1:N7=0
222 GOSUB 354
224 CLS:GOTO 206
226 GOSUB 354
228 PRINT:PRINT"   WANT A SCREEN PLOT OF THE RESPONSE?(Y OR N)"
```

```
230 INPUT X$
232 IF X$="N" OR X$="n" THEN 240
234 IF X$="Y" OR X$="y" THEN 238
236 GOTO 228
238 GOSUB 242
240 KEY ON:SCREEN 0,0,0:END
242 REM PLOT ROUTINE
244 LO=1E+08:HI=-1E+08
246 FOR K=0 TO 60
248 IF Y(K)<=LO THEN LO=Y(K):KL=K
250 IF Y(K)>=HI THEN HI=Y(K):KH=K
252 NEXT K
254 HI1=ABS(HI):LO1=ABS(LO)
256 SCREEN 2:CLS
258 LOCATE 2,10
260 PRINT "MAX VALUE=";HI;" Tmx=";KH*T
262 LOCATE 3,10
264 PRINT "MIN VALUE=";LO;" Tmn=";KL*T
266 IF (HI>0) AND (LO>=0) THEN 288
268 IF (HI>0) AND (LO<0)  THEN 306
270 D5=LO
272 Z=40:ZP=7
274 X7=(100*Y(0))/D5+40
276 FOR K=0 TO 59
278 T5=10*K+40:T6=T5+10
280 X8=(100*Y(K+1))/D5+40
282 LINE (T5,X7)-(T6,X8)
284 X7=X8:NEXT K
286 GOTO 326
288 D5=HI:Z=140:ZP=20
290 X7=140-(100*Y(0))/D5
292 FOR K=0 TO 59
```

```
294 T5=10*K+40
296 T6=T5+10
298 X8=140-(100*Y(K+1))/D5
300 LINE (T5,X7)-(T6,X8):X7=X8
302 NEXT K
304 GOTO 326
306 Z=100:ZP=15
308 IF HI1>LO1 THEN D5=HI
310 IF LO1>HI1 THEN D5=LO
312 X7=100-(60*Y(0))/D5
314 FOR K=0 TO 59
316 T5=10*K+40
318 T6=T5+10
320 X8=100-(60*Y(K+1))/D5
322 LINE (T5,X7)-(T6,X8)
324 X7=X8:NEXT K
326 LINE (40,Z)-(639,Z)
328 LINE (340,Z+10)-(340,Z)
330 LINE (40,40)-(40,140)
332 LOCATE ZP,42
334 PRINT 30*T
336 LOCATE ZP,70
338 PRINT"TIME,t"
340 LOCATE 6,1
342 PRINT"y(t)"
344 LOCATE 21,1
346 LOCATE ZP,2
348 PRINT 0
350 GOSUB 354
352 CLS:RETURN
354 LOCATE 24,2:PRINT"PRESS RETURN TO CONTINUE":INPUT"", S$:CLS:RETURN
```

INDEX

E

F

G

H

I

K

L

M

N

O

P